SCHAUM'S OUTLINE OF

THEORY AND PROBLEMS

OF

MECHANICAL
VIBRATIONS

•

BY

WILLIAM W. SETO, B.S. in M.E., M.S.

Associate Professor of Mechanical Engineering
San Jose State College

•

SCHAUM'S OUTLINE SERIES

McGRAW-HILL BOOK COMPANY
New York, St. Louis, San Francisco, Toronto, Sydney

ISBN 07-056327-6

10 11 12 13 14 15 SH SH 7 5 4 3 2 1 0 6 9

Preface

This book is designed primarily to supplement standard texts in elementary Mechanical Vibrations, based on the belief that numerous solved problems constitute one of the best means for clarifying and fixing in mind basic principles. Moreover, the statements of theory and principle are sufficiently complete that, with suitable handling of lecture-recitation time, the book could be used as a text by itself. Because of its extensive coverage, graduate students taking further course work in vibrations will find the last few chapters especially helpful. It should also be of considerable value as a reference book to the practicing engineer.

Throughout the book, emphasis is placed on the fundamentals, with discussions and problems extending into many phases and applications of mechanical vibrations. The subject matter is divided into chapters covering duly-recognized areas of theory and study. Each chapter begins with statements of pertinent definitions, principles and theorems. This is followed by graded sets of solved and supplementary problems. The solved problems illustrate and amplify the theory, present methods of analysis, provide practical examples, and bring into sharp focus those fine points which enable the student to apply the basic principles correctly and confidently. Numerous proofs of theorems and derivations of basic results are included among the solved problems. The large number of supplementary problems with answers serve as a complete review of the material of each chapter.

Topics covered include the fundamental single-degree-of-freedom systems and the complex multiple-degree-of-freedom systems using Newton's law of motion, the energy method, Lagrange's equations, influence coefficients, matrix iteration, the Holzer method, the Stodola method, and the mechanical impedance method. Special advanced topics include longitudinal and transverse vibration of uniform beams and circular shafts, nonlinear and self-excited vibration, and the vibrating string. Added features are the chapters on electrical analogies and the electronic analog computer which are powerful tools used extensively in vibration analysis.

Considerably more material has been included here than can be covered in most first courses. This has been done to make the book more flexible, to provide a more useful book of reference and to stimulate further interest in the topics.

I wish to take this opportunity to thank the staff of the Schaum Publishing Company for their valuable suggestions and helpful cooperation.

W. W. Seto

San Jose State College
March, 1964

CONTENTS

Symbols and Abbreviations

The following tabulation lists the symbols used in this book. Because the alphabet is limited, the same letter is sometimes used to represent more than one concept. Since each symbol is defined when it is first used, no confusion should result.

a acceleration in in/sec², velocity of propagation of wave in in/sec

A area in in²

b length or width in in.

B length or width in in.

c linear damping coefficient in lb-sec/in

C capacitance in microfarads

d diameter in in.

D.E.   dissipation energy in lb-in

e eccentricity in in., base of natural logarithm

e_i input voltage in volts

e_0 output voltage in volts

E Young's modulus of elasticity in lb/in²

E_0 initial voltage in volts

f coefficient of Coulomb damping

f_d damped natural frequency in cyc/sec

f_n natural frequency in cyc/sec

F force in lb

F_0 magnitude of applied force in lb

$f(t)$ function of time

g gravitational acceleration (32.2 ft/sec² or 386 in/sec²)

G shear modulus of elasticity in lb/in²

h height or thickness in in.

i $\sqrt{-1}$, loop current in amperes

I moment of inertia in in⁴

Im   imaginary number

I_p polar moment of inertia in in⁴

j integer labeling normal modes of vibration

J mass moment of inertia in in-lb-sec²/rad

k linear spring stiffness in lb/in

K torsional spring stiffness in in-lb/rad

K.E.   kinetic energy in lb-in

L inductance in henrys or length in inches

ln   natural logarithm

log   logarithm to base 10

m mass in lb-sec²/in

M mass in lb-sec²/in, moment in in-lb

n gear reduction ratio

P.E.   potential energy in lb-in

p_i natural frequencies of beams in rad/sec

q charge in coulombs

Q generalized force in lb, shear force in lb

r radius in in., root of equation

R radius in in., resistance in megohms

Re   real number

R_i input resistor in megohms

R_0 output resistor in megohms

s root of characteristic equation

S tensile force in lb, scale factor

t thickness in in., time in seconds

T machine time in seconds, period in seconds, tensile force in pounds

T_0 magnitude of applied torque

TR   transmissibility

u longitudinal elongation of bars

v velocity in in/sec, voltage in volts

V volume in in³

V_0 magnitude of applied velocity in in/sec

w load intensity in lb/in

W	weight in lb	$\ddot{x}$	rectilinear acceleration in in/sec^2 or ft/sec^2
x	rectilinear displacement in in. or ft	X	normal function
x_c	complementary solution	$x(t)$	x is a function of t
x_p	particular solution	y	deflection of beams in in. or ft
$\dot{x}$	rectilinear velocity in in/sec or ft/sec	Z	mechanical impedance

---

α (alpha)	angular acceleration in rad/sec^2
α_{ij}	influence coefficient in in/lb
β (beta)	any angle
γ (gamma)	specific weight in lb/in^3
δ (delta)	logarithmic decrement
δ_{st}	static deflection in in.
ϵ (epsilon)	strain
ζ (zeta)	damping factor
η (eta)	torsional damping coefficient in lb-in-sec/rad
θ (theta)	any angle
λ (lambda)	amplitude ratio
μ (mu)	coefficient of friction
ν (nu)	Poisson's ratio
π (pi)	3.14159
ρ (rho)	mass per unit length, mass per unit volume in lb-sec^2/in^4
σ (sigma)	stress in lb/in^2
τ (tau)	period in sec
ϕ (phi)	any angle
ψ (psi)	any angle
ω (omega)	natural angular frequency in rad/sec
ω_d	damped natural angular frequency in rad/sec
ω_e	natural angular frequency of electrical system in rad/sec

Chapter 1

Single-Degree-of-Freedom System

INTRODUCTION

Engineering systems possessing mass and elasticity are capable of relative motion. If the motion of such systems repeats itself after a given interval of time, the motion is known as *vibration*. Vibration, in general, is a form of wasted energy and undesirable in many cases. This is particularly true in machinery; for it generates noise, breaks down parts, and transmits unwanted forces and movements to close-by objects.

EQUATION OF MOTION

To eliminate the adverse effects of most vibration, one of the approaches is to make a complete study of the equation of motion of the system in question. The system is first idealized and simplified in terms of *mass, spring*, and *dashpot*, which represent the body, the elasticity, and the friction of the system respectively. The *equation of motion*, then, expresses displacement as a function of time or will give the distance between any instantaneous position of the mass during its motion and the equilibrium position. The important property of a vibrating system, the natural frequency, is then obtained from the equation of motion.

FREQUENCY AND PERIOD

In both the *rectilinear* and *torsional* types of vibration analysis, the *period* is the time required for a periodic motion to repeat itself; and the *frequency* is the number of cycles per unit time. Due to the similarities between rectilinear and torsional types of vibration, the discussion and analysis of one type apply equally well to the other type.

Natural frequency is the frequency of the system having free vibration without friction, while *damped natural frequency* is the frequency of the system having free vibration and with friction.

FREE VIBRATION

Free vibration is the periodic motion observed as the system is displaced from its static equilibrium position. The forces acting are the spring force, the friction force, and the weight of the mass. Due to the presence of friction, the vibration will diminish with time. This is *free vibration* or sometimes called *transient*.

$$x_c = e^{-\zeta \omega_n t}(A \cos \omega_d t + B \sin \omega_d t)$$

where x_c = amplitude of free vibration,

ζ = damping factor,

ω_n = natural circular frequency,

ω_d = natural damped circular frequency,

A, B = arbitrary constants.

(See Problem 28)

1

FORCED VIBRATION

When external forces, usually as $F(t) = F_0 \sin \omega t$ or $F_0 \cos \omega t$, are acting on the system during its vibratory motion, it is termed *forced vibration*. At forced vibration, the system will tend to vibrate at its own natural frequency as well as to follow the frequency of the excitation force. In the presence of friction, that portion of motion not sustained by the sinusoidal excitation force will gradually die out. As a result, the system will vibrate at the frequency of the excitation force regardless of the initial conditions or the natural frequency of the system. That part of sustained vibration is called *steady state vibration* or *response* of the system. Very often, the steady state response is required in vibration analysis because of its continuous effects.

$$x_p = \frac{F_0}{\sqrt{(k - m\omega^2)^2 + (c\omega)^2}} \sin(\omega t - \phi), \qquad \phi = \tan^{-1} \frac{c\omega}{k - m\omega^2}$$

where x_p = amplitude of steady state vibration,

   F_0 = magnitude of the excitation force,

   k = spring constant,

   m = mass of the system,

   c = damping coefficient,

   ω = frequency of the excitation force,

   ϕ = phase angle.   (See Problem 28)

DAMPING

In actuality, most engineering systems during their vibratory motion encounter friction or resistance in the form of damping. Damping, in its various forms such as air damping, fluid friction, Coulomb dry friction, magnetic damping, internal damping, etc., will always slow down the motion, and cause the eventual dying out of the oscillation. If the damping is heavy, oscillatory motion will not occur; the system is said to be *over-damped*. If the damping is light, oscillation is possible; the system is said to be *under-damped*. A *critically-damped* system is one in which the amount of damping is such that the resultant motion is on the borderline between the two cases just mentioned. The mass upon being released will simply return to its static equilibrium position. In most problems in vibration, air damping is so small that it is neglected except for special cases.

RESONANCE

Resonance occurs when the frequency of the excitation is equal to the natural frequency of the system. When this happens, the amplitude of vibration will increase without bound and is governed only by the amount of damping present in the system. Therefore, in order to avoid disastrous effects resulting from very large amplitude of vibration at resonance, the natural frequency of a system must be known and properly taken care of.

SINGLE-DEGREE-OF-FREEDOM SYSTEM

Many systems can vibrate in more than one manner and direction. If a system is constrained so that it can vibrate in only one mode or manner, or if only one independent coordinate is required to specify completely the geometric location of the masses of the system in space, it is a single-degree-of-freedom system. The following four systems are of the single-degree-of-freedom.

In the spring-mass system shown in Fig. 1-1 below, if the mass m is constrained to move vertically, only one coordinate, $x(t)$, is required to define the location of the mass at any time from the static equilibrium position. Thus the system is said to possess one degree of freedom.

Similarly, if the torsional pendulum shown in Fig. 1-2 is constrained to oscillate about the longitudinal axis of the shaft, the configuration of the system can be specified by one coordinate, $\theta(t)$. This is also a single-degree-of-freedom system.

The spring-mass-pulley system of Fig. 1-3 is of single degree of freedom because either $x(t)$ or $\theta(t)$ can be used to determine the relative position of the masses. But $x(t)$ and $\theta(t)$ are not independent of each other.

By attaching the base to the body whose motion is to be measured, as shown in Fig. 1-4, the vibration pickup will be able to measure the oscillatory motion of the body. This is possible by finding the relative motion of the base and the mass. Hence only one coordinate is needed to specify the configuration of the system.

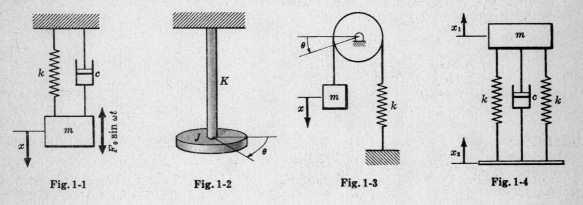

Fig. 1-1   Fig. 1-2   Fig. 1-3   Fig. 1-4

SIMPLE HARMONIC MOTION

For a particle in rectilinear motion, if its acceleration is always proportional to the distance of the particle from a fixed point on the path and is directed toward the fixed point, then the particle is said to have *simple harmonic motion* or simply SHM. SHM is the simplest form of periodic motion. The periodic motion of vibration, whether simple or complex, may be considered to be composed of SHM, or a number of SHM of various amplitudes and frequencies by means of a Fourier series. In differential equation form, SHM is represented as

$$a = -Kx \quad \text{or} \quad \ddot{x} + Kx = 0$$

and
$$x = A \sin \sqrt{K}\, t + B \cos \sqrt{K}\, t \quad \text{or} \quad x = C \sin(\sqrt{K}\, t + \phi)$$

NEWTON'S LAW OF MOTION

The equation of motion is just another form of Newton's law of motion, $\Sigma F = ma$ (total forces in the same direction as motion). Equations of motion for many systems are conveniently determined by Newton's law of motion. Some of them, however, are more easily found by other methods such as Energy method, Lagrange's equation, etc.

ENERGY METHOD

For a conservative system, the total energy of the system is unchanged at all time. If the total energy of the system is expressed as *potential* and *kinetic energy*, then the following is true:

$$\text{K.E.} + \text{P.E.} = \text{constant} \qquad \text{or} \qquad \frac{d}{dt}(\text{K.E.} + \text{P.E.}) = 0$$

where   K.E. = kinetic energy,   P.E. = potential energy.

The resulting equation is the equation of motion of the system under consideration. This is, then, the Energy method.

RAYLEIGH'S METHOD

Again, if the given system is a conservative one, the total kinetic energy of the system is zero at the maximum displacement, but is a maximum at the static equilibrium point. For the total potential energy of the system, on the other hand, the reverse is true. Hence,

$$(\text{K.E.})_{max} = (\text{P.E.})_{max} = \text{total energy of the system}$$

This is known as Rayleigh's method. The resulting equation will readily yield the natural frequency of the system.

MECHANICAL IMPEDANCE METHOD

In the determination of the steady state vibration of a system, the Mechanical Impedance method is simple and straightforward compared to other methods. This method is based on the vectorial representation of harmonic functions. Let the force vector be $F = Fe^{i\omega t}$. Since the steady state response must lag the excitation force, the displacement vector is $x = Xe^{i(\omega t - \phi)}$. The velocity vector then is $\dot{x} = Xe^{i(\omega t - \phi + \pi/2)}$ or $\dot{x} = i\omega x$. Similarly, the acceleration vector is $\ddot{x} = -\omega^2 x$. Therefore the mechanical impedances of the three elements are as follow:

$$\text{mass} = -m\omega^2$$
$$\text{damper} = ic\omega$$
$$\text{spring} = k$$

as shown in the diagram in Fig. 1-5.

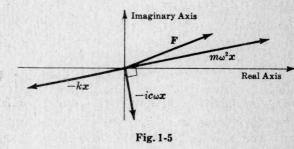

Fig. 1-5

UNBALANCE

Rotating unbalance of a machine exists if the center of gravity of the rotating part of the machine does not coincide with the axis of rotation. Usually, the amount of rotating unbalance is expressed by me, where m is an equivalent eccentric mass and e is the eccentricity. The centrifugal force, $me\omega^2$, as a result of this unbalance me will produce the unwanted excitation. For *reciprocating unbalance*, the same reasoning will apply.

$$m\ddot{x} + c\dot{x} + kx = me\omega^2 \sin \omega t$$

(See Problem 33)

CRITICAL SPEED OF SHAFT

When the rotating speeds of the shaft coincide with one of the natural frequencies of the system of rotors or disks mounted on the elastic shaft, violent vibration will take place. These are commonly known as 'critical speeds' of the shaft, and must be avoided.

TRANSMISSIBILITY

In order to reduce as much as possible the amount of force transmitted to the foundation due to the vibration of machinery, machines are usually isolated from the foundation by mounting them on springs and dampers. As a result, the force transmitted to the foundation is the sum of the spring and damper force, i.e. $F_t = kx + c\dot{x}$. *Transmissibility* is thus defined as the amplitude ratio of the transmitted force to the impressed force.

$$\text{TR} = F_t/F_0 = \sqrt{1 + (2\zeta r)^2}\Big/\sqrt{(1 - r^2)^2 + (2\zeta r)^2}$$

where r = frequency ratio, ζ = damping factor.   (See Problem 40)

For the same reason, it is frequently desirable to isolate the motion of the surroundings from a delicate instrument. The effectiveness of such an isolator would be the ratio of the amplitude of vibration of the body to that of the supporting part. This ratio is the same as that for force isolator. Hence the same isolator can be used for both force and *motion isolation*.

SEISMIC INSTRUMENTS

Seismic instruments are essentially vibratory systems consisting of the support or the base and the mass with spring attached. The support or the base is attached to the body whose motion is to be measured. The relative motion between the mass and the base, recorded by a rotating drum or some other devices inside the instrument, will indicate the motion of the body. For measuring the displacement of a machine part, a *vibrometer* should be used, whose natural frequency is low compared to the frequency of the vibration to be measured. An *accelerometer* is used to measure acceleration because its natural frequency is high compared to that of the vibration to be measured. *Seismographs*, the oldest seismic instruments, are used for the recording of earthquake vibration. The more elaborate modern types of seismic instruments such as the *torsiograph*, are used to record torsional vibration.

Solved Problems

EQUATION OF MOTION AND NATURAL FREQUENCY

1. Determine the equation of motion and natural frequency of vibration of the simple spring-mass system shown in Fig. 1-6.

    Applying Newton's law of motion, $\Sigma F = ma$:

    For vertical oscillations, the forces acting are the spring force, $k(\delta_{st} + x)$, and the weight mg of the mass. Therefore the equation of motion is

    $$m\ddot{x} = -k(\delta_{st} + x) + mg$$

    where $\ddot{x} = d^2x/dt^2$, and δ_{st} is the static deflection due to the weight of the mass acting on the spring. Then $mg = \delta_{st}k$, and the equation of motion becomes

    $$m\ddot{x} + kx = 0$$

    which is the differential equation for SHM. The most general form of solution for this equation is

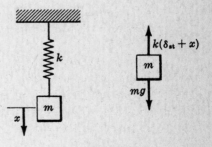

Fig. 1-6

$$x = A \sin \sqrt{k/m}\, t + B \cos \sqrt{k/m}\, t$$
or
$$x = C \cos (\sqrt{k/m}\, t + \phi)$$

where $A, B, C,$ and ϕ are arbitrary constants depending on initial conditions $x(0)$ and $\dot{x}(0)$. Two constants must appear in the general solution because this is a second order differential equation.

For an initial displacement x_0, we have $A = 0$, $B = x_0$ and hence

$$x = x_0 \cos (\sqrt{k/m}\, t)$$

Physically, this equation represents an undamped free vibration, one cycle of which occurs when $\sqrt{k/m}\,t$ varies through 360 degrees; therefore

$$\text{period } T = \frac{2\pi}{\sqrt{k/m}} \quad\text{and}\quad \text{natural frequency } f_n = 1/T \text{ cps}$$

where $\sqrt{k/m} = \omega_n$ rad/sec is the angular natural frequency.

2. Find the equivalent springs for the systems shown in Fig. 1-7a and Fig. 1-7b where the springs are in parallel and in series respectively.

For springs in parallel,

$$F_1 = k_1 x, \quad F_2 = k_2 x, \quad\text{and}\quad F = F_1 + F_2 = (k_1 + k_2)x$$

Then $\qquad k_{eq} = F/x = k_1 + k_2$

and in general, $\qquad k_{eq} = \sum_{i=1}^{n} k_i$

For springs in series, force is the same in each spring, but the total displacement is the sum of the individual displacements. Thus

$$F = k_1 x_1 = k_2 x_2 \quad\text{and}\quad x = x_1 + x_2 = F/k_1 + F/k_2$$

Then $\qquad k_{eq} = F/x = \dfrac{1}{1/k_1 + 1/k_2}$

and in general, $\qquad k_{eq} = \dfrac{1}{\displaystyle\sum_{i=1}^{n} 1/k_i}$

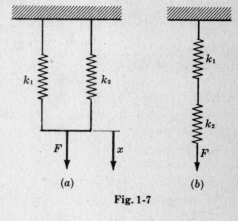

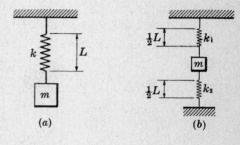

Fig. 1-7

3. Assuming that the spring constant is inversely proportional to the number of coils in the spring, find and compare the natural frequencies of the systems shown in Fig. 1-8.

For case (a), the equation of motion is $m\ddot{x} + kx = 0$, so $\omega_n = \sqrt{k/m}$.

For case (b), the equation of motion is

$$m\ddot{x} + (k_1 + k_2)x = 0$$

corresponding to springs in parallel, so $\omega_n = \sqrt{(k_1 + k_2)/m}$. But since $k_1 = k_2 = 2k$, $\omega_n = \sqrt{4k/m}$ and the required ratio is $1:2$.

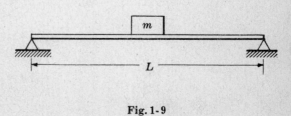

Fig. 1-8

4. A simply-supported beam with a concentrated load acting on the midspan is shown in Fig. 1-9. If the mass of the beam is negligible compared to the mass acting, find the natural frequency of the system.

From Strength of Materials, the deflection at the midspan of a simply-supported beam due to a concentrated load P at the center of the beam is given by $\delta = PL^3/48EI$ where E and I take on the usual meaning. For small deflections, $k = P/\delta = 48EI/L^3$; hence the equation of motion for this free undamped vibration is

$$m\ddot{x} + kx = 0$$

and $\qquad \omega_n = \sqrt{k/m} = \sqrt{48EI/mL^3}$ rad/sec

Fig. 1-9

5. The string is under tension T which can be assumed to remain constant for small displacements. For small oscillations, find the natural frequency of the vertical vibration of the string.

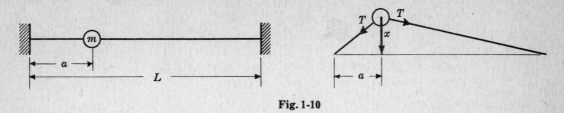

Fig. 1-10

This is free undamped vibration; the restoring force is the tension in the string. Since the tension is essentially constant, this is $T[x/a + x/(L-a)]$. Applying Newton's second law of motion, the equation of motion is

$$m\ddot{x} + T[x/a + x/(L-a)] = 0$$

or $\qquad m\ddot{x} + T[L/a(L-a)]x = 0,$ and $\qquad \omega_n = \sqrt{\dfrac{TL}{ma(L-a)}}$ rad/sec

6. A steel cantilever beam of length 10 in. has a square cross-section of $\frac{1}{4} \times \frac{1}{4}$ in. A mass of 10 lb is attached to the free end of the beam as shown in Fig. 1-11. Determine the natural frequency of the system if the mass is displaced slightly and released.

Fig. 1-11

Assume the mass of the cantilever is small. From Strength of Materials, the deflection at the free end of the cantilever due to mass m is $\delta = PL^3/3EI$.

For small oscillations, the cantilever behaves elastically; hence the spring scale $k = F/\delta = 3EI/L^3$ lb/in.

The moment of inertia I of the beam is $I = b^3h/12 = (\frac{1}{4})^3(\frac{1}{4})/12 = 1/3072$ in^4, and the modulus of elasticity of the steel is $E = 30(10)^6$ lb/in^2.

The equation of motion for free undamped vibration is $m\ddot{x} + kx = 0$, and

$$\omega_n = \sqrt{k/m} = \sqrt{\frac{3(30)(10)^6(32.2)(12)}{10(3072)(10)^3}} = 33.7 \text{ rad/sec}$$

7. A manometer used in a fluid mechanics laboratory has a uniform bore of cross-section area A. If a column of liquid of length L and density ρ is set into motion as shown in Fig. 1-12, find the frequency of the resulting motion.

Newton's law of motion:

$$\Sigma F_x = m\ddot{x} \qquad \text{or} \qquad -2Axg\rho = LA\rho\ddot{x}$$

from which $\quad \ddot{x} + (2g/L)x = 0 \quad$ and $\quad \omega_n = \sqrt{2g/L}$ rad/sec.

Energy method:

$$\text{K.E.} = \tfrac{1}{2}m\dot{x}^2 = \tfrac{1}{2}(LA\rho)\dot{x}^2$$
$$\text{P.E.} = \tfrac{1}{2}kx^2 = \tfrac{1}{2}(2Ax\rho g/x)x^2 = Ax^2\rho g$$

and $\qquad \dfrac{d}{dt}(\text{K.E.} + \text{P.E.}) = LA\rho\dot{x}\ddot{x} + 2A\rho gx\dot{x} = 0$

from which $\quad \ddot{x} + (2g/L)x = 0 \quad$ and $\quad \omega_n = \sqrt{2g/L}$ rad/sec.

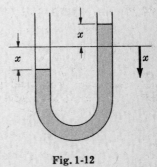

Fig. 1-12

8. An electric motor is supported by 4 springs, each having spring constant k lb/in as shown in Fig. 1-13. If the moment of inertia of the motor about the central axis of rotation is J_0, find its natural frequency of oscillation.

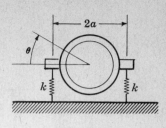

Fig. 1-13

Newton's law of motion in the form of the torque equation is $\Sigma M = J_0 \ddot{\theta}$ or

$$\text{twisting torque} = \text{restoring torque}$$
$$-4ka^2\theta = J_0 \ddot{\theta}$$

from which $\ddot{\theta} + (4ka^2/J_0)\theta = 0$ and $\omega_n = 2a\sqrt{k/J_0}$ rad/sec.

9. A semicircular homogeneous disk of radius r and mass m is pivoted freely about its center as shown in Fig. 1-14. Determine its natural frequency of oscillation for small displacements.

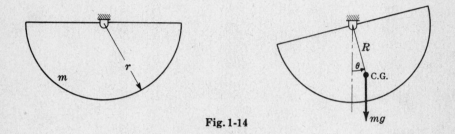

Fig. 1-14

Employ Newton's torque equation, $\Sigma M = J_0 \ddot{\theta}$.

Here $J_0 = \frac{1}{2}mr^2$ and the distance of the mass center from the center of the disk is $R = 4r/3\pi$ as shown. Hence the restoring torque is $mgR \sin\theta$ and the equation of motion is

$$(\tfrac{1}{2}mr^2)\ddot{\theta} = -mgR \sin\theta$$

where the minus sign is required because the force mg is acting in the decreasing θ direction. For small values of θ, $\sin\theta \doteq \theta$ and the equation of motion becomes

$$\ddot{\theta} + (8g/3r\pi)\theta = 0$$

which is the differential equation of a system having simple harmonic motion; hence

$$\omega_n = \sqrt{\frac{8g}{3r\pi}} \text{ rad/sec}$$

10. Determine the natural frequency of the spring-mass-pulley system shown in Fig. 1-15.

Newton's law of motion:

For the mass m,

$$mg - T = m\ddot{x} \qquad (1)$$

For the pulley M,

$$J_0 \ddot{\theta} = Tr - kr^2(\theta + \theta_0) \qquad (2)$$

where $J_0 = \frac{1}{2}Mr^2$ is the moment of inertia of the pulley.

But at static equilibrium, $mgr = kr^2\theta_0$. Hence equation (2) becomes

$$\tfrac{1}{2}Mr^2 \ddot{\theta} = r(mg - m\ddot{x}) - kr^2\theta - mgr$$

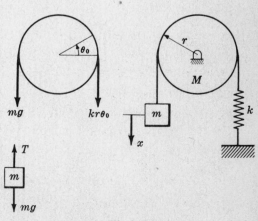

Fig. 1-15

Replacing x by $r\theta$, or $\ddot{x}$ by $r\ddot{\theta}$, the differential equation of motion becomes

$$(\tfrac{1}{2}Mr^2 + mr^2)\,\ddot{\theta} + kr^2\theta = 0, \quad \text{and} \quad \omega_n = \sqrt{\frac{k}{M/2 + m}} \text{ rad/sec}$$

Energy method:

$$\text{K.E.} = \text{K.E. of mass} + \text{K.E. of pulley}$$
$$= \tfrac{1}{2}m\dot{x}^2 + \tfrac{1}{2}J_0\dot{\theta}^2 = \tfrac{1}{2}mr^2\dot{\theta}^2 + \tfrac{1}{2}J_0\dot{\theta}^2$$
$$\text{P.E.} = \tfrac{1}{2}kx^2 = \tfrac{1}{2}kr^2\theta^2$$

Since the total energy of the system remains unchanged,

$$\frac{d}{dt}(\text{K.E.} + \text{P.E.}) = 0 \quad \text{or} \quad mr^2\dot{\theta}\ddot{\theta} + J_0\dot{\theta}\ddot{\theta} + kr^2\theta\dot{\theta} = 0$$

and

$$\dot{\theta}(mr^2\ddot{\theta} + J_0\ddot{\theta} + kr^2\theta) = 0$$

Since $\dot{\theta}$ is not always zero, $(mr^2\ddot{\theta} + J_0\ddot{\theta} + kr^2\theta)$ is equal to zero.   Then

$$\ddot{\theta} + \frac{kr^2}{J_0 + mr^2}\theta = 0 \quad \text{and} \quad \omega_n = \sqrt{\frac{k}{M/2 + m}} \text{ rad/sec}$$

11. Determine the natural frequency of the system shown in Fig. 1-16 where the mass of the spring is not small.

If the mass of the spring is not to be ignored,

$$\text{K.E. of the system} = \text{K.E. of mass} + \text{K.E. of spring}$$
$$= \tfrac{1}{2}m\dot{x}^2 + \int_0^L \tfrac{1}{2}(\rho\, de)[(e/L)\dot{x}]^2$$

where $(e/L)x$ is the displacement at an intermediate point of the spring at a distance e from the upper end, and ρ is the mass of the spring per unit length.

Rayleigh's method: $(\text{K.E.})_{max} = (\text{P.E.})_{max}$

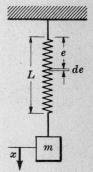

$$(\text{K.E.})_{max} = \tfrac{1}{2}m\dot{x}_{max}^2 + \tfrac{1}{2}\rho(\dot{x}_{max}/L)^2\int_0^L e^2\, de$$
$$= \tfrac{1}{2}m\dot{x}_{max}^2 + \tfrac{1}{2}\rho(\dot{x}_{max}/L)^2\,(L^3/3) = \tfrac{1}{2}(m + \tfrac{1}{3}\rho L)\dot{x}_{max}^2$$
$$(\text{P.E.})_{max} = \tfrac{1}{2}kx_{max}^2$$

For sinusoidal oscillation, let $x = A\sin\omega_n t$; then, equating the two maximum energy expressions,

$$\tfrac{1}{2}(m + \tfrac{1}{3}\rho L)(\omega_n A)^2 = \tfrac{1}{2}kA^2$$

which gives

$$\omega_n = \sqrt{k/(m + \tfrac{1}{3}\rho L)} \text{ rad/sec}$$

If ρ is negligible compared to m, then $\omega_n = \sqrt{k/m}$, which is the natural angular frequency of the simple spring-mass system shown in Problem 1.

Fig. 1-16

12. A simple pendulum is shown in Fig. 1-17. Determine the natural frequency of the swing, (*a*) if the mass of the rod is small compared to the mass at the end, (*b*) if the mass of the rod is not negligible.

(*a*) *Energy method:* $\dfrac{d}{dt}(\text{K.E.} + \text{P.E.}) = 0$

For small rotation of the mass about the pivot,

$$\text{K.E.} = \tfrac{1}{2}m\dot{x}^2 = \tfrac{1}{2}m(L\dot{\theta})^2, \quad \text{P.E.} = mgL(1 - \cos\theta)$$

and

$$\frac{d}{dt}(\text{K.E.} + \text{P.E.}) = mL^2\dot{\theta}\ddot{\theta} + mgL\sin\theta\,\dot{\theta} = 0$$

or

$$\ddot{\theta} + (g/L)\sin\theta = 0$$

For small angle of oscillation, $\sin\theta \doteq \theta$; the equation of motion then becomes

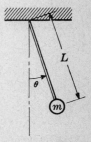

Fig. 1-17

$$\ddot{\theta} + (g/L)\theta = 0 \quad \text{with} \quad \omega_n = \sqrt{g/L} \text{ rad/sec}$$

(b) Let M be the mass of the uniform rod; it acts through the center of the rod halfway from both ends.

$$\text{K.E.} = \text{K.E. of mass} + \text{K.E. of rod}$$
$$= \tfrac{1}{2}m(L\dot{\theta})^2 + \tfrac{1}{2}(1/3)M(L\dot{\theta})^2$$
$$\text{P.E.} = \text{P.E. of mass} + \text{P.E. of rod}$$
$$= mgL(1 - \cos\theta) + Mg(L/2)(1 - \cos\theta)$$

$$\frac{d}{dt}(\text{K.E.} + \text{P.E.}) = (m + M/3)L^2\dot{\theta}\ddot{\theta} + gL(m + M/2)\sin\theta\,\dot{\theta} = 0$$

or

$$\ddot{\theta} + \left[\frac{m + M/2}{m + M/3}\sin\theta\right](g/L) = 0$$

For small angle of oscillation, $\sin\theta \doteq \theta$ and the equation of motion becomes

$$\ddot{\theta} + \frac{m + M/2}{m + M/3}(g/L)\theta = 0 \quad\text{with}\quad \omega_n = \sqrt{\frac{m + M/2}{m + M/3}(g/L)}\ \text{rad/sec}$$

Note that if M is much smaller than m, then $\dfrac{1 + M/2m}{1 + M/3m}$ is essentially 1 and ω_n has approximately the same value as in part (a).

13. The Compound Pendulum is a rigid body of mass m and is pivoted at a point a distance d from its center of mass G. It is free to rotate under its own gravitational force. Find the frequency of oscillation of such a pendulum.

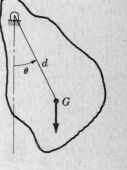

At any instant under consideration, the weight of the body is the only force acting. Then

$$J\ddot{\theta} = -mgd\sin\theta$$

is the equation of motion, where J is the moment of inertia of the body about the axis of rotation.

Considering only small oscillations, $\sin\theta \doteq \theta$ and

$$\ddot{\theta} + (mgd/J)\theta = 0, \quad \omega_n = \sqrt{mgd/J}\ \text{rad/sec}$$

But $\omega_n = \sqrt{g/d}$ for a simple pendulum of length d. Hence J/md can be defined as the length of the equivalent simple pendulum.

Fig. 1-18

14. The uniform stiff rod is restrained to move vertically by both linear and torsional springs as shown in Fig. 1-19. Calculate the frequency of the vertical oscillation of the rod.

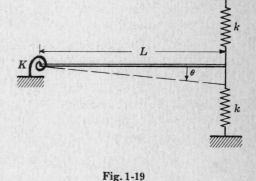

Applying the torque equation $\Sigma M = J\ddot{\theta}$,

$$-K\theta - 2kL^2\sin\theta = (mL^2/3)\ddot{\theta}$$

where m is the mass of the rod, and J is the moment of inertia of the rod with axis of rotation at the end. Thus

$$(mL^2/3)\ddot{\theta} + K\theta + 2kL^2\theta = 0$$

and

$$\omega_n = \sqrt{\frac{3K + 6kL^2}{mL^2}}\ \text{rad/sec}$$

Fig. 1-19

15. The mass M at the end of the cord of the conical pendulum is revolving around the vertical axis as shown in Fig. 1-20 below. The plane of the circular path is horizontal, and will rise when the angular velocity of rotation is increased. Determine the frequency of the system.

From the free-body diagram of mass M,

$$\Sigma F_n = Ma_n \qquad \text{or} \qquad T \sin \theta = M(L \sin \theta)\omega^2$$

where ω is the angular velocity of rotation of the mass at the instant under consideration. Also,

$$\Sigma F_v = Ma_v \qquad \text{or} \qquad T \cos \theta - Mg = Ma_v$$

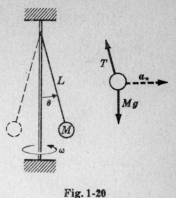

But $a_v = 0$; hence $T \cos \theta = Mg$ or $T = Mg/(\cos \theta)$. Substituting this value of T into the first equation yields

$$\frac{Mg}{\cos \theta} \sin \theta = M(L \sin \theta)\omega^2 \qquad \text{and} \qquad \omega_n = \sqrt{g/(L \cos \theta)} \text{ rad/sec}$$

If $\theta = 0$, then $\cos \theta = 1$, and $\omega_n = \sqrt{g/L}$ which is the natural frequency of a simple pendulum. This means that the conical (sometimes called spherical) pendulum is reduced to a simple pendulum if θ is equal to zero.

Fig. 1-20

16. Determine the equation of motion of the simple pendulum as shown in Fig. 1-21 below, taking into account the effect of amplitude of oscillation.

Employing Newton's law of motion,

$$-mg \sin \theta = mL\ddot{\theta} \qquad \text{or} \qquad \ddot{\theta} + (g/L) \sin \theta = 0$$

where the minus sign indicates that the force is in the direction of decreasing θ.

For small oscillations, $\sin \theta \doteq \theta$ and the equation is simplified to $\ddot{\theta} + (g/L)\theta = 0$, the usual form for a simple pendulum.

For slightly larger values of θ, $\sin \theta \doteq \theta - \theta^3/3! + \theta^5/5! - \cdots$ and the equation of motion becomes, if only the first two terms are taken for the second approximation,

$$\ddot{\theta} + (g/L)\theta - (g/6L)\theta^3 = 0$$

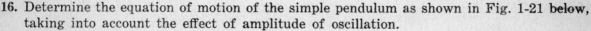

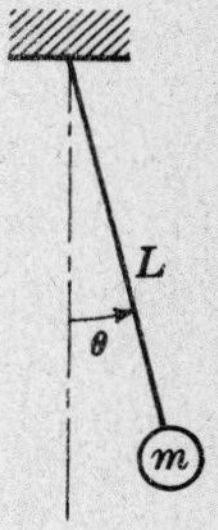

Fig. 1-21

which is a specific form of the general Duffing's equation in non-linear vibration. (Non-linear vibration is dealt with separately and much more in detail in a separate chapter.)

Hence the assumption of small angles of oscillation not only simplifies the problem but also restricts us to the linear case.

17. A solid wooden cylinder of radius r is partially immersed in a bath of distilled water as shown in Fig. 1-22 below. The cylinder is depressed slightly and released. Find the natural frequency of oscillation of the cylinder if it stays upright all the time. What will the frequency be if salt water of specific gravity 1.2 is used instead of distilled water?

Let the displacement of the cylinder be x ft; then the weight of water displaced is $\pi r^2 x(64.4)$ lb, and this is the restoring force. The mass of the cylinder $= \pi r^2 hs(64.4)/g$, where s is the specific gravity of the wood.

From Newton's law of motion, $\Sigma F = ma$. Since the only force acting is the weight of water displaced, the equation of motion is

$$m\ddot{x} + kx = 0 \qquad \text{or} \qquad \frac{64.4\pi r^2 hs}{g}\ddot{x} + 64.4\pi r^2 x = 0,$$

and

$$\omega_n = \sqrt{g/hs} \text{ rad/sec}$$

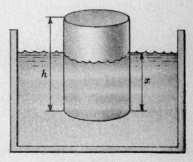

If salt water is used, the restoring force will be $1.2(64.4\pi r^2 x)$ while the mass of the cylinder is unchanged. The equation of motion is now

$$\frac{64.4\pi r^2 hs}{g}\ddot{x} + 1.2(64.4\pi r^2 x) = 0$$

and

$$\omega_n = \sqrt{1.2g/hs} \text{ rad/sec}$$

Fig. 1-22

Thus the frequency is increased as the restoring force is increased.

18. A circular cylinder of mass m and radius r is connected by a spring of modulus k as shown in Fig. 1-23. If it is free to roll on the rough horizontal surface without slipping, find its frequency.

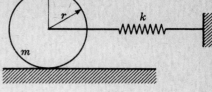

Fig. 1-23

Energy method:

The total energy of the system consists of kinetic energy (rotational and translational) and potential energy; and should remain the same all the time.

Translational K.E. $= \frac{1}{2}m\dot{x}^2$,   Rotational K.E. $= \frac{1}{2}J_0\dot{\theta}^2$

where the moment of inertia of the cylinder is $J_0 = \frac{1}{2}mr^2$. Also $r\theta = x$, or $r\dot{\theta} = \dot{x}$. Thus, for the system at any time,

$$\text{K.E.} = \frac{1}{2}m\dot{x}^2 + \frac{1}{2}(\frac{1}{2}mr^2)(\dot{x}/r)^2 = \frac{3}{4}m\dot{x}^2$$
$$\text{P.E.} = \frac{1}{2}kx^2$$

and $$\frac{d}{dt}(\text{K.E.} + \text{P.E.}) = 0   \text{   or   }   (\frac{3}{2}m\ddot{x} + kx)\dot{x} = 0$$

Since $\dot{x}$ is not always zero, the equation of motion becomes

$$\frac{3}{2}m\ddot{x} + kx = 0,   \text{   and so   }   \omega_n = \sqrt{2k/3m} \text{ rad/sec}$$

Newton's law of motion:

Applying Newton's law of motion to the cylinder,

$$\Sigma F = ma   \text{   or   }   m\ddot{x} = -kx + F_f$$

where F_f is the friction force.

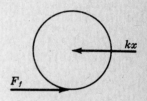

Using the torque equation, $\Sigma M = J_0\ddot{\theta}$,

$$J_0\ddot{\theta} = -F_f r   \text{   or   }   (\frac{1}{2}mr^2)(\ddot{x}/r) = -F_f r$$

and hence $F_f = -\frac{1}{2}m\ddot{x}$. Substitute this expression for F_f into the force equation to obtain

$$m\ddot{x} = -kx - \frac{1}{2}m\ddot{x}   \text{   or   }   \frac{3}{2}m\ddot{x} + kx = 0$$

and so $$\omega_n = \sqrt{2k/3m} \text{ rad/sec}$$

Fig. 1-24

19. The homogeneous circular disc has a moment of inertia about its center equal to 10 lb-in-sec^2. At the static equilibrium position, both springs are stretched one inch. Find the natural angular frequency of oscillation of the disc when it is given a small angular displacement and released. $k = 10$ lb/in.

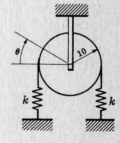

Using the torque equation $\Sigma M_0 = I\ddot{\theta}$, the initial tension in the spring is 10 lb, and the change in tension is $10(10\theta)$ lb, and

$$I_0\ddot{\theta} = [(10 - 100\theta) - (10 + 100\theta)]10$$

or $$\ddot{\theta} + 200\theta = 0$$

from which $\omega_n = \sqrt{200} = 14.2$ rad/sec.

Fig. 1-25

20. The mass of the slender uniform rod shown in Fig. 1-26 below is small compared to the mass attached to it. For small oscillations, calculate the natural frequency of the swing of the mass.

Using the torque equation $\Sigma M_0 = I\ddot{\theta}$, we obtain

$$mL^2\ddot{\theta} = -mgL\sin\theta - (a\cos\theta)k(a\sin\theta)$$

where $a\sin\theta$ is the amount the spring stretched. Assuming small oscillations, $\sin\theta \doteq \theta$, $\cos\theta \doteq 1$, and the equation of motion becomes

$$mL^2\ddot{\theta} + (mgL + ka^2)\theta = 0$$

or

$$\ddot{\theta} + \left(\frac{mgL + ka^2}{mL^2}\right)\theta = 0$$

which gives

$$\omega_n = \sqrt{\frac{mgL + ka^2}{mL^2}} \text{ rad/sec.}$$

If the spring constant k is zero, the expression for the angular natural frequency of the pendulum is reduced to the familiar expression for a simple pendulum, $\omega_n = \sqrt{g/L}$ rad/sec.

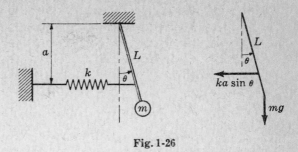

Fig. 1-26

21. Use Rayleigh's method to find the natural frequency of the semi-circular shell of mass m and radius r which rolls from side to side without slipping as shown in Fig. 1-27 below.

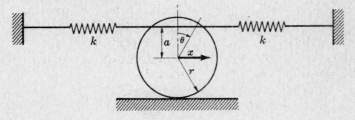

Fig. 1-27

$$(\text{K.E.})_{max} = \tfrac{1}{2}I_A\omega_n^2 = mr(r-a)\omega_n^2$$

where $I_A = I_{c.g.} + m(r-a)^2 = I_0 - ma^2 + m(r-a)^2 = 2mr(r-a)$. Then

$$(\text{K.E.})_{max} = (\text{P.E.})_{max}$$

or

$$mr(r-a)\omega_n^2 = mga(1-\cos\theta)$$

and

$$\omega_n = \sqrt{\frac{ga(1-\cos\theta)}{r(r-a)}} \text{ rad/sec}$$

22. Use the Energy method to find the natural frequency of oscillation of the homogeneous cylinder as shown in Fig. 1-28 below.

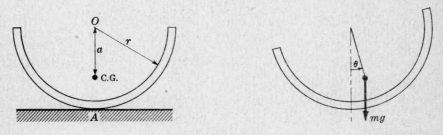

Fig. 1-28

$$\text{K.E.} = \tfrac{1}{2}m\dot{x}^2 + \tfrac{1}{2}J_0\dot{\theta}^2 = \tfrac{1}{2}mr^2\dot{\theta}^2 + \tfrac{1}{4}mr^2\dot{\theta}^2 = \tfrac{3}{4}mr^2\dot{\theta}^2$$

where m is the mass of cylinder and J_0 its mass moment of inertia.

$$\text{P.E.} = \tfrac{1}{2}kx^2 = 2\cdot\tfrac{1}{2}k(r+a)^2\theta^2$$

where $x = r\theta$ and $\sin\theta \doteq \theta$. Then

$$\frac{d}{dt}(\text{K.E.} + \text{P.E.}) = 0 \quad \text{or} \quad \tfrac{3}{2}mr^2\dot{\theta}\ddot{\theta} + 2k(r+a)^2\theta\dot{\theta} = 0 \quad \text{or} \quad \ddot{\theta} + \frac{4k(r+a)^2}{3mr^2}\theta = 0$$

from which

$$\omega_n = \sqrt{\frac{4k(r+a)^2}{3mr^2}} \text{ rad/sec}$$

23. The cord can be assumed inextensible in the spring-mass-pulley system as shown in Fig. 1-29. Find the natural frequency of vibration if the mass m is displaced slightly and released. Use the Energy method.

$$\begin{aligned}
\text{K.E.}_{\text{system}} &= \text{K.E.}_{\text{mass}} + \text{K.E.}_{\text{pulley}} \\
&= \tfrac{1}{2}m\dot{x}^2 + \tfrac{1}{2}M\dot{y}^2 + \tfrac{1}{2}J_0\dot{\theta}^2 \\
&= \tfrac{1}{2}m\dot{x}^2 + \tfrac{1}{2}M(\dot{x}^2/4) + \tfrac{1}{2}(\tfrac{1}{2}Mr^2)(\dot{x}/2r)^2 \\
&= \tfrac{1}{2}m\dot{x}^2 + \tfrac{3}{16}M\dot{x}^2
\end{aligned}$$

$$\begin{aligned}
\text{P.E.}_{\text{system}} &= \text{elastic energy of spring} \\
&= \tfrac{1}{2}ky^2 = \tfrac{1}{2}k(\tfrac{1}{2}x)^2 = \tfrac{1}{8}kx^2
\end{aligned}$$

Then $\dfrac{d}{dt}(\text{K.E.} + \text{P.E.}) = 0$ or $m\dot{x}\ddot{x} + \tfrac{3}{8}M\dot{x}\ddot{x} + \tfrac{1}{4}kx\dot{x} = 0$

and $\dot{x}(4m\ddot{x} + \tfrac{3}{2}M\ddot{x} + kx) = 0$

Since $\dot{x}$ is not always zero,

$$\ddot{x}(4m + \tfrac{3}{2}M) + kx = 0, \quad \text{and} \quad \omega_n = \sqrt{\frac{k}{4m + \tfrac{3}{2}M}} \text{ rad/sec}$$

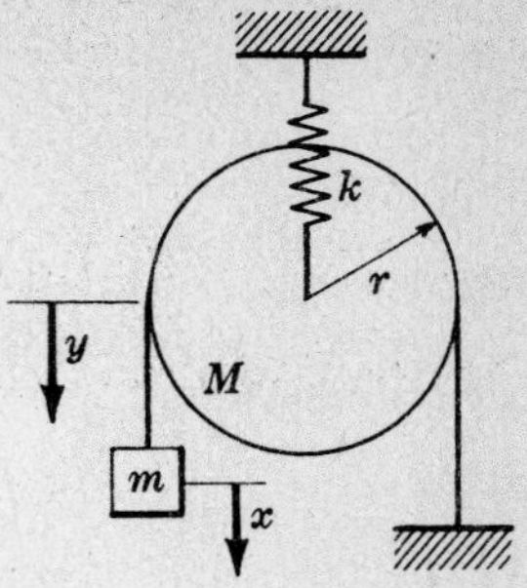

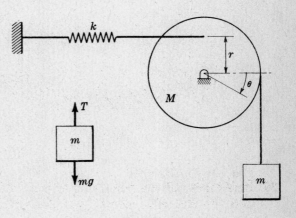

Fig. 1-29

24. The mass m is hanging from a cord attached to the circular homogeneous disc of mass M and radius R ft, as shown in Fig. 1-30. The disc is restrained from rotating by a spring attached at radius r ft from the center. If the mass is displaced downward from the rest position, determine the frequency of oscillation.

Applying Newton's second law of motion to the mass,

$$\Sigma F = ma \quad \text{or} \quad T - mg = -ma$$

and tension $T = m(g - a)$.

Let α be the angular acceleration of the disc; then $a = R\alpha$ and

$$T = m(g - R\alpha) = m(g - R\ddot{\theta})$$

When the mass is at rest, the torque due to the weight of the mass is balanced by the moment due to the restoring force in the spring, i.e. $mgR = T_0 r$ where T_0 is the initial tension in the spring; then $T_0 = mgR/r$.

Additional torque due to further stretching of the spring is $kr(r\theta)$, where k is the modulus of the spring. Also, the total torque due to the weight of the mass is $mR(g - R\ddot{\theta})$.

Fig. 1-30

Using the torque equation, $\Sigma M = J_0\alpha$,

$$J_0\ddot{\theta} = -(kr^2\theta + mgR) + (mgR - mR^2\ddot{\theta})$$

or

$$(J_0 + mR^2)\ddot{\theta} + kr^2\theta = 0$$

which is the differential equation of motion of a system having simple harmonic motion; therefore,

$$\omega_n = \sqrt{\frac{kr^2}{J_0 + mR^2}} = \sqrt{\frac{kr^2}{R^2(M/2 + m)}} \text{ rad/sec}$$

where $J_0 = \tfrac{1}{2}MR^2$ is the mass moment of inertia of the disc with respect to the central longitudinal axis.

25. The cylinder of mass m and radius r rolls without slipping on a circular surface of radius R as shown in Fig. 1-31 below. Determine the frequency of oscillation when the cylinder is displaced slightly from its equilibrium position. Use the Energy method.

The total energy of the system consists of kinetic energy (rotational and translational) and potential energy; and should remain the same all the time.

Translational K.E. $= \frac{1}{2}m\,[(R-r)\dot{\theta}]^2$.   Rotational K.E. $= \frac{1}{2}J_0\,(\dot{\phi}-\dot{\theta})^2$, where the moment of inertia of the cylinder is $J_0 = \frac{1}{2}mr^2$. Also, arc length $AB = R\theta = r\phi$; hence $\phi = R\theta/r$. Thus, for the system at any time,

$$\text{K.E.} = \tfrac{1}{2}m[(R-r)\dot{\theta}]^2 + \tfrac{1}{2}(\tfrac{1}{2}mr^2)(R/r-1)^2\,\dot{\theta}^2$$
$$\text{P.E.} = mg(R-r)(1-\cos\theta)$$

and

$$\frac{d}{dt}(\text{K.E.} + \text{P.E.}) = 0 \quad\text{or}\quad \tfrac{3}{2}(R-r)^2m\,\ddot{\theta} + mg(R-r)\sin\theta = 0$$

For small angles of oscillation, $\sin\theta \doteq \theta$ and the equation of motion becomes

$$\ddot{\theta} + \frac{2g}{3(R-r)}\theta = 0, \quad\text{and so}\quad \omega_n = \sqrt{\frac{2g}{3(R-r)}}\ \text{rad/sec}$$

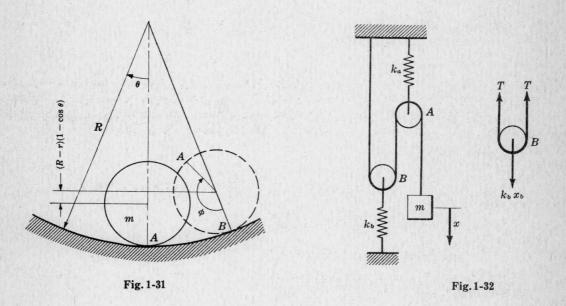

Fig. 1-31   Fig. 1-32

26. If the mass of the pulleys shown in Fig. 1-32 above is small and the cord is inextensible, find the natural frequency of the system.

Let T be the tension in the cord, and x_a and x_b the displacements of pulleys A and B respectively. Then, if pulley B is fixed, $x = 2x_a$; if instead pulley A is fixed, $x = 2x_b$. But since neither pulley A nor pulley B is fixed,

$$x = 2x_a + 2x_b$$

Consider pulley B as a free-body, and sum forces:

$$2T - k_b x_b = 0 \quad\text{or}\quad 2T = k_b x_b$$

Similarly for pulley A:

$$2T = k_a x_a \quad\text{or}\quad 2T = k_a x_a = k_b x_b$$

Then

$$x = 2x_a + 2x_b = 4T(1/k_a + 1/k_b) = T/k_{eq}$$

where $k_{eq} = \dfrac{1}{4(1/k_a + 1/k_b)}$ is the equivalent spring constant. This has reduced the problem to a simple spring-mass system with spring constant equal to k_{eq}.

But $m\ddot{x} + kx = 0$ is the equation of motion for a simple spring-mass system with natural frequency $\sqrt{k/m}$; hence

$$\omega_n = \sqrt{k_{eq}/m} = \sqrt{\frac{k_a k_b}{4m(k_a + k_b)}}\ \text{rad/sec}$$

27. The mass m is attached to one end of a weightless stiff rod which is rigidly connected to the center of a homogeneous cylinder of radius r as shown in Fig. 1-33. If the cylinder rolls without slipping, what is the natural frequency of oscillation of the system?

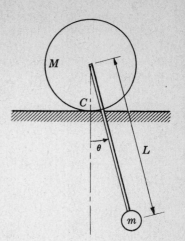

By Newton's method:

    At any instant under consideration, the masses M and m are rotating around point C, the instantaneous center of zero velocity. Hence the equation of motion takes the form $\Sigma M_C = J_C\ddot{\theta}$, where J_C is the moment of inertia of masses M and m with respect to point C. Then

$$J_C = (\tfrac{1}{2}Mr^2 + Mr^2) + mR^2$$

where $R^2 = L^2 + r^2 - 2rL\cos\theta$. For small oscillations, we have $\sin\theta \doteq \theta$, $\cos\theta \doteq 1$, and

$$J_C = 3Mr^2/2 + m(L-r)^2$$

Thus the equation of motion becomes

$$[3Mr^2/2 + m(L-r)^2]\,\ddot{\theta} = -mgL\sin\theta = -mgL\theta$$

or $\quad \ddot{\theta} + \dfrac{mgL}{3Mr^2/2 + m(L-r)^2}\,\theta = 0, \quad$ and so $\quad \omega_n = \sqrt{\dfrac{mgL}{3Mr^2/2 + m(L-r)^2}}$ rad/sec

Fig. 1-33

By Rayleigh's method:

$$\text{(K.E.)}_{max} = \text{(P.E.)}_{max} \quad \text{or} \quad \tfrac{1}{2}J_C(\dot{\theta})^2_{max} = mg(\Delta h)_{max}$$

Assume simple harmonic motion; then $\theta = A\sin\omega t$ and

$$(\dot{\theta})_{max} = (\omega A\cos\omega t)_{max} = \omega A$$
$$(\Delta h) = L - L\cos\theta = L(1-\cos\theta) = L\left[1 - \left(1 - \frac{\theta^2}{2!} + \frac{\theta^4}{4!} - \cdots\right)\right]$$
$$(\Delta h)_{max} = L\left[1 - \left(1 - \frac{\theta^2_{max}}{2!} + \frac{\theta^4_{max}}{4!}\right)\right] = \tfrac{1}{2}L\,\theta^2_{max} = \tfrac{1}{2}LA^2$$

Equating the two energy expressions yields

$$\tfrac{1}{2}[3Mr^2/2 + m(L-r)^2]\omega^2A^2 = mg(\tfrac{1}{2}LA^2), \quad \text{and} \quad \omega_n = \sqrt{\dfrac{mgL}{3Mr^2/2 + m(L-r)^2}} \text{ rad/sec}$$

FORCED VIBRATION WITH DAMPING

28. A generalized single-degree-of-freedom system having forced vibration with damping is shown in Fig. 1-34 below. Investigate its general motion.

    Employing Newton's law of motion,

$$\Sigma m\ddot{x} = \text{sum of forces in the } x\text{-direction}$$
$$= -k(x + \delta_{st}) + mg - c\dot{x} + F_0\sin\omega t$$

But $k\delta_{st} = mg$, the weight of the mass; hence the equation of motion takes its most general form

$$m\ddot{x} + c\dot{x} + kx = F_0\sin\omega t$$

The general solution for this second order differential equation with constant coefficients is

$$x = x_c + x_p$$

where x_c is called the complementary solution, or the solution for part of the equation, $m\ddot{x} + c\dot{x} + kx = 0$. x_p is the particular solution for the given equation.

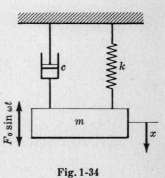

Fig. 1-34

The usual form for the complementary solution is

$$x_c = A e^{r_1 t} + B e^{r_2 t}$$

where A and B are arbitrary constants determined by the initial conditions imposed on the system, and r_1 and r_2 are the roots of the auxiliary equation $mr^2 + cr + k = 0$. Then

$$r_1 = \frac{-c + \sqrt{c^2 - 4mk}}{2m} \quad \text{and} \quad r_2 = \frac{-c - \sqrt{c^2 - 4mk}}{2m}$$

If $\omega_n^2 = k/m$ and $\zeta = c/2m\omega_n$, where ζ is called the damping factor, then

$$r_1 = \omega_n(-\zeta + \sqrt{\zeta^2 - 1}), \qquad r_2 = \omega_n(-\zeta - \sqrt{\zeta^2 - 1})$$

These values of r may be real and distinct, real and equal, or complex conjugates depending on the magnitude of ζ, i.e. whether it is greater than, equal to, or less than unity.

If ζ is greater than unity, the values of r are real and negative. Therefore no oscillatory motion is possible from the complementary solution of the equation of motion regardless of the initial conditions imposed on the system. This is overdamped, where

$$x_c = A e^{-r_1 t} + B e^{-r_2 t}$$

If ζ is equal to unity, the values of r are equal to $-\omega_n$. The motion is again not oscillatory, and its amplitude will eventually diminish to zero. This is critically-damped, where

$$x_c = (C + Dt) e^{-\omega_n t}$$

If ζ is less than unity, the values of r are complex conjugates. They then are

$$r_1 = \omega_n(-\zeta + i\sqrt{1 - \zeta^2}), \qquad r_2 = \omega_n(-\zeta - i\sqrt{1 - \zeta^2})$$

And if the damped natural angular frequency $\omega_d = \sqrt{1 - \zeta^2}\,\omega_n$, the equation finally takes the following form for the underdamped case

$$x_c = e^{-\zeta \omega_n t}(A \cos \omega_d t + B \sin \omega_d t)$$

or

$$x_c = C e^{-\zeta \omega_n t} \sin(\omega_d t + \phi)$$

where $C = \sqrt{A^2 + B^2}$, and $\phi = \tan^{-1}(A/B)$. This is a harmonic motion of angular frequency ω_d, the amplitude $C e^{-\zeta \omega_n t}$ of which decreases exponentially with time.

It is clear that no matter which one of the three values ζ takes, the complementary solution $x_c(t)$, usually called the transient, will eventually die out. Also, the rate of decay and natural frequency of the system depend on the system parameters only, while the amplitude of vibration and phase angle are determined by the initial conditions.

The particular solution x_p has the form $(A \sin \omega t + B \cos \omega t)$; therefore,

$$x_p = \frac{F_0}{(k - m\omega^2)^2 + (c\omega)^2} [(k - m\omega^2) \sin \omega t - c\omega \cos \omega t]$$

or

$$x_p = \frac{F_0}{\sqrt{(k - m\omega^2)^2 + (c\omega)^2}} \sin(\omega t - \psi)$$

where ω = angular frequency of the excitation, $\psi = \tan^{-1}\dfrac{c\omega}{k - m\omega^2} = \tan^{-1}\dfrac{2\zeta(\omega/\omega_n)}{1 - (\omega/\omega_n)^2}$.

Hence it can be concluded that the particular solution is of the same frequency as that of the excitation and is the steady state vibration of the system. The amplitude of the steady state vibration depends on the amplitude and frequency of the excitation. At resonance, i.e. when the forcing frequency is equal to the natural frequency or $\omega/\omega_n = 1$, the amplitude of vibration is limited only by the damping factor ζ. The steady state vibration is not in phase with the excitation; its variation in phase angle with excitation frequency is due to the presence of damping in the system.

The general solution is therefore

$$x = D e^{-\zeta \omega_n t} \sin(\omega_d t + \phi) + E \sin(\omega t - \psi)$$

where the first part of the solution is the transient, and the second part is the steady state response.

The detailed analysis of this generalized single-degree-of-freedom system is very useful. It brings out the concept of natural frequency, the role of damping on the oscillatory motion of the system, and the response to excitation. The results and conclusions from the model study can be advantageously applied to physical problems of this class of single-degree-of-freedom.

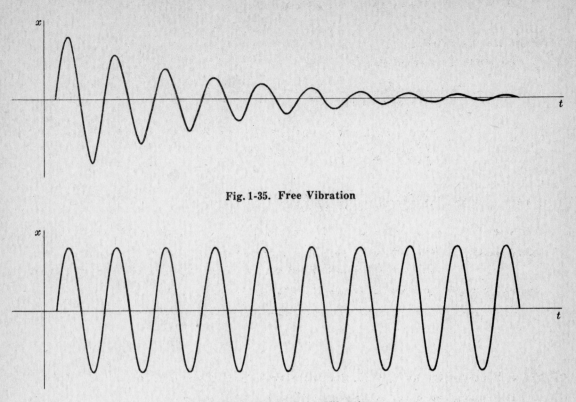

Fig. 1-35. Free Vibration

Fig. 1-36. Steady State Vibration

29. The mass shown in Fig. 1-37 below is initially at rest when a velocity of 4 in/sec is given to it. Find the subsequent displacement and velocity of the mass. Given: $c = 0.85$ lb-sec/in, $k = 25$ lb/in, and $W = 40$ lb.

As in Problem 28, the equation of motion is

$$m\ddot{x} + c\dot{x} + 2kx = 0$$

and

$$x = e^{-\zeta\omega_n t}(A\cos\omega_d t + B\sin\omega_d t)$$

where $\omega_n = \sqrt{k/m} = \sqrt{50/m} = 22$ rad/sec, $m = 40/(32.2 \times 12)$

$$\zeta = \frac{c}{2m\omega_n} = \frac{0.85}{2[40/(32.2 \times 12)]22} \doteq 0.181$$

$$\omega_d = \sqrt{1 - \zeta^2}\,\omega_n = 22\sqrt{1 - 0.181^2} = 21.6 \text{ rad/sec.}$$

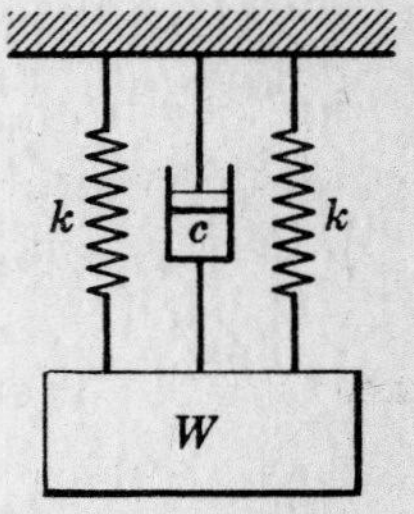

Substituting these values into the equation gives

$$x = e^{-3.98t}(A\cos 21.6t + B\sin 21.6t)$$

and $\dot{x} = -3.98\,e^{-3.98t}(A\cos 21.6t + B\sin 21.6t) + 21.6\,e^{-3.98t}(-A\sin 21.6t + B\cos 21.6t)$

At $t = 0$, $x = 0$: then $A = 0$. At $t = 0$, $\dot{x} = 4$; then $B = 4/21.6 = 0.185$. Hence

$$x = 0.185\,e^{-3.98t}\sin 21.6t$$

$$\dot{x} = e^{-3.98t}(4\cos 21.6t - 0.737\sin 21.6t)$$

or

$$\dot{x} = 4.08\,e^{-3.98t}\cos(21.6t + 9.5°)$$

Fig. 1-37

30. Calculate the transient and the steady state response of the mass in Problem 29 if an excitation force $F_0\sin\omega t = 10\sin 15t$ is acting on the mass.

As is given in Problem 29, the transient is

$$x_c = e^{-3.98t} \sin(21.6t + \phi)$$

where $\phi = \tan^{-1} A/B$.

The steady state response is

$$x_p = \frac{F_0}{\sqrt{(k - m\omega^2)^2 + (c\omega)^2}} \sin(\omega t - \psi)$$

$$= \frac{10}{\sqrt{\left(50 - \frac{40 \times 15^2}{32.2 \times 12}\right)^2 + (0.85 \times 15)^2}} \sin(15t - \psi) = 0.337 \sin(15t - 28°)$$

where $\psi = \tan^{-1} \dfrac{2\zeta\omega/\omega_n}{1 - (\omega/\omega_n)^2} = \dfrac{2(0.18)(15/22)}{1 - (15/22)^2} = 28°$.

Now $x = x_c + x_p = E e^{-3.98t} \sin(21.6t + \phi) + 0.337 \sin(15t - 28°)$

Initial conditions: At $t = 0$, $x = 0$ and $E \sin \phi + 0.337 \sin(-28°) = 0$.

At $t = 0$, $\dot{x} = 4$ and $E[-3.98 \sin \phi + 21.6 \cos \phi] + (0.337)(15) \cos(-28°) = 4$.

Solving these two equations, we obtain $E = 0.176$ and $\phi = 65°$. Hence

$$x = 0.176 e^{-3.98t} \sin(21.6t + 65°) + 0.345 \sin(15t - 28°)$$

31. Solve Problem 28 by the Mechanical Impedance method.

The equation of motion is given as $m\ddot{x} + c\dot{x} + kx = F_0 \sin \omega t$.

Since $\mathbf{F} = F_0 e^{i\omega t} = F_0 (\cos \omega t + i \sin \omega t)$, if $\mathbf{F}$ is used to represent $F_0 \sin \omega t$, the answer must be $\text{Im}(F_0 e^{i\omega t})$, i.e. the imaginary part of the solution.

Let displacement vector $= \mathbf{X} = X e^{i(\omega t - \psi)}$, velocity vector $= i\omega\mathbf{X}$, acceleration vector $= -\omega^2\mathbf{X}$. Substituting these values into the equation of motion gives

$$(k - m\omega^2 + ic\omega)X e^{-i\psi} = F_0 \qquad \text{or} \qquad X e^{-i\psi} = \frac{F_0}{(k - m\omega^2 + ic\omega)}$$

But $X e^{-i\psi} = X(\cos \psi - i \sin \psi)$, where $\psi = \tan^{-1} \omega c/(k - m\omega^2)$. Hence

$$X = \frac{F_0}{\sqrt{(k - m\omega^2)^2 + (c\omega)^2}}$$

and $$x_p = \text{Im}(\mathbf{X}) = \frac{F_0}{\sqrt{(k - m\omega^2)^2 + (c\omega)^2}} \sin(\omega t - \psi)$$

32. Find the steady state vibration of Problem 30 by the Mechanical Impedance method.

As in Problem 31, substituting the corresponding values into the equation of motion yields

$$[(k - m\omega^2) + icw]X e^{i(\omega t - \psi)} = F_0 e^{i\omega t}$$

$$[50 - (40/386)(15)^2 + i(0.85)(15)]X e^{-i\psi} = 10$$

$$(26.7 + i12.8)X^{-i\psi} = 10, \quad X e^{-i\psi} = \frac{10}{26.7 + i12.8}$$

and $X = 0.345$ with $\psi = \tan^{-1}(12.8/26.7) = 28°$. Then

$$\mathbf{X} = X e^{i(\omega t - \psi)} = 0.345 e^{i(15t - 28°)}$$

and $x = \text{Im}(\mathbf{X}) = X \sin(\omega t - \psi) = 0.345 \sin(15t - 28°)$

ROTATING UNBALANCE

33. A machine having a rotating part is simplified to having a mass M of its own and an unbalance me as shown in Fig. 1-38 below. Determine the steady state response of the machine.

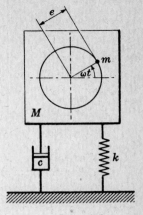

Assume the machine is constrained to move in the vertical direction only. The centrifugal force due to unbalance in the rotating part of the machine is $me\omega^2$. Since the vertical component of the centrifugal force affects the motion of the machine, the equivalent excitation is $me\omega^2 \sin \omega t$, where ω is the angular velocity of the rotating part of the machine. Hence the differential equation of motion is

$$M\ddot{x} + c\dot{x} + kx = me\omega^2 \sin \omega t$$

which is damped forced vibration. Solving this differential equation as before,

$$x = \frac{(me\omega^2)/k}{\sqrt{[1 - (\omega/\omega_n)^2]^2 + [2\zeta(\omega/\omega_n)]^2}} \sin(\omega t - \psi)$$

where $\omega_n = \sqrt{k/M}$ and $\psi = \tan^{-1} \dfrac{2\zeta\omega/\omega_n}{1 - (\omega/\omega_n)^2}$

Thus $me\omega^2$ may be regarded as the amplitude F_0 of the harmonic excitation.

Fig. 1-38

34. The rotor of mass m is mounted on an elastic shaft whose mass is negligible compared to that of the rotor. The rotor has an eccentricity of e from the center. Find the critical speed of the shaft if the natural frequency of the rotor is $\sqrt{k/m}$, where k is the equivalent spring stiffness of the shaft.

Let R be the center of rotation, O the geometric center of the rotor, and G the center of mass of the rotor as shown in Fig. 1-39.

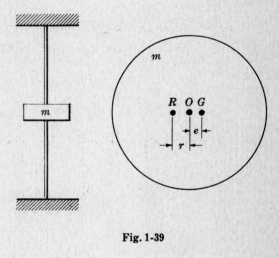

At any instant during rotation, the restoring force due to the elastic shaft is kr, while the centrifugal force due to the unbalance is $m(e + r)\omega^2$. These two forces must equal each other, i.e.,

$$kr = m(e + r)\omega^2 \quad \text{or} \quad k/m = \omega^2(e + r)/r$$

But $k/m = \omega_n^2$; therefore the equation becomes

$$\omega_n^2 r \left(1 - \frac{\omega^2}{\omega_n^2}\right) = e\omega^2 \quad \text{or} \quad r = \frac{e(\omega^2/\omega_n^2)}{1 - (\omega^2/\omega_n^2)}$$

If $\omega^2 = \omega_n^2$, the value of r will become very large, i.e. the shaft is unstable. Thus the critical speed of the shaft is $\omega = \sqrt{k/m} = \omega_n$; at this special speed of rotation resonance will occur.

Fig. 1-39

35. The armature of an electric motor weighs 20 lb and its center of gravity is 0.01 in. off center from the bearing axis. The motor, weighing a total of 60 lb, is resting on four springs of stiffness of 150 lb/in each. Find the critical speed of the motor, and the vertical amplitude of vibration of the motor when running at three times this critical speed.

Total spring constant $= 4(150) = 600$ lb/in.

Natural frequency of motor is $\omega_n = \sqrt{k/m} = \sqrt{600(32.2)(12)/60} = 62.4$ rad/sec or 600 rpm. Hence the critical speed of motor is at 600 rpm (see Problem 34). At three times this critical speed, the force due to rotating unbalance is $F = me\omega^2 = (20/32.2)(0.01/12)(62.4)^2 = 18.2$ lb.

As discussed in Problem 33, the steady state response without damping is

$$x_p = \frac{F/k}{r^2 - 1} = \frac{me\omega^2/k}{(\omega/\omega_n)^2 - 1} = \frac{18.2/600}{9 - 1} = 0.03 \text{ in.}$$

36. The amplitude of vibration of the system shown in Fig. 1-40 is observed to decrease to 25% of the initial value after five consecutive cycles of motion as shown in Fig. 1-41 below. Determine the damping coefficient c of the system if $k = 20$ lb/in and $m = 10$ lb.

For damped free vibration, as discussed in Problem 28,

$$x_c = C e^{-\zeta \omega_n t} \sin(\omega_d t + \phi)$$

The maximum amplitude in a cycle occurs when $\sin(\omega_d t + \phi)$ is equal to unity. Therefore the maximum amplitudes are

$$x_1 = C e^{-\zeta \omega_n t_1} \quad \text{and} \quad x_2 = C e^{-\zeta \omega_n t_2}$$

The ratio

$$x_1/x_2 = e^{\zeta \omega_n (t_2 - t_1)} = e^{\zeta \omega_n (2\pi/\omega_d)} = e^{2\pi\zeta/\sqrt{1-\zeta^2}}$$

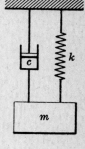

since $(t_2 - t_1)$, the time difference between two consecutive amplitudes, is also the period of oscillation, and $\omega_d = \sqrt{1 - \zeta^2}\,\omega_n$. The logarithm of this ratio x_1/x_2 is $\ln(x_1/x_2) = 2\pi\zeta/\sqrt{1-\zeta^2} = \delta$, where δ is called the *logarithmic decrement*. Since the damping coefficient $c = 2\zeta\sqrt{km}$, by knowing the ratio of two consecutive amplitudes or the logarithmic decrement δ, the damping factor ζ is known and hence c.

Fig. 1-40

In this problem, $x_1/x_6 = 1/0.25$. But $x_1/x_6 = (x_1/x_2)(x_2/x_3)(x_3/x_4)(x_4/x_5)(x_5/x_6)$, and $\ln(x_1/x_2) = \delta$. Taking the logarithm of both sides of the equation gives

$$\ln 4 = \ln(x_1/x_2) + \ln(x_2/x_3) + \ln(x_3/x_4) + \ln(x_4/x_5) + \ln(x_5/x_6)$$

or $\ln 4 = 5\delta$, $\delta = 0.28$.

From $\delta = 0.28 = 2\pi\zeta/\sqrt{1-\zeta^2}$, $\zeta = 0.044$. Then

$$c = 2\zeta\sqrt{km} = 2(0.044)\sqrt{(20/12)(10/32.2)} = 0.063 \text{ lb-sec/in.}$$

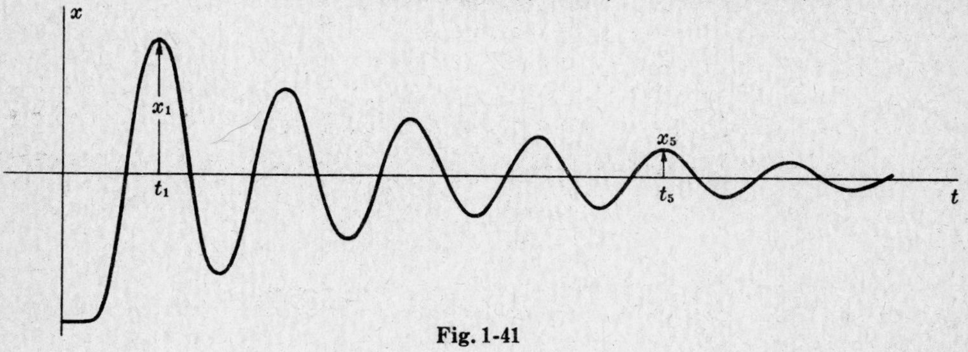

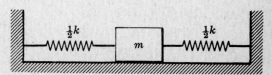

Fig. 1-41

37. The coefficient of friction between the dry surfaces of the block and the plane as shown in Fig. 1-42 below is f, a constant value. The constant friction force is always acting against motion to produce what is known as *Coulomb damping*. Investigate the motion of the block if it is given a displacement x_0 from the center position where the springs are unstressed.

Let the block be displaced to the right and released; the forces acting are the spring force kx and the friction force fmg. Then $\Sigma F = ma$ gives

$$-kx + fmg = m\ddot{x} \quad \text{or} \quad \ddot{x} + (k/m)x = fg$$

with solution

$$x = A \cos \omega_n t + B \sin \omega_n t + fmg/k$$

Fig. 1-42

From initial conditions: $t = 0$, $x = x_0$, $x_0 = A + fmg/k$, $A = x_0 - fmg/k$.

$t = 0$, $\dot{x} = 0$, $\omega_n B = 0$.

Since ω_n is not always zero, $B = 0$. Hence the solution takes the form

$$x = (x_0 - fmg/k) \cos \omega_n t + fmg/k$$

At $t = \pi/\omega_n$, i.e. at the end of half a cycle, at the extreme left position,

$$x = -(x_0 - fmg/k) + fmg/k = -x_0 + 2fmg/k$$

Due to the presence of damping in the system, however, the amplitude of motion is continuously diminishing. Therefore, at the extreme left position at the end of the first half cycle, the distance of the block from the central unstressed position is $x_0 - 2(fmg/k)$.

From the symmetry of the problem, applying the same reasoning for motion of the block to the right with an initial displacement of $x_0 - 2(fmg/k)$ from the center to the left, it can be concluded that the block will reach the extreme right position. The distance from this position to the center unstressed position will be $x_0 - 4(fmg/k)$.

Thus during each half cycle the amplitude of vibration is diminished by $2(fmg/k)$. Finally, the block will come to rest at one of its extreme positions. This will happen as soon as the amplitude becomes less than fmg/k, because here the friction force is large enough to balance the force exerted by the springs. It is true, therefore, that the motion of the block is not simple harmonic; instead, the shape of the displacement-time curve changes at each half cycle.

In order to determine the damped natural frequency of the block, the equation of motion is written as

$$m\ddot{x} + k(x - fmg/k) = 0$$

Let $x' = x - fmg/k$; then since fmg/k is a constant, $\ddot{x}' = \ddot{x}$ and the equation of motion becomes

$$m\ddot{x}' + kx' = 0, \quad \text{from which} \quad \omega_n = \sqrt{k/m} \text{ rad/sec}$$

It is clear, therefore, that the damped natural frequency of the block is equal to the undamped natural frequency. In short, the frequency of vibration of a system is not affected by a constant damping.

38. Two harmonic motions of the same amplitude but of slightly different frequencies are imposed on a vibrating body. Analyze the motion of the body.

Let $x_1 = A \cos \omega t$, $x_2 = A \cos (\omega + \Delta\omega)t$. The motion of the body, then, is the superposition of the two vibrations:

$$x = x_1 + x_2 = A \cos \omega t + A \cos (\omega + \Delta\omega)t$$
$$= A[\cos \omega t + \cos (\omega + \Delta\omega)t]$$

From trigonometry, $\cos x + \cos y = 2 \cos \frac{1}{2}(x + y) \cdot \cos \frac{1}{2}(x - y)$. Then

$$x = A[2 \cos \frac{1}{2}(\omega t + \omega t + \Delta\omega\, t) \cdot \cos (\Delta\omega/2)t]$$
$$= [2A \cos (\Delta\omega/2)t] \cos (\omega + \Delta\omega/2)t$$

The amplitude of x is seen to fluctuate between zero and $2A$ according to the $2A \cos (\Delta\omega/2)t$ term, while the general motion of x is a cosine wave of angular frequency equal to $(\omega + \Delta\omega/2)$. This special pattern of motion is known as the *beating phenomenon*. Whenever the amplitude reaches a maximum, there is said to be a *beat*. The beat frequency as determined by two consecutive maximum amplitude is equal to

$$f_b = \frac{\Delta\omega + \omega}{2\pi} - \frac{\omega}{2\pi} = \frac{\Delta\omega}{2\pi} \text{ cycles/sec}$$

and the period $T = 1/f_b = 2\pi/\Delta\omega$ sec.

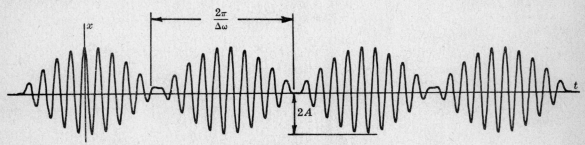

Fig. 1-43. The Beating Phenomenon

39. A periodic excitation as shown in Fig. 1-44(b) is applied to the base of the spring-dashpot-mass system. Determine the resulting motion of the mass m if $k = 40$ lb/in, $m = 10$ lb-sec^2/in, and $c = 20$ lb-sec/in.

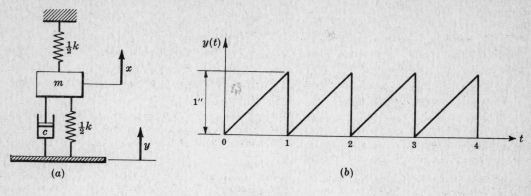

Fig. 1-44

General periodic motions can be represented by Fourier series as the sum of two harmonic functions.

$$F(t) = \sum_{n=0}^{\infty} (a_n \cos n\omega t + b_n \sin n\omega t)$$

or in an expanded form,

$$F(t) = a_0 + a_1 \cos \omega t + a_2 \cos 2\omega t + \cdots + b_1 \sin \omega t + b_2 \sin 2\omega t + \cdots$$

where $a_0 = (1/\tau) \int_0^\tau F(t)\, dt$, $a_n = (2/\tau) \int_0^\tau F(t) \cos n\omega t\, dt$, $b_n = (2/\tau) \int_0^\tau F(t) \sin n\omega t\, dt$, and $\tau = 2\pi/\omega$ is the period of $F(t)$.

For this given saw-tooth periodic motion, $F(t) = t$ and $\tau = 1$. Then $y(t) = F(t) = t$ and

$$a_0 = (1/\tau) \int_0^\tau F(t)\, dt = \int_0^1 t\, dt = \tfrac{1}{2}$$

$$a_n = (2/\tau) \int_0^\tau F(t) \cos n\omega t\, dt = 2 \int_0^1 t \cos n\omega t\, dt = 0$$

$$b_n = (2/\tau) \int_0^\tau F(t) \sin n\omega t\, dt = 2 \int_0^1 t \sin n\omega t\, dt = -1/n\pi$$

Thus the Fourier series expansion of $y(t)$ is

$$y(t) = \tfrac{1}{2} - (1/\pi) \sum_{n=1}^{\infty} (1/n) \sin n\omega t$$

Applying $\Sigma F = ma$ to this system gives

$$m\ddot{x} = -c(\dot{x} - \dot{y}) - k(x - y) \quad \text{or} \quad m\ddot{x} + c\dot{x} + kx = c\dot{y} + ky$$

where $\dot{y} = -\sum_{n=1}^{\infty} (\omega/\pi) \cos n\omega t$. Hence the equation of motion becomes

$$m\ddot{x} + c\dot{x} + kx = c\left[-\sum_{n=1}^{\infty} (\omega/\pi) \cos n\omega t \right] + k\left[\tfrac{1}{2} - (1/\pi) \sum_{n=1}^{\infty} (1/n) \sin n\omega t \right]$$

$$= k/2 - \sum_{n=1}^{\infty} \left[\frac{k}{\pi n} \sin n\omega t + \frac{c\omega}{\pi} \cos n\omega t \right]$$

But $A \sin n\omega t + B \cos n\omega t = \sqrt{A^2 + B^2} \sin(n\omega t + \phi)$ where $\phi = \tan^{-1}(B/A)$. Then

$$c\dot{y} + ky = k/2 - (1/\pi) \sum_{n=1}^{\infty} (1/n) \sqrt{k^2 + c^2 n^2 \omega^2} \sin(n\omega t + \phi)$$

where $\phi = \tan^{-1}(c\omega/k)$; and since $\omega_n^2 = k/m$, $\zeta = c/2m\omega_n$,

$$c\dot{y} + ky = k/2 - (1/\pi) \sum_{n=1}^{\infty} (k/n) \sqrt{1 + (2rn\zeta)^2} \sin(n\omega t + \phi)$$

This is the impressed force, and the steady state response, according to earlier discussions, is

$$x_p = \frac{1}{2} - (1/\pi) \sum_{n=1}^{\infty} (1/n^2) \sqrt{\frac{1 + (2rn\zeta)^2}{(1 - n^2r^2)^2 + (2rn\zeta)^2}} \sin(n\omega t - \psi)$$

where $\psi = \tan^{-1}\left(\frac{2nr\zeta}{1 - n^2r^2} - \phi\right)$.

For this particular case, $\omega_n = \sqrt{k/m} = \sqrt{40/10} = 2$ rad/sec, $r = \omega/\omega_n = 1/2 = 0.5$, $\zeta = c/2m\omega_n = 20/(2 \times 10 \times 2) = 0.5$. Substituting these values into the equation of motion and simplifying yields the steady state response of the mass m:

$$x_p = 0.5 - 0.396 \sin(6.24t - 7.1°) - 0.225 \sin(12.48t - 26.5°) - 0.096 \sin(18.9t - 74.3°) \ldots$$

TRANSMISSIBILITY

40. A centrifugal fan weighs 100 lb and has a rotating unbalance of 20 lb-in. When dampers having damping factor $\zeta = 0.2$ are used, specify the springs for mounting such that only 10% of the unbalance force is transmitted to the floor. Also, determine the magnitude of the transmitted force. The fan is running at constant speed of 1000 rpm.

The total force transmitted is the sum of the reactions at the fixed ends of the spring and dashpot.

$$F_t = kx + c\dot{x}$$

Under steady state vibration condition as discussed earlier, the amplitude of vibration is

$$x_p = \frac{F_0/k}{\sqrt{(1 - r^2)^2 + (2\zeta r)^2}} \sin(\omega t - \psi)$$

For convenience, let $x_p = A \sin(\omega t - \psi)$; then $\dot{x}_p = A\omega \cos(\omega t - \psi)$, and

$$F_t = kA \sin(\omega t - \psi) + cA\omega \cos(\omega t - \psi)$$

But spring force is a maximum when velocity is equal to zero (or displacement is a maximum), while damping force is a maximum when the displacement is equal to zero (or velocity is a maximum). Since the spring force is at right angle to the damping force, the resultant maximum transmitted force is $A\sqrt{k^2 + (c\omega)^2}$. Then

$$F_t = A\sqrt{k^2 + (c\omega)^2} \cos(\omega t + \phi), \qquad \phi = \tan^{-1}(c\omega/k)$$

Transmissibility TR, which is the ratio of the maximum transmitted force to the maximum impressed force, takes the form

$$\text{TR} = \frac{A\sqrt{k^2 + (c\omega)^2}}{F_0} \cos(\omega t + \phi)$$

But $x = x_p$ for steady state vibration; therefore

$$\text{TR} = \frac{\sqrt{1 + (2\zeta r)^2}}{\sqrt{(1 - r^2)^2 + (2\zeta r)^2}} \qquad \text{or} \qquad 0.1 = \frac{\sqrt{1 + 0.16r^2}}{\sqrt{1 + r^4 - 2r^2 + 0.16r^2}}$$

where $r = \omega/\omega_n$, $\zeta = c/2m\omega_n$, and TR = 0.1. Then

$$\frac{1 + 0.16r^2}{r^4 - 1.84r^2 + 1} = 0.01 \qquad \text{or} \qquad r^4 - 1.84r^2 - 16r - 99 = 0$$

which gives $r = 3.7$. Then $\omega_n = \omega/r = 105/3.7 = 28.4$ rad/sec and

$$k = m\omega_n^2 = (100)(28.4)(28.4)/(32.2)(12) = 210 \text{ lb/in}$$

Amplitude of force transmitted: $F_t = (0.1)(me\omega^2) = \dfrac{(0.1)(20)(105)^2}{(32.2)(12)} = 57$ lb.

41. In order to cut down the vibratory motion transmitted to the instruments, the instrument boards are mounted on isolators as in aircrafts. If the isolator, having very little damping, deflects $\frac{1}{8}$ in. under a weight of 50 lb, find the percentage of motion transmitted to the instrument board if the vibration of the aircraft is at 2000 rpm.

$$TR = \frac{\sqrt{1 + (2\zeta r)^2}}{\sqrt{(1 - r^2)^2 + (2\zeta r)^2}}$$

where $r = \omega/\omega_n$ and $\zeta = c/2m\omega_n$. But since the damping present is small, $c = 0$ and $\zeta = 0$.

Using $\omega_n = \sqrt{k/m}$, where $k = F/\delta_{st}$, we have

$$\omega_n^2 = \frac{F}{m\delta_{st}} = \frac{50}{(50/386)(1/8)} = 3056 \ (\text{rad/sec})^2$$

or $f_n = 55.4/2\pi = 8.85$ cycles/sec or 530 cpm. Then $r = f/f_n = 2000/530 = 3.77$ and

$$TR = \frac{1}{r^2 - 1} = \frac{1}{14.2 - 1} = 0.076$$

Hence, by using isolators, only 7.6% of the vibratory motion of the aircraft is transmitted to the instrument board.

SEISMIC INSTRUMENTS

42. A generalized model for vibration measurement is shown in Fig. 1-45. The base is attached to the body having an unknown vibration $A \sin \omega t$. Investigate the motion of the system and its application in vibration instruments.

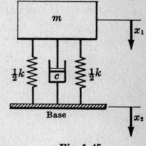

The forces acting on the mass are the spring force $= k(\frac{1}{2} + \frac{1}{2})(x_1 - x_2)$ and damping force $= c(\dot{x}_1 - \dot{x}_2)$, assuming x_1 is greater than x_2. Using $\Sigma F = ma$, the equation of motion is

$$-k(x_1 - x_2) - c(\dot{x}_1 - \dot{x}_2) = m\ddot{x}_1$$

Let the relative motion $x_1 - x_2 = x$; then

$$x_1 = x + x_2 \quad \text{and} \quad \dot{x}_1 = \dot{x} + \dot{x}_2$$

Fig. 1-45

and the equation of motion takes the form

$$m\left(\frac{d^2x}{dt^2} + \frac{d^2x_2}{dt^2}\right) + c\frac{dx}{dt} + kx = 0 \quad \text{or} \quad m\frac{d^2x}{dt^2} + c\frac{dx}{dt} + kx = -m\frac{d^2x_2}{dt^2}$$

But $x_2 = A \sin \omega t$ is the vibration of the body; substituting,

$$m\frac{d^2x}{dt^2} + c\frac{dx}{dt} + kx = mA\omega^2 \sin \omega t.$$

which is the standard differential equation of motion for forced vibration with F_0 replaced by $mA\omega^2$. The steady state response is, as before,

$$x_p = \frac{F_0}{\sqrt{(k - m\omega^2)^2 + (c\omega)^2}} \sin(\omega t - \phi), \qquad \phi = \tan^{-1}\frac{c\omega}{k - m\omega^2}.$$

Substituting $F_0 = \omega^2 mA$, $\omega_n^2 = k/m$, $\zeta = c/2m\omega_n$, and $r = \omega/\omega_n$, we obtain

$$x_p = \frac{Ar^2}{\sqrt{(1 - r^2)^2 + (2\zeta r)^2}} \sin(\omega t - \phi)$$

Here ω_n is the natural frequency of the vibration pickup while ω is the frequency of the unknown vibration.

If ω_n is small, (i.e. $\omega_n^2 = k/m$, by having small spring constant or by having a large mass), the ratio $r = \omega/\omega_n$ is quite large. The expression for x_p, after dividing out by r^2, is

$$x_p = \frac{A}{\sqrt{1/r^4 - (2 - 4\zeta)/r^2 + 1}} \sin(\omega t - \phi)$$

Since r is large, the denominator of the above expression is approximately equal to unity; then

$$x_p = A \quad \text{and} \quad \phi = \tan^{-1} 2r\zeta/(1 - r^2) = \tan^{-1}\frac{2\zeta/r}{(1/r^2) - 1}$$

Hence $\phi = \tan^{-1} 0 = 180°$, which means that the two values of motion are 180° out of phase from each other, or $x_p = -A$.

One type of seismic instrument, the vibrometer, makes use of this principle to measure the amplitude of vibration. The relative motion between the mass and the base is usually recorded by a pen pressing against a rotating drum. Because the natural frequency of the vibrometer is made to have a low value, the amplitude of vibration is seen equal to the relative motion recorded with a phase difference of $180°$.

Another type of seismic instrument, the accelerometer, which is used to measure acceleration, uses the same principle. Instead of using soft spring, very hard springs are used to give a high natural frequency. Consequently the ratio of frequency r will be very small. The expression for x_p becomes

$$x_p = \frac{Ar^2}{\sqrt{(1-r^2)^2 + (2r\zeta)^2}} = \frac{Ar^2}{\sqrt{(1-0)^2 + 0}} = Ar^2 = A\omega^2/\omega_n^2$$

But $A\omega^2$ is the amplitude of the acceleration of the body vibrating with $x_2 = A \sin \omega t$. Therefore the relative motion is a measure of the acceleration.

43. A simplified spring-mass vibration pickup as shown in Fig. 1-46 is used to measure the vertical acceleration of a train which has vertical frequency of 10 rad/sec. The mass weighs 3.86 lb and the modulus of the spring is 100 lb/in. The amplitude of the relative motion of the mass is 0.05 in. as recorded by the instrument. Find the maximum vertical acceleration of the train. What is the amplitude of vibration of the train?

From Problem 42, the amplitude of the relative motion of the mass is given by

$$x_p = \frac{Ar^2}{\sqrt{(1-r^2)^2 + (2r\zeta)^2}}$$

Since there is no damping in this system,

$$x_p = \frac{Ar^2}{1-r^2} = \frac{A\omega^2}{\omega_n^2 - \omega^2}$$

where $r = \omega/\omega_n$. Hence the maximum vertical acceleration of the train is

$$A\omega^2 = x_p(\omega_n^2 - \omega^2) = 0.05[(100)(386)/3.86 - 100] = 495 \text{ in/sec}^2$$

The amplitude of vibration is

$$A = x_p\frac{\omega_n^2 - \omega^2}{\omega^2} = 0.05\frac{(10,000 - 100)}{100} = 4.95 \text{ in.}$$

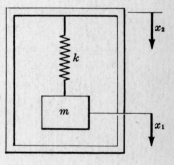

Fig. 1-46

44. A mass weighing 3.86 lb is suspended in a box by a vertical spring whose constant $k = 50$ lb/in as shown in Fig. 1-47. The box is placed on top of a shake table producing a vibration $x = 0.09 \sin 3t$. Find the absolute amplitude of the mass.

The amplitude of the relative motion of the mass is given by x_p as in Problem 42.

$$x_p = \frac{Ar^2}{\sqrt{(1-r^2)^2 + (2r\zeta)^2}}$$

As in most engineering problems, damping due to the presence of air is neglected; hence there is no effective damping in this case. Then

$$x_p = \frac{Ar^2}{1-r^2}$$

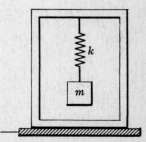

where $r = \omega/\omega_n$. But $\omega_n^2 = k/m = 50(32.2)(12)/3.86 = 500$ (rad/sec)^2, $\omega = 3$ rad/sec, and $\omega^2 = 9$ (rad/sec)^2; hence

Fig. 1-47

$$x_p = \frac{Ar^2}{1 - r^2} = \frac{A\omega^2}{\omega_n^2 - \omega^2} = \frac{0.09(9)}{500 - 9} = 0.0016 \text{ in.}$$

The absolute amplitude of the mass is $x_1 = x_p + A = 0.0016 + 0.09 = 0.0916$ in.

45. A vibrometer, whose damping is negligible, is employed to find the magnitude of vibration of a machine structure. It gives a reading of the relative displacement of 0.002 in. The natural frequency of the vibrometer is given as 300 cpm and the machine is running at 100 rpm. What will be the magnitude of displacement, velocity, and acceleration of the vibrating machine part?

As discussed earlier, the amplitude of the relative motion of the mass is given as

$$x_p = \frac{Ar^2}{\sqrt{(1 - r^2)^2 + (2r\zeta)^2}}$$

where A = amplitude, $r = \omega/\omega_n$, ζ = damping factor = $c/2m\omega_n$. Since there is very little damping, $c = 0$ and $\zeta = 0$; then

$$x_p = \frac{Ar^2}{1 - r^2}$$

where $r = \omega/\omega_n = 100/300 = 0.333$. Therefore the magnitude of the displacement is

$$A = x_p(1 - r^2)/r^2 = 0.002(1 - 0.333^2)/(1/9) = 0.016 \text{ in.}$$

the magnitude of the velocity is

$$A\omega = x_p(1 - r^2)\omega/r^2 = 0.016[100(2\pi/60)] = 0.17 \text{ in/sec}$$

and the magnitude of the acceleration is

$$A\omega^2 = x_p(1 - r^2)\omega_n^2 = 0.016\omega^2 = 0.016(108) = 1.74 \text{ in/sec}^2$$

Supplementary Problems

46. Show that the sum of two harmonic motions of the same frequency but with different phase angles is also a harmonic motion of the same frequency, i.e. $A \cos \omega t + B \cos (\omega t + \phi) = C \cos (\omega t + \psi)$.

47. Show that the equations of motion of the two simple spring-mass systems as shown in Fig. 1-48(a) and Fig. 1-48(b) are the same, and are equal to that of the system as shown in Fig. 1-48(c).

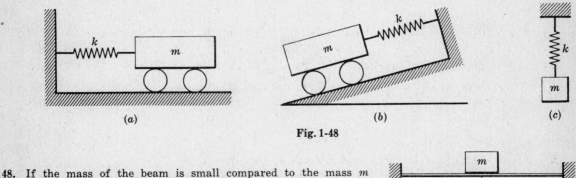

(a)   (b)   (c)

Fig. 1-48

48. If the mass of the beam is small compared to the mass m itself, derive an expression for the frequency of the mass m. See Fig. 1-49.   *Ans.* $\omega_n = \sqrt{192EI/mL^3}$ rad/sec

Fig. 1-49

49. Assuming that the pendulum in a clock follows simple pendulum theory, what will be its length if it has a period of 1 second on the clock?   *Ans.* $L = 9.83$ in.

50. A homogeneous square plate of side L ft and mass m is suspended from the midpoint of one of the sides as shown in Fig. 1-50 below. Find its frequency of vibration.
     Ans. $\omega_n = \sqrt{6g/5L}$ rad/sec

51. The rigid weightless rod is restrained to oscillate in a vertical plane as shown in Fig. 1-51 below. Determine the natural frequency of the mass m.   *Ans.* $\omega_n = \sqrt{k/9m}$ rad/sec

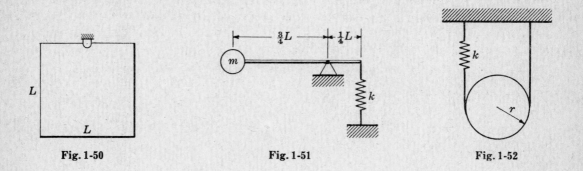

| Fig. 1-50 | Fig. 1-51 | Fig. 1-52 |

52. A homogeneous cylinder of mass m is suspended by a spring of constant k lb/in and an inextensible cord as shown in Fig. 1-52 above. Find the natural frequency of vibration of the cylinder.
     Ans. $\omega_n = \sqrt{8k/3m}$ rad/sec

53. Solve Problem 5 by the Energy method.

54. A homogeneous solid cylinder of mass m is linked by a spring of constant k lb/in and is resting on an inclined plane as shown in Fig. 1-53. If it rolls without slipping, show that the frequency of oscillation is $\sqrt{2k/3m}$ rad/sec.

55. Solve Problem 14 by the Energy method.

56. Solve Problem 20 by the Energy method.

57. Solve Problem 23 by Newton's second law of motion.

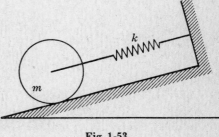

Fig. 1-53

58. A homogeneous sphere of radius r and mass m is free to roll without slipping on a spherical surface of radius R. If the motion of the sphere is restricted to a vertical plane as shown in Fig. 1-54 below, determine the natural frequency of oscillation of the sphere.   *Ans.* $\omega_n = \sqrt{5g/7(R-r)}$ rad/sec

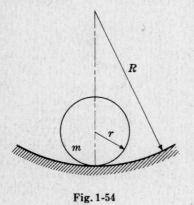

Fig. 1-54

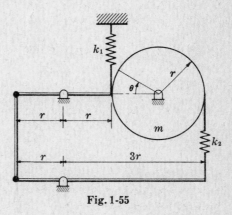

Fig. 1-55

59. For small angles of oscillation, find the frequency of oscillation of the system shown in Fig. 1-55 above.
     Ans. $\omega_n = \sqrt{(2k_1 + 8k_2)/m}$ rad/sec

60. Solve Problem 24 by the Energy method.

61. Use Newton's law of motion to solve Problem 25.

62. Solve Problem 26 by the Energy method.

63. The simple pendulum is pivoted at point O as shown in Fig. 1-56 below. If the mass of the rod is negligible, and for small oscillations, find the damped natural frequency of the pendulum.

Ans. $\omega_n = \sqrt{\dfrac{kL_1^2 + mgL}{mL^2} - \left[\dfrac{cL_1L_2^2}{2mL^2}\right]^2}$ rad/sec

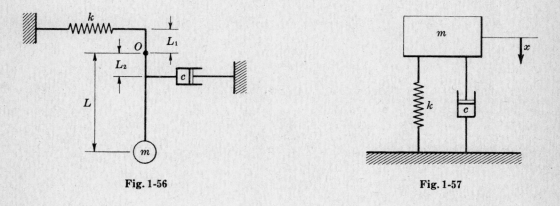

Fig. 1-56   **Fig. 1-57**

64. The 50 lb mass is resting on a spring of 25 lb/in and dashpot of 0.75 lb-sec/in as shown in Fig. 1-57 above. If a velocity of 4 in/sec is applied to the mass at the rest position, what will be its displacement at the end of the first second?   *Ans.* 0.0013 in.

65. Show that the mass of an overdamped system will never pass through the static equilibrium position
(a) if it is given an initial displacement only.
(b) if it is given an initial velocity only.

66. A simply-supported beam has a concentrated mass M acting on the midspan. Find the natural frequency of the system if the mass of the beam is m.

Ans. $\omega_n = \sqrt{\dfrac{48EI}{(M + 0.486m)L^3}}$ rad/sec

67. Determine the natural frequency of vibration of the mass M, attached to the end of a cantilever beam of length L and mass m, when the mass of the beam is not negligible.

Ans. $\omega_n = \sqrt{\dfrac{3EI}{L^3(M + 0.236m)}}$ rad/sec

68. Fig. 1-58 shows a rectangular block of mass m resting on top of a semi-cylindrical surface. If the block is slightly tipped at one end, find its frequency of oscillation.

Ans. $\omega_n = 3.47\sqrt{\dfrac{(r - d/2)g}{4d^2 + L^2}}$ rad/sec

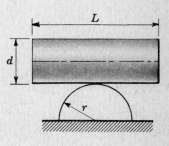

Fig. 1-58

69. The block of mass m is supported by a spring of constant k which is mounted on a weightless base having up and down harmonic motion $A_0 \sin \omega t$ as shown in Fig. 1-59 below. Determine the motion of the block.

Ans. $x = A \sin \omega_n t + B \cos \omega_n t + \dfrac{A_0 \sin \omega t}{1 - (\omega/\omega_n)^2}$

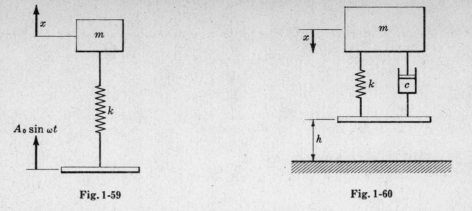

Fig. 1-59 Fig. 1-60

70. If the system shown in Fig. 1-60 above is dropped from an elevation h to a hard surface, what will be the resulting motion of the mass m?

Ans. $x = \dfrac{\sqrt{2gh}\, e^{-(c/2m)t}}{\omega_d} \sin \omega_d t$

71. A simple one-degree-of-freedom spring-mass system is having steady state forced vibration. Show that the principle of linear superposition holds, i.e. the solutions for components of the forcing function can be added together to form a solution for the complete forcing function.

72. What will be the steady state response of the mass in Fig. 1-61 below if the forcing function is

$$f(t) = 10 \sin 0.5t + 10 \cos 1.5t + 20 \sin t + 20 \cos 2t$$

and $k = 10$ lb/in, $m = 1$ lb-sec^2/in.

Ans. $x = 1.03 \sin 0.5t + 2.22 \sin t + 1.29 \cos 1.5t + 3.33 \cos 2t$

73. The static deflection of the spring due to the mass m in Fig. 1-61 below is 1.2 in, and the amplitude of vibration due to a harmonic excitation $10 \cos 20t$ is 0.02 in. What is the weight of the mass?
Ans. 15.12 lb

74. The piston shown in Fig. 1-62 below oscillates with a harmonic motion $x = A \cos \omega t$ in a cylinder of mass m supported by a spring of constant k. If there is viscous damping between the piston and the cylinder wall of magnitude c, find the amplitude of motion of the cylinder and its phase relationship with the piston.

Ans. $|x| = \dfrac{cA\omega}{\sqrt{(k - m\omega^2)^2 + (c\omega)^2}}$, $\phi = \tan^{-1} \dfrac{(k - m\omega^2)}{c\omega}$

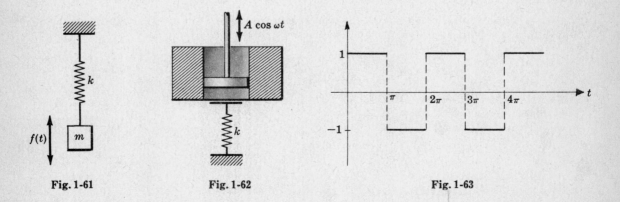

Fig. 1-61 Fig. 1-62 Fig. 1-63

75. Find the first four terms of the Fourier series representation of the square wave or meander function as shown in Fig. 1-63 above.

Ans. $f(t) = \dfrac{4}{\pi}\left(\sin t + \dfrac{1}{3} \sin 3t + \dfrac{1}{5} \sin 5t + \dfrac{1}{7} \sin 7t + \cdots \right)$

76. Find the first four terms of the Fourier series representation of the triangular wave as shown in Fig. 1-64 below.

   Ans. $f(t) = -\dfrac{8}{\pi^2}\left(\cos t + \dfrac{1}{3^2}\cos 3t + \dfrac{1}{5^2}\cos 5t + \dfrac{1}{7^2}\cos 7t + \cdots\right)$

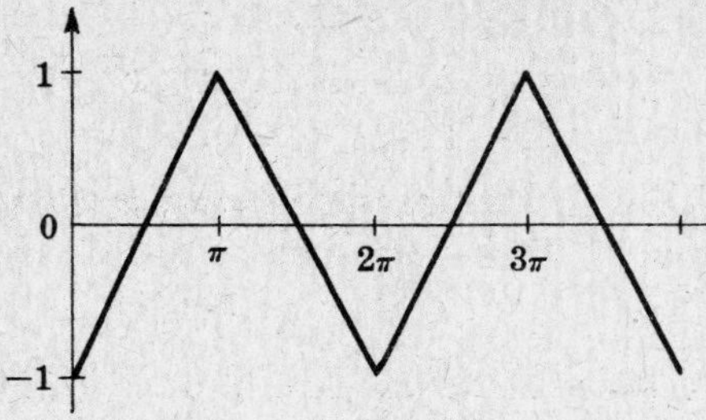

Fig. 1-64

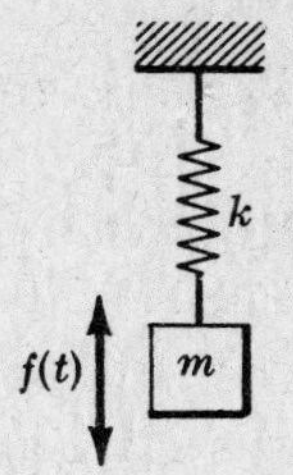

Fig. 1-65

77. A periodic excitation as given in Problem 75 is acting on the simple vibrating spring-mass system as shown in Fig. 1-65 above. Determine the steady state vibration of the system if the magnitude of k is 20 and $m = 5$, $\omega = 1$.

   Ans. $x_p = 1.69 \sin t - 0.34 \sin 3t + 0.048 \sin 5t - \cdots$

78. A motor weighs 200 lb and is running at a constant speed 1800 rpm. If the transmissibility of force between the motor and the floor should be 0.1 or 10%, what should be the spring constant for such mounting of the motor?   *Ans.* $k = 1700$ lb/in

79. If a loaded automobile weighing 2000 lb is running at 60 mi/hr over a rough road whose surface varies sinusoidally with 16 ft/cycle and amplitude of $\frac{1}{2}$ ft, determine the amplitude ratio of the automobile when it is loaded and empty. It weighs 500 lb when it is empty, and the damping factor ζ is 0.5 when it is loaded.   *Ans.* Amplitude ratio = 0.68

80. A vibrometer having a natural frequency of 31.4 rad/sec is employed to measure the amplitude of vibration of a machine part. If it gives a reading of 0.06 in, what is the amplitude of vibration of the machine part?   *Ans.* 0.045 in.

Chapter 2

Two Degrees of Freedom

INTRODUCTION

Systems that require two independent coordinates to specify their position are called two-degree-of-freedom systems. The following three systems are of two degrees of freedom.

(a) In the spring-mass system shown in Fig. 2-1 below, if the masses m_1 and m_2 are constrained to move vertically, at least one coordinate $x(t)$ is required to define the location of each mass at any time. Thus the system requires altogether two coordinates to specify their positions; it is a two-degree-of-freedom system.

(b) If the mass m supported by two equal springs as shown in Fig. 2-2 below is constrained to move in a vertical plane, two coordinates are required to specify the configuration of the system. One of these should be a rectilinear displacement, such as the displacement of the mass, $x(t)$. The other coordinate should be the angular displacement, $\theta(t)$, to account for the rotation of the mass. These two coordinates are independent of each other; hence the system is of two degrees of freedom.

(c) For the double-pendulum shown in Fig. 2-3 below, it is clear that two coordinates are required to specify the position of the masses m_1 and m_2 at any time, and the system is of two degrees of freedom. But either x_1 and x_2, or y_1 and y_2, or θ_1 and θ_2 is a possible set of coordinates for this system.

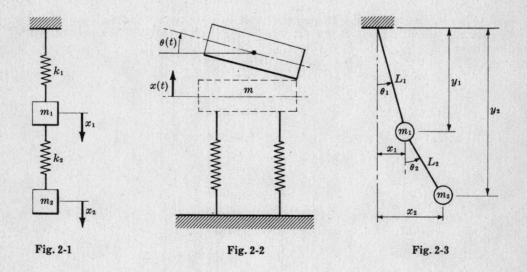

Fig. 2-1   Fig. 2-2   Fig. 2-3

GENERALIZED COORDINATES

As discussed earlier, it is sometimes possible to specify the configuration of a system by more than one set of independent coordinates or parameters such as length, angle, or some other physical parameters; any set of such coordinates may be called *generalized coordinates*.

NORMAL MODES

There are two equations of motion for a two-degree-of-freedom system, one for each mass. As a result, there are two natural frequencies for a two-degree-of-freedom system. The natural frequencies are found by solving the *frequency equation* of an undamped system or the *characteristic equation* of a damped system.

When the masses of a system are oscillating in such a manner that they reach maximum displacements simultaneously and pass their equilibrium points simultaneously, or all moving parts of the system are oscillating in phase with one frequency, such a state of motion is called *normal mode* or *principal mode of vibration*.

PRINCIPAL COORDINATES

It is always possible to find a particular set of coordinates such that each equation of motion contains only one unknown quantity. Then the equations of motion can be solved independently of each other. Such a particular set of coordinates is called *principal coordinates*.

COORDINATE COUPLING

This is a concept of coupling action where a vibration in one part of the system induces vibration in another part of the same system due to the force transmitted through the coupling spring or dashpot. In other words, the displacement of one mass will be felt by another mass in the same system since they are coupled together. There are two types of coupling: the *static coupling* due to static displacements, and the *dynamic coupling* due to inertia forces.

LAGRANGE'S EQUATIONS

Lagrange's equation, in its fundamental form for generalized coordinates q_i, is

$$\frac{d}{dt}\frac{\partial(\text{K.E.})}{\partial \dot{q}_i} - \frac{\partial(\text{K.E.})}{\partial q_i} + \frac{\partial(\text{P.E.})}{\partial q_i} + \frac{\partial(\text{D.E.})}{\partial \dot{q}_i} = Q_i$$

where K.E. = kinetic energy of the system = $\frac{1}{2}m\dot{x}^2$
        P.E. = potential energy of the system = $\frac{1}{2}kx^2$
        D.E. = dissipation energy of the system = $\frac{1}{2}c\dot{x}^2$
        Q_i = generalized external force acting on the system.

For a conservative system, Lagrange's equation can be written as

$$\frac{d}{dt}\frac{\partial L}{\partial \dot{q}_i} - \frac{\partial L}{\partial q_i} = 0$$

where $L = \text{K.E.} - \text{P.E.}$ is called the Lagrangian.

The use of Lagrange's equation will yield directly as many equations of motion as the number of degrees of freedom of the system when the basic energy expressions of the system are known.

DYNAMIC VIBRATION ABSORBER

A dynamic vibration absorber is simply a single-degree-of-freedom system, usually in the form of a simple spring-mass system. When added to another single-degree-of-freedom system as an auxiliary system, it will transform the whole system into a two-degree-of-freedom with two natural frequencies of vibration. One of the natural frequencies is set above the excitation frequency while the other is set below it so that the main mass of the entire system will have very small amplitude of vibration instead of very large amplitude under the given excitation. (See Problems 36 and 37.)

ORTHOGONALITY PRINCIPLE

The principal modes of vibration for systems having two or more degrees of freedom are orthogonal. This is known as the orthogonality principle. This important property that the principal modes are vibrations along mutually perpendicular straight lines is very useful for the calculation of natural frequencies. Though the principal modes for systems with more than three degrees of freedom may not be literally perpendicular to one another, the orthogonality principle still holds.

The orthogonality principle for two degrees of freedom systems can be written as

$$m_1 A_1 A_2 + m_2 B_1 B_2 = 0$$

where A_1, A_2, B_1, B_2 are the amplitudes of the two coordinates for first and second modes of vibration. (See Problems 41 and 42.)

SEMI-DEFINITE SYSTEMS

Sometimes, when one of the roots of the frequency equation of a vibrating system is equal to zero, this indicates that one of the natural frequencies of the system is equal to zero. Such systems are known as semi-definite systems. Physically, this simply means that the system will move as a rigid body without distortion of the springs and dashpots that connect the different parts of the system together. (See Problems 43, 44, 45.)

Solved Problems

1. Determine the equations of motion and the natural frequencies of the two-degree-of-freedom spring-mass system shown in Fig. 2-4.

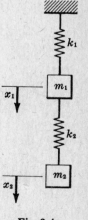

Employing $\Sigma F = ma$, we have

$$m_1 \ddot{x}_1 = -k_1 x_1 - k_2(x_1 - x_2)$$
$$m_2 \ddot{x}_2 = -k_2(x_2 - x_1)$$

Rearranging, the equations of motion become

$$m_1 \ddot{x}_1 + (k_2 + k_1)x_1 - k_2 x_2 = 0$$
$$m_2 \ddot{x}_2 + k_2 x_2 - k_2 x_1 = 0$$

Assume that the motion is periodic, and is composed of harmonic motions of various amplitudes and frequencies. Let one of these components be

$$x_1 = A \sin(\omega t + \psi), \qquad x_2 = B \sin(\omega t + \psi)$$

where A, B, and ψ are arbitrary constants and ω one of the natural frequencies of the system. Substituting these values into the equations of motion, we obtain

$$-m_1 A \omega^2 \sin(\omega t + \psi) + (k_1 + k_2)A \sin(\omega t + \psi) - k_2 B \sin(\omega t + \psi) = 0$$
$$-m_2 B \omega^2 \sin(\omega t + \psi) + k_2 B \sin(\omega t + \psi) - k_2 A \sin(\omega t + \psi) = 0$$

Fig. 2-4

Cancelling out $\sin(\omega t + \psi)$, the equations of motion become

$$(k_1 + k_2 - m_1 \omega^2)A - k_2 B = 0$$
$$-k_2 A + (k_2 - m_2 \omega^2)B = 0$$

These are homogeneous linear algebraic equations in A and B. The solution $A = B = 0$ simply defines the equilibrium condition of the system. The other solution is obtained by equating to zero the determinant of the coefficients of A and B, i.e.,

$$\begin{vmatrix} k_1 + k_2 - m_1\omega^2 & -k_2 \\ -k_2 & k_2 - m_2\omega^2 \end{vmatrix} = 0$$

Since the periodic motion is assumed to be composed of harmonic motions, either the sine or cosine function can be used to represent the motion. The resulting algebraic equations in A and B will be the same.

Expanding the determinant gives

$$\omega^4 - \left[\frac{k_1 + k_2}{m_1} + \frac{k_2}{m_2}\right]\omega^2 + \frac{k_1 k_2}{m_1 m_2} = 0$$

which is the frequency equation of the system. Solving,

$$\omega^2 = \frac{k_1 + k_2}{2m_1} + \frac{k_2}{2m_2} \pm \sqrt{\frac{1}{4}\left[\frac{k_1 + k_2}{m_1} + \frac{k_2}{m_2}\right]^2 - \frac{k_1 k_2}{m_1 m_2}}$$

Thus the general solution of the equations of motion is composed of two harmonic motions of frequencies ω_1 and ω_2; they are the fundamental and first harmonic.

$$x_1 = A_1 \sin(\omega_1 t + \psi_1) + A_2 \sin(\omega_2 t + \psi_2)$$
$$x_2 = B_1 \sin(\omega_1 t + \psi_1) + B_2 \sin(\omega_2 t + \psi_2)$$

where A's, B's, ψ's are arbitrary constants. But the amplitude ratios are

$$\frac{A_1}{B_1} = \frac{k_2}{k_1 + k_2 - m_1\omega_1^2} = \frac{k_2 - m_2\omega_1^2}{k_2} = \frac{1}{\lambda_1}$$

$$\frac{A_2}{B_2} = \frac{k_2}{k_1 + k_2 - m_1\omega_2^2} = \frac{k_2 - m_2\omega_2^2}{k_2} = \frac{1}{\lambda_2}$$

Hence the general solutions finally become

$$x_1 = A_1 \sin(\omega_1 t + \psi_1) + A_2 \sin(\omega_2 t + \psi_2)$$
$$x_2 = \lambda_1 A_1 \sin(\omega_1 t + \psi_1) + \lambda_2 A_2 \sin(\omega_2 t + \psi_2)$$

and the four constants A_1, A_2, ψ_1, ψ_2 are to be evaluated by the four initial conditions $x_1(0), \dot{x}_1(0),$ $x_2(0), \dot{x}_2(0)$.

2. Two equal masses are attached to a string having high tension as shown in Fig. 2-5. Determine the natural frequencies of the system.

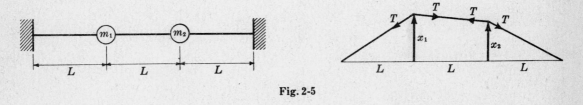

Fig. 2-5

For small oscillations, the tension in the string can be assumed to remain unchanged. Applying Newton's equation of motion to mass m_1 yields

$$m_1 \ddot{x}_1 = -(x_1/L)T - [(x_1 - x_2)/L]T$$

or

$$m_1 \ddot{x}_1 + x_1 T/L + (x_1 - x_2)T/L = 0$$

Similarly, for mass m_2,

$$m_2 \ddot{x}_2 + x_2 T/L + (x_2 - x_1)T/L = 0$$

As discussed earlier, let $x_1 = A \cos \omega t$, and $\ddot{x}_1 = -\omega^2 A \cos \omega t$
$x_2 = B \cos \omega t$, and $\ddot{x}_2 = -\omega^2 B \cos \omega t$.

Using these values, the equations of motion become

$$-m_1\omega^2 A + TA/L + TA/L - TB/L = 0$$
$$-m_2\omega^2 B + TB/L + TB/L - TA/L = 0$$

Rearranging,

$$(2T/L - m_1\omega^2)A - (T/L)B = 0$$
$$-(T/L)A + (2T/L - m_2\omega^2)B = 0$$

The solution for these two algebraic equations other than the trivial one is found by equating to zero the determinant of the coefficients A and B:

$$\begin{vmatrix} (2T/L - m_1\omega^2) & -(T/L) \\ -(T/L) & (2T/L - m_2\omega^2) \end{vmatrix} = 0$$

from which the frequency equation is

$$\omega^4 - (4T/Lm)\omega^2 + 3T^2/L^2m^2 = 0$$

and

$$\omega^2 = \frac{4T/Lm \pm \sqrt{16T^2/L^2m^2 - 12T^2/L^2m^2}}{2} = 3T/Lm,\ T/Lm$$

Then $\omega_1 = \sqrt{T/Lm}$ rad/sec and $\omega_2 = 1.73\sqrt{T/Lm}$ rad/sec, where T is the tensile force in the string.

3. The equations of motion representing the tuned pendulum as shown in Fig. 2-6 are:

$$\ddot{x} + 1000x - 100\theta = 0$$
$$\ddot{\theta} + 1000\theta - 100x = 0$$

If the pendulum is turned one radian and released, find the resultant motion of the system.

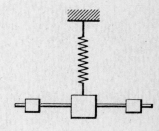

Fig. 2-6

Assume that the motion is periodic and is composed of harmonic motions of various amplitudes and frequencies. Let one of these harmonic components be

$$x = A\cos(\omega t + \psi), \qquad \theta = B\cos(\omega t + \psi)$$

where A, B, and ψ are arbitrary constants and ω is one of the natural frequencies of the system.

After substitution and simplification, the equations of motion yield the frequency equation of the system as

$$\begin{vmatrix} (1000 - \omega^2) & -100 \\ -100 & (1000 - \omega^2) \end{vmatrix} = 0$$

or $\omega^4 - 2000\omega^2 + 1000^2 - 10,000 = 0$ from which $\omega_1 = 30$ and $\omega_2 = 33.1$ rad/sec.

The amplitude ratios are $A/B = 100/(1000 - \omega^2)$, i.e.,

$$A_1/B_1 = 100/100 = 1, \qquad A_2/B_2 = 100/(-100) = -1$$

and the equations of motion become

$$x = A_1\cos(30t + \psi_1) + A_2\cos(33.1t + \psi_2)$$
$$\theta = A_1\cos(30t + \psi_1) - A_2\cos(33.1t + \psi_2)$$

where the four unknowns A_1, A_2, ψ_1, ψ_2 are to be evaluated by the four initial conditions.

The initial displacements are $x(0) = 0$, $\theta(0) = 1$, or

$$x(0) = A_1\cos\psi_1 + A_2\cos\psi_2 = 0$$
$$\theta(0) = A_1\cos\psi_1 - A_2\cos\psi_2 = 1$$

which altogether give $A_1 = 1/(2\cos\psi_1)$ and $A_2 = 1/(-2\cos\psi_2)$.

The initial velocities are $\dot{x}(0) = 0$, $\dot{\theta}(0) = 0$, or

$$\dot{x}(0) = -\omega_1 A_1\sin\psi_1 - \omega_2 A_2\sin\psi_2 = 0$$
$$\dot{\theta}(0) = -\omega_1 A_1\sin\psi_1 + \omega_2 A_2\sin\psi_2 = 0$$

Adding these two equations yields

$$2A_1\omega_1 \sin\psi_1 = 0, \qquad 2A_2\omega_2 \sin\psi_2 = 0$$

But $\omega_1, \omega_2, A_1, A_2$ cannot be zero all the time; hence $\sin\psi_1 = \sin\psi_2 = 0$ or $\psi_1 = \psi_2 = 0$. Then $A_1 = \frac{1}{2}, A_2 = -\frac{1}{2}$; and so the resultant motion of the system is expressed by the equations

$$x(t) = \tfrac{1}{2}\cos 30t - \tfrac{1}{2}\cos 33.1t$$
$$\theta(t) = \tfrac{1}{2}\cos 30t + \tfrac{1}{2}\cos 33.1t$$

4. The two-degree-of-freedom spring-mass system shown in Fig. 2-7 is constrained to have vertical oscillations. Determine the frequency equation and the amplitude ratios of the system.

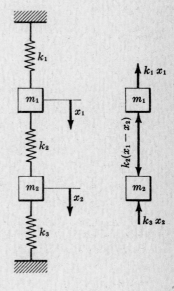

Fig. 2-7

Applying $\Sigma F = ma$ to the two masses, we obtain

$$m_1\ddot{x}_1 = -k_1 x_1 - k_2(x_1 - x_2)$$
$$m_2\ddot{x}_2 = -k_3 x_2 - k_2(x_2 - x_1)$$

Rearranging, the equations of motion become

$$m_1\ddot{x}_1 + (k_1 + k_2)x_1 - k_2 x_2 = 0$$
$$m_2\ddot{x}_2 + (k_2 + k_3)x_2 - k_2 x_1 = 0$$

Assume that the motion is periodic and is composed of harmonic motions of various amplitudes and frequencies. Let one of these components be

$$x_1 = A\sin(\omega t + \phi), \qquad x_2 = B\sin(\omega t + \phi)$$

where A, B, and ϕ are arbitrary constants and ω one of the natural frequencies of the system. Substituting these expressions into the equations of motion and simplifying, we have

$$(k_1 + k_2 - m_1\omega^2)A - k_2 B = 0$$
$$-k_2 A + (k_2 + k_3 - m_2\omega^2)B = 0$$

These are homogeneous linear algebraic equations in A and B; the solution that $A = B = 0$ simply defines the equilibrium condition of the system. The other solution is obtained by equating to zero the determinant of the coefficients of A and B, i.e.,

$$\begin{vmatrix} (k_1 + k_2 - m_1\omega^2) & -k_2 \\ -k_2 & (k_2 + k_3 - m_2\omega^2) \end{vmatrix} = 0$$

Expand the determinant to obtain the frequency equation

$$\omega^4 - \left[\frac{k_1 + k_2}{m_1} + \frac{k_2 + k_3}{m_2}\right]\omega^2 + \frac{k_1 k_2 + k_2 k_3 + k_3 k_1}{m_1 m_2} = 0$$

The amplitude ratios are found from the algebraic equations of coefficients A and B:

$$\frac{A_1}{B_1} = \frac{k_2}{k_1 + k_2 - m_1\omega_1^2} = \frac{k_2 + k_3 - m_2\omega_1^2}{k_2}$$

$$\frac{A_2}{B_2} = \frac{k_2}{k_1 + k_2 - m_1\omega_2^2} = \frac{k_2 + k_3 - m_2\omega_2^2}{k_2}$$

5. If mass m_1 is displaced 1 in. from its static equilibrium position and released, determine the resulting displacements $x_1(t)$ and $x_2(t)$ of the masses shown in Fig. 2-8 below.

Applying $\Sigma F = ma$ to masses m_1 and m_2, we obtain

$$m_1\ddot{x}_1 = -kx_1 - k(x_1 - x_2)$$
$$m_2\ddot{x}_2 = -kx_2 - k(x_2 - x_1)$$

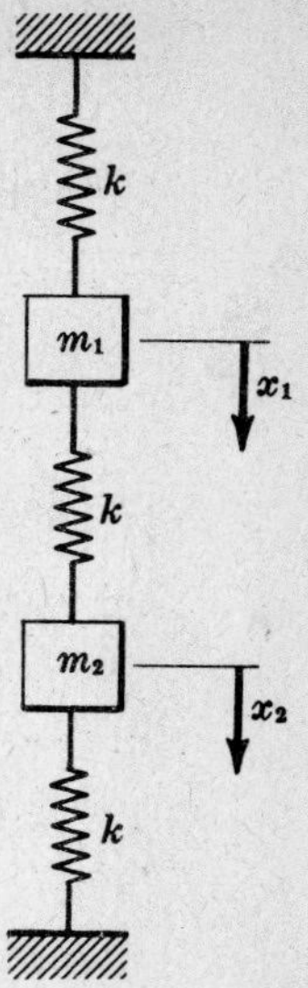

Assume that the motion is periodic and is composed of harmonic components of various amplitudes and frequencies. Let one of these components be

$$x_1 = A \cos(\omega t + \phi), \qquad x_2 = B \cos(\omega t + \phi)$$

Substituting these values into the equations of motion yields

$$(2k - m\omega^2)A - kB = 0$$
$$-kA + (2k - m\omega^2)B = 0$$

These are homogeneous linear algebraic equations in A and B. The solution that $A = B = 0$ simply defines the equilibrium condition of the system. The other solution is obtained by equating to zero the determinant of the coefficients of A and B, i.e.,

$$\begin{vmatrix} (2k - m\omega^2) & -k \\ -k & (2k - m\omega^2) \end{vmatrix} = 0$$

Expand the determinant to obtain the frequency equation

$$\omega^4 - (4k/m)\omega^2 + 3(k/m)^2 = 0$$

from which $\omega_1 = \sqrt{k/m}$ and $\omega_2 = \sqrt{3k/m}$ rad/sec.

The amplitude ratios are given by

$$A_1/B_1 = k/(2k - m\omega_1^2) = 1$$
$$A_2/B_2 = k/(2k - m\omega_2^2) = -1$$

Hence the motions of the masses are expressed by

$$x_1(t) = A_1 \cos(\sqrt{k/m}\, t + \phi_1) + A_2 \cos(\sqrt{3k/m}\, t + \phi_2)$$
$$x_2(t) = A_1 \cos(\sqrt{k/m}\, t + \phi_1) - A_2 \cos(\sqrt{3k/m}\, t + \phi_2)$$

where the four constants of integration are to be evaluated by the four initial conditions: $x_1(0) = 1$, $x_2(0) = 0$, $\dot{x}_1(0) = 0$, $\dot{x}_2(0) = 0$.

$$1 = A_1 \cos \phi_1 + A_2 \cos \phi_2 \qquad (1)$$
$$0 = A_1 \cos \phi_1 - A_2 \cos \phi_2 \qquad (2)$$
$$0 = -\omega_1 A_1 \sin \phi_1 - \omega_2 A_2 \sin \phi_2 \qquad (3)$$
$$0 = -\omega_1 A_1 \sin \phi_1 + \omega_2 A_2 \sin \phi_2 \qquad (4)$$

Solving equations (1) and (2) together, $A_1 = 1/(2 \cos \phi_1)$; and from equations (3) and (4), $\sin \phi_1 = \sin \phi_2 = 0$ or $\phi_1 = \phi_2 = 0$. Hence, $A_1 = A_2 = \frac{1}{2}$. Thus the motions of the masses are

$$x_1(t) = \tfrac{1}{2} \cos \sqrt{k/m}\, t + \tfrac{1}{2} \cos \sqrt{3k/m}\, t$$
$$x_2(t) = \tfrac{1}{2} \cos \sqrt{k/m}\, t - \tfrac{1}{2} \cos \sqrt{3k/m}\, t$$

For the first principal mode of vibration, the two masses move in the same direction with equal amplitudes. The coupling spring in this case is unstressed. For the second mode of vibration, the two masses move in opposite directions with equal amplitudes. The midpoint of the coupling spring is stationary due to the symmetry of the system. This stationary point is called a node.

Fig. 2-8

6. What are the appropriate initial conditions such that the system in Problem 5 is vibrating at (a) first principal mode, and (b) second principal mode?

The general motion of the system is given by

$$x_1(t) = A_1 \cos(\omega_1 t + \phi_1) + A_2 \cos(\omega_2 t + \phi_2)$$
$$x_2(t) = A_1 \cos(\omega_1 t + \phi_1) - A_2 \cos(\omega_2 t + \phi_2)$$

Applying the initial conditions $x_1(0)$, $\dot{x}_1(0)$, $x_2(0)$ and $\dot{x}_2(0)$, we obtain

$$x_1(0) = A_1 \cos \phi_1 + A_2 \cos \phi_2$$
$$\dot{x}_1(0) = -A_1 \omega_1 \sin \phi_1 - A_2 \omega_2 \sin \phi_2$$
$$x_2(0) = A_1 \cos \phi_1 - A_2 \cos \phi_2$$
$$\dot{x}_2(0) = -A_1 \omega_1 \sin \phi_1 + A_2 \omega_2 \sin \phi_2$$

Solving,

$$A_1 = \frac{-x_1(0) - x_2(0)}{-2 \cos \phi_1} = \frac{\dot{x}_2(0) + \dot{x}_1(0)}{-2\omega_1 \sin \phi_1}, \qquad A_2 = \frac{x_2(0) - x_1(0)}{-2 \cos \phi_2} = \frac{\dot{x}_1(0) - \dot{x}_2(0)}{-2\omega_2 \sin \phi_2}$$

(a) For first principal mode of vibration, $A_2 = 0$ and hence

$$x_1(t) = A_1 \cos(\omega_1 t + \phi_1), \qquad x_2(t) = A_1 \cos(\omega_1 t + \phi_1)$$

which requires $x_1(0) = x_2(0)$ and $\dot{x}_1(0) = \dot{x}_2(0)$.

(b) For second principal mode of vibration, $A_1 = 0$ and hence

$$x_1(t) = A_2 \cos(\omega_2 t + \phi_2), \qquad x_2(t) = -A_2 \cos(\omega_2 t + \phi_2)$$

which requires $x_1(0) = -x_2(0)$ and $\dot{x}_1(0) = -\dot{x}_2(0)$.

7. Two uniform slender rods, each weighing 5 lb/ft, are suspended at their upper ends and connected by a spring of stiffness 5 lb/in as shown in Fig. 2-9. The system is displaced slightly and released. What are the natural frequencies of oscillation?

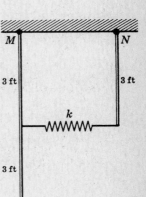

Taking moments with respect to the suspension points M and N gives

$$J_1 \ddot{\theta}_1 = -m_1 g(36 \sin \theta_1) - (36)^2 (\theta_1 - \theta_2)k$$
$$J_2 \ddot{\theta}_2 = -m_2 g(18 \sin \theta_2) - (36)^2 (\theta_2 - \theta_1)k$$

where J_1 and J_2 are the moments of inertia of the two rods with respect to M and N. For small oscillations, $\sin \theta \doteq \theta$. Substituting $J_1 = \frac{1}{3}m_1 L_1^2 = 134$ and $J_2 = \frac{1}{3}m_2 L_2^2 = 16.8$, the equations of motion become

$$\ddot{\theta}_1 + 56.4\theta_1 - 48.4\theta_2 = 0$$
$$\ddot{\theta}_2 + 401.1\theta_2 - 385\theta_1 = 0$$

Let $\quad \theta_1 = A \sin(\omega t + \psi), \qquad \ddot{\theta}_1 = -\omega^2 A \sin(\omega t + \psi)$
$\qquad \theta_2 = B \sin(\omega t + \psi), \qquad \ddot{\theta}_2 = -\omega^2 B \sin(\omega t + \psi)$

Substituting these values into the equations of motion, we have

$$(56.4 - \omega^2)A - 48.4B = 0$$
$$-385A + (401.1 - \omega^2)B = 0$$

and the frequency equation, obtained by equating to zero the determinant of the coefficients of A and B, is

$$\omega^4 - 457.5\omega^2 + 2000 = 0$$

Fig. 2-9

from which $\omega_1 = 1.95$ and $\omega_2 = 21.3$ rad/sec.

8. Find the natural frequencies of oscillation of the double-pendulum as shown in Fig. 2-10 below where $m_1 = m_2 = m$, and $L_1 = L_2 = L$.

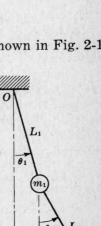

Taking moments with respect to mass m_1 and the pivoted point O, we obtain the following equations of motion:

$$m_2 L_2^2 \ddot{\theta}_2 = -m_2 g L_2 \sin \theta_2 - m_2 L_1 L_2 \ddot{\theta}_1$$
$$m_1 L_1^2 \ddot{\theta}_1 = -m_1 g L_1 \theta_1 - m_2 g L_1 \theta_1 - m_2(L_1 \ddot{\theta}_1 + L_2 \ddot{\theta}_2)L_1$$

For small angles of oscillation, $\sin \theta \doteq \theta$, and the equations of motion become

$$\ddot{\theta}_1 + (g/L)\theta_1 + \ddot{\theta}_2/2 = 0$$
$$\ddot{\theta}_2 + \ddot{\theta}_1 + (g/L)\theta_2 = 0$$

Assume that the motion is periodic, and is composed of harmonic motion of various amplitudes and frequencies. Let one of these harmonic components be

$$\theta_1 = A \cos(\omega t + \psi) \quad \text{and} \quad \theta_2 = B \cos(\omega t + \psi)$$

Substitute these values into the equations of motion and obtain

$$(g/L - \omega^2)A - (\omega^2/2)B = 0$$
$$-\omega^2 A + (g/L - \omega^2)B = 0$$

Fig. 2-10

and the frequency equation, found by equating to zero the determinant of the coefficients of A and B, is

$$\omega^4 - (4g/L)\omega^2 + 2g^2/L^2 = 0$$

from which $\omega_1 = 0.75\sqrt{g/L}$ and $\omega_2 = 1.86\sqrt{g/L}$ rad/sec.

9. Derive the frequency equation for the system as shown in Fig. 2-11. Assume the cord passing over the cylinder does not slip.

Employing the force equation $\Sigma F = ma$ for mass m_1 and the torque equation for the cylinder of mass m_2, we obtain

$$m_1 \ddot{x} = -k_1(x - r\theta)$$
$$J_0 \ddot{\theta} = -k_2 r^2 \theta - k_1(r\theta - x)r$$

where $J_0 = \frac{1}{2}m_2 r^2$ is the moment of inertia of the cylinder of radius r. Rearranging, the equations of motion become

$$m_1 \ddot{x} + k_1 x - k_1 r\theta = 0$$
$$J_0 \ddot{\theta} + (k_2 r^2 + k_1 r^2)\theta - k_1 x r = 0$$

Assume that the motion is periodic and is composed of harmonic motion of various amplitudes and frequencies. Let

$$x = A \sin(\omega t + \psi), \qquad \ddot{x} = -\omega^2 A \sin(\omega t + \psi)$$
$$\theta = B \sin(\omega t + \psi), \qquad \ddot{\theta} = -\omega^2 B \sin(\omega t + \psi)$$

Substituting these values into the equations of motion, we have

$$(k_1 - m_1\omega^2)A - k_1 r B = 0$$
$$-k_1 r A + (k_2 r^2 + k_1 r^2 - J_0\omega^2)B = 0$$

The frequency equation, obtained by equating to zero the determinant of the coefficients of A and B, is

$$\omega^4 - \left[\frac{2(k_1 + k_2)}{m_2} + \frac{k_1}{m_1}\right]\omega^2 + \frac{2k_1 k_2}{m_1 m_2} = 0$$

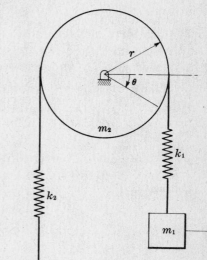

Fig. 2-11

10. Two simple pendulums are connected by a spring as shown in Fig. 2-12 below. Determine the natural frequency of each pendulum.

Taking moments about the two hinged points M and N, we have

$$mL^2 \ddot{\theta}_1 = -mgL\theta_1 - ka^2(\theta_1 - \theta_2)$$
$$mL^2 \ddot{\theta}_2 = -mgL\theta_2 - ka^2(\theta_2 - \theta_1)$$

Let $\theta_1 = A \cos \omega t$, $\theta_2 = B \cos \omega t$. Substituting these values into the equations of motion, we obtain

$$(-\omega^2 mL^2 + mgL + ka^2)A - ka^2 B = 0$$
$$(-\omega^2 mL^2 + mgL + ka^2)B - ka^2 A = 0$$

The frequency equation is obtained by equating to zero the determinant of the coefficients of A and B, i.e.,

$$\begin{vmatrix} -\omega^2 mL^2 + mgL + ka^2 & -ka^2 \\ -ka^2 & -\omega^2 mL^2 + mgL + ka^2 \end{vmatrix} = 0$$

Expand the determinant to obtain

$$\omega^4 - 2(g/L + ka^2/mL^2)\omega^2 + (g^2/L^2 + 2ka^2 g/mL^3) = 0$$

which gives

$$\omega_1 = \sqrt{g/L} \text{ and } \omega_2 = \sqrt{g/L + 2ka^2/mL^2} \text{ rad/sec}$$

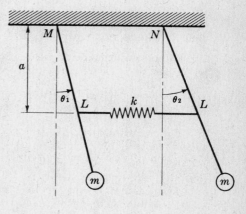

Fig. 2-12

11. Determine the frequency equation of the system as shown in Fig. 2-13, if $k_1 = k_2 = k_3 = k$, $m_1 = m_2 = m$, $r_1 = r_2 = r$, and $J_1 = J_2 = J$.

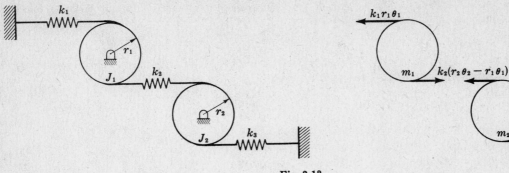

Fig. 2-13

Applying the torque equation $\Sigma T = J\ddot{\theta}$, we have

$$J_1\ddot{\theta}_1 = -k_1(r_1\theta_1)r_1 - k_2(r_1\theta_1 - r_2\theta_2)r_1$$
$$J_2\ddot{\theta}_2 = -k_3(r_2\theta_2)r_2 - k_2(r_2\theta_2 - r_1\theta_1)r_2$$

where $J_1 = J_2 = \frac{1}{2}mr^2$ are the moments of inertia of the cylinders. Rearranging,

$$J_1\ddot{\theta}_1 + (k_1r_1^2 + k_2r_1^2)\theta_1 - k_2r_1r_2\theta_2 = 0$$
$$J_2\ddot{\theta}_2 + (k_3r_2^2 + k_2r_2^2)\theta_2 - k_2r_1r_2\theta_1 = 0$$

Assume that the motion is periodic and is composed of harmonic motions of various amplitudes and frequencies. Let

$$\theta_1 = A\sin(\omega t + \psi), \qquad \ddot{\theta}_1 = -\omega^2 A\sin(\omega t + \psi)$$
$$\theta_2 = B\sin(\omega t + \psi), \qquad \ddot{\theta}_2 = -\omega^2 B\sin(\omega t + \psi)$$

Substitute these values into the equations of motion and obtain

$$(k_1r_1^2 + k_2r_1^2 - \omega^2 J_1)A - k_2r_1r_2B = 0$$
$$-(k_2r_1r_2)A + (k_3r_2^2 + k_2r_2^2 - \omega^2 J_2)B = 0$$

Putting $k_1 = k_2 = k_3 = k$, $m_1 = m_2 = m$, $r_1 = r_2 = r$, and $J_1 = J_2 = J$, the equations of motion become

$$(2kr^2 - \omega^2 J)A - kr^2B = 0$$
$$-kr^2A + (2kr^2 - \omega^2 J)B = 0$$

The solution for these two algebraic equations, other than the trivial solution that A and B are both equal to zero, is obtained by equating to zero the determinant of the coefficients A and B, i.e.,

$$\begin{vmatrix} (2kr^2 - \omega^2 J) & -kr^2 \\ -kr^2 & (2kr^2 - \omega^2 J) \end{vmatrix} = 0$$

Expand the determinant to obtain

$$\omega^4 - (8k/m)\omega^2 + 6k^2/m^2 = 0$$

from which $\omega_1 = 0.92\sqrt{k/m}$ and $\omega_2 = 2.68\sqrt{k/m}$ rad/sec.

12. For small angles of oscillation, derive the equations of motion of the system as shown in Fig. 2-14.

Employing $\Sigma F = ma$,

$$m_1\ddot{x}_1 = -m_2\ddot{x}_2 - 2kx_1$$

where $x_2 = x_1 + L\sin\theta$ and L is the length of the pendulum. For small angles of oscillation, $\ddot{x}_2 = \ddot{x}_1 + L\cos\theta\,\ddot{\theta}$. Hence the equation of motion becomes

$$(m_1 + m_2)\ddot{x}_1 + 2kx_1 + m_2L\ddot{\theta} = 0$$

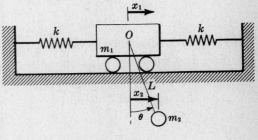

Fig. 2-14

The second equation of motion is obtained by considering the moments about point O:

$$-m_2 \ddot{x}_2 L \cos \theta \;=\; (m_2 g - m_2 \ddot{y}) L \sin \theta$$

where $y = L \cos \theta$ and $\ddot{y} = -L \sin \theta \, \ddot{\theta}$. Assuming small angles of oscillation, $\sin \theta \doteq \theta$ and $\cos \theta \doteq 1$, the second equation of motion becomes

$$\ddot{\theta} + (g/L)\theta + \ddot{x}_1/L \;=\; 0$$

Note. If m_1 is held stationary, the system is reduced to a single-degree-of-freedom system having simple pendulum action, i.e., $\ddot{\theta} + (g/L)\theta = 0$.

13. A double physical pendulum of mass m_1 and m_2 is shown in Fig. 2-15. a_1 and a_2 are the distances from the mass centers to the corresponding pivoted points respectively. Derive the equations of motion.

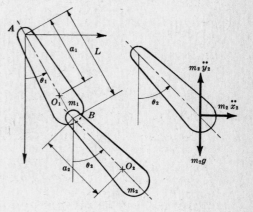

Let (x_1, y_1) and (x_2, y_2) be the coordinates of the mass centers O_1 and O_2 and J_1 and J_2 be the moments of inertia of the pendulums with respect to the axis of rotation. Taking moments with respect to points B and A, we have

$$J_2 \ddot{\theta}_2 = -m_2 g a_2 \sin \theta_2 - m_2 \ddot{x}_2 a_2 \cos \theta_2 + (m_2 a_2 \sin \theta_2) \ddot{y}_2$$
$$J_1 \ddot{\theta}_1 = -m_1 g x_1 + m_1 \ddot{y}_1 x_1 - m_1 \ddot{x}_1 y_1 + m_2 \ddot{y}_2 L \sin \theta_1$$
$$\qquad - m_2 \ddot{x}_2 y_2 - m_2 \ddot{y}_2 x_2 - m_2 g L \sin \theta_1$$

where
$$x_1 = a_1 \sin \theta_1$$
$$x_2 = L \sin \theta_1 + a_2 \sin \theta_2$$
$$y_1 = a \cos \theta_1$$
$$y_2 = L \cos \theta_1 + a_2 \cos \theta_2$$

Fig. 2-15

Hence
$$\ddot{x}_1 = a_1 \cos \theta_1 \, \ddot{\theta}_1 - a_1 \sin \theta_1 \, \dot{\theta}_1^2$$
$$\ddot{x}_2 = L \cos \theta_1 \, \ddot{\theta}_1 - L \sin \theta_1 \, \dot{\theta}_1^2 + a_2 \cos \theta_2 \, \ddot{\theta}_2 - a_2 \sin \theta_2 \, \dot{\theta}_2^2$$
$$\ddot{y}_1 = -a_1 \sin \theta_1 \, \ddot{\theta}_1 - a_1 \cos \theta_1 \, \dot{\theta}_1^2$$
$$\ddot{y}_2 = -L \sin \theta_1 \, \ddot{\theta}_1 - L \cos \theta_1 \, \dot{\theta}_1^2 - a_2 \sin \theta_2 \, \ddot{\theta}_2 - a_2 \cos \theta_2 \, \dot{\theta}_2^2$$

Substituting these values into the moment equations, we obtain

$$J_2 \ddot{\theta}_2 + m_2 [a_2 L \cos (\theta_2 - \theta_1) \ddot{\theta}_1 + a_2 L \sin (\theta_2 - \theta_1) \dot{\theta}_1^2 + a_2^2 \ddot{\theta}_2 + a_2 g \sin \theta_2] \;=\; 0$$
$$J_1 \ddot{\theta}_1 + m_1 (a_1^2 \ddot{\theta}_1 + a_1 g \sin \theta_1) + m_2 [L^2 \ddot{\theta}_1 + a_2 L \cos (\theta_2 - \theta_1) \ddot{\theta}_2]$$
$$\qquad + m_2 [a_2 L \sin (\theta_1 - \theta_2) \dot{\theta}_2^2 + gL \sin \theta_1] \;=\; 0$$

Assuming small oscillations $(\sin \theta \doteq \theta, \cos \theta \doteq 1)$ and neglecting higher order terms, the equations of motion become

$$(J_2 + m_2 a_2^2) \ddot{\theta}_2 + m_2 g a_2 \theta_2 + m_2 a_2 L \ddot{\theta}_1 \;=\; 0$$
$$(J_1 + m_1 a_1^2 + m_2 L^2) \ddot{\theta}_1 + (m_1 g a_1 + m_2 g L)\theta_1 + m_2 a_2 L \ddot{\theta}_2 \;=\; 0$$

By setting m_2 equal to zero in the last expression, it then represents the equation of motion of a single physical pendulum as discussed in Problem 13, Chapter 1.

PRINCIPAL COORDINATES

14. For the two-degree-of-freedom spring-mass system as shown in Fig. 2-16 below, find the principal coordinates. Let $k = m = 1$.

As discussed in Problem 1, the general motion of the system is found to be

$$x_1(t) = A_1 \sin (0.63t + \psi_1) + A_2 \sin (1.62t + \psi_2)$$
$$x_2(t) = 1.6 A_1 \sin (0.63t + \psi_1) - 0.63 A_2 \sin (1.62t + \psi_2)$$

where $\lambda_1 = 1.6$ and $\lambda_2 = -0.63$.

Define a new set of coordinates y_1 and y_2 such that

$$y_1 = A_1 \sin(0.63t + \psi_1)$$
$$y_2 = A_2 \sin(1.62t + \psi_2)$$

Since y_1 and y_2 are harmonic motions, their corresponding equations of motion are given by

$$\ddot{y}_1 + 0.4y_1 = 0$$
$$\ddot{y}_2 + 2.62y_2 = 0$$

This set of equations of motion represents a two-degree-of-freedom vibratory system with natural frequencies $\omega_1 = 0.63$ and $\omega_2 = 1.62$ rad/sec. Because there is neither a static nor a dynamic coupling term in the equations of motion, y_1 and y_2 are principal coordinates. Now

$$x_1 = y_1 + y_2$$
$$x_2 = 1.6y_1 - 0.63y_2$$

Hence

$$y_1 = 0.28x_1 + 0.45x_2$$
$$y_2 = 0.72x_1 - 0.45x_2$$

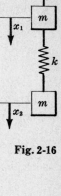

Fig. 2-16

15. Determine the principal coordinates for the system as shown in Fig. 2-17.

As discussed in Problem 4, the general motion of the system is found to be

$$x_1(t) = A_1 \sin(\sqrt{k/m}\,t + \psi_1) + A_2 \sin(\sqrt{3k/m}\,t + \psi_2)$$
$$x_2(t) = A_1 \sin(\sqrt{k/m}\,t + \psi_1) - A_2 \sin(\sqrt{3k/m}\,t + \psi_2)$$

Define a new set of coordinates y_1 and y_2 such that

$$y_1 = A_1 \sin(\sqrt{k/m}\,t + \psi_1)$$
$$y_2 = A_2 \sin(\sqrt{3k/m}\,t + \psi_2)$$

Since y_1 and y_2 are harmonic motions, their corresponding equations of motion are given by

$$\ddot{y}_1 + (k/m)y_1 = 0$$
$$\ddot{y}_2 + (3k/m)y_2 = 0$$

This set of equations of motion represents a two-degree-of-freedom vibratory system with natural frequencies $\omega_1 = \sqrt{k/m}$ and $\omega_2 = \sqrt{3k/m}$ rad/sec. Because there is neither a static nor a dynamic coupling term in the equations of motion, y_1 and y_2 are principal coordinates. Now

$$x_1 = y_1 + y_2$$
$$x_2 = y_1 - y_2$$

Hence

$$y_1 = \tfrac{1}{2}(x_1 + x_2)$$
$$y_2 = \tfrac{1}{2}(x_1 - x_2)$$

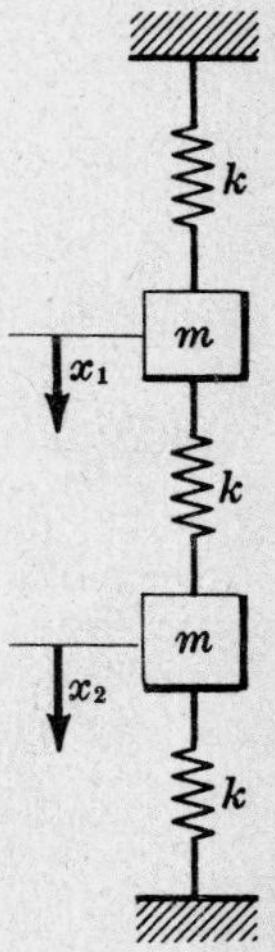

Fig. 2-17

COORDINATE COUPLING

16. A two-degree-of-freedom vibrating system consisting of a mass m and two springs of stiffness k_1 and k_2 is shown in Fig. 2-18. Investigate the coupling action of the system.

As shown in Fig. 2-18, two independent coordinates $x(t)$ and $\theta(t)$ are required to specify the configuration of the system. Using $\Sigma F = ma$,

$$m\ddot{x} = -k_1(x - L_1\theta) - k_2(x + L_2\theta)$$
$$J\ddot{\theta} = k_1(x - L_1\theta)L_1 - k_2(x + L_2\theta)L_2$$

or

$$m\ddot{x} + (k_1 + k_2)x - (k_1L_1 - k_2L_2)\theta = 0$$
$$J\ddot{\theta} + (k_1L_1^2 + k_2L_2^2)\theta - (k_1L_1 - k_2L_2)x = 0$$

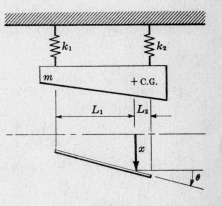

Fig. 2-18

which will be independent of each other if the coupling term $(k_1 L_1 - k_2 L_2)$ is equal to zero, i.e. $k_1 L_1 = k_2 L_2$. If not, the resultant motion of the mass will consist of both translational and rotational motion when either a displacement or torque is applied through the center of gravity of the body as initial condition. In other words, the mass will rotate in a vertical plane and will have vertical motion as well, unless $k_1 L_1 = k_2 L_2$. This, then, is known as static or elastic coupling.

Referring to Fig. 2-19, where $y(t)$ and $\phi(t)$ are used as the coordinates for the system, the equations of motion are given by

$$m\ddot{y} = -k_1(y - L_1\phi) - k_2(y + L_2\phi) - mL\ddot{\phi}$$
$$J\ddot{\phi} = k_1(y - L_1\phi)L_1 - k_2(y + L_2\phi)L_2 - m\ddot{y}L$$

or

$$m\ddot{y} + (k_1 + k_2)y + mL\ddot{\phi} + (k_2 L_2 - k_1 L_1)\phi = 0$$
$$J\ddot{\phi} + (k_1 L_1^2 + k_2 L_2^2)\phi + mL\ddot{y} + (k_2 L_2 - k_1 L_1)y = 0$$

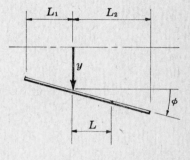

This set of equations of motion then contains static as well as dynamic coupling terms. If $k_1 L_1 = k_2 L_2$, the system is having dynamic or inertia coupling only. In this case, if the mass is moving up and down in the y-direction, the inertia force $m\ddot{y}$, which acts through the center of gravity of the mass, will induce a motion in the ϕ-direction, by virtue of the moment $m\ddot{y}L$. On the other hand, a motion in the ϕ-direction will similarly create a motion of the mass in the y-direction due to the force $mL\ddot{\phi}$.

Fig. 2-19

17. A schematic representation of an automobile is shown in Fig. 2-20 below. If the automobile weighs 4000 lb, and has a radius of gyration about the center of gravity of 4.5 ft, find the principal modes of vibration of the automobile. The combined front springs k_1 is 250 lb/in and k_2 is 270 lb/in.

An automobile has more than two degrees of freedom. In order to bring out the coupling action, only motion in the vertical plane is considered. Therefore let x and θ be the coordinates. The equations of motion are given by

$$m\ddot{x} = -k_1(x - L_1\theta) - k_2(x + L_2\theta)$$
$$J\ddot{\theta} = k_1(x - L_1\theta)L_1 - k_2(x + L_2\theta)L_2$$

where $J = mk^2$ is the moment of inertia of the automobile. Rearranging,

$$m\ddot{x} + (k_1 + k_2)x - (k_1 L_1 - k_2 L_2)\theta = 0$$
$$J\ddot{\theta} + (k_1 L_1^2 + k_2 L_2^2)\theta - (k_1 L_1 - k_2 L_2)x = 0$$

which contain the static coupling term $(k_1 L_1 - k_2 L_2)$. Substituting the values for k_1, k_2 and $J = mk^2$ into the equations of motion, we obtain

$$\ddot{x} + 50.12x + 532\theta = 0$$
$$\ddot{\theta} + 64.22\theta + 0.54x = 0$$

Let $x = A \sin(\omega t + \psi)$ and $\theta = B \sin(\omega t + \psi)$. Substituting these values into the equations of motion and simplifying,

$$(50.12 - \omega^2)A + 532B = 0$$
$$0.54A + (64.22 - \omega^2)B = 0$$

Fig. 2-20

The frequency equation, obtained by equating to zero the determinant of the coefficients A and B, is

$$\omega^4 - 114.2\omega^2 + 2923 = 0$$

which gives $\omega_1 = 5.5$ and $\omega_2 = 8.9$ rad/sec.

The principal modes of vibration are found from the amplitude ratios

$$A_1/B_1 = 532/(50.12 - 5.5^2) = 26.1$$
$$A_2/B_2 = 532/(50.12 - 8.9^2) = -18.6$$

as shown in Fig. 2-21(a) and 2-21(b) below.

First Mode

Second Mode

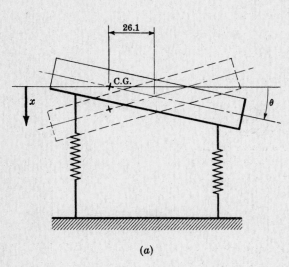
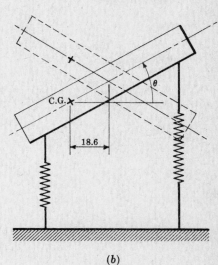

(a)   (b)

Fig. 2-21

18. A weightless stiff rod with two equal masses m at each end is attached to a cantilever with torsional stiffness K and linear stiffness k as shown in Fig. 2-22. What are the coupling terms of the system?

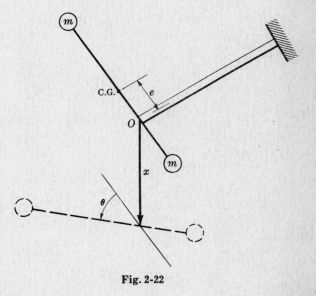

Assume the center of gravity of the rod is at a distance e from the end of the cantilever. Using $\Sigma F = ma$, the first equation of motion is

$$2m\,\ddot{x} = -kx - 2me\,\ddot{\theta}$$

where kx is the restoring force due to the elastic behavior of the cantilever, and $2me\,\ddot{\theta}$ is the inertia force due to the eccentricity of the center of gravity of the stiff rod.

The second equation of motion of the system, found by employing the moment equation $\Sigma M = J\,\ddot{\theta}$, is

$$J\,\ddot{\theta} = -K\theta - 2m\ddot{x}e$$

Fig. 2-22

where J is the moment of inertia of the rod with respect to the point O, $K\theta$ is the restoring moment of the cantilever which acts as a shaft, and $2m\ddot{x}e$ is the moment due to the eccentricity of the center of gravity of the rod. Rearranging, the equations of motion become

$$\ddot{x} + (k/2m)x + e\,\ddot{\theta} = 0$$
$$\ddot{\theta} + (K/J)\theta + (2me/J)\ddot{x} = 0$$

and hence $e\,\ddot{\theta}$ and $(2me/J)\,\ddot{x}$ are the dynamic coupling terms of the system.

LAGRANGE'S EQUATION

19. Use Lagrange's equation to derive the equations of motion for the coupled-pendulum as shown in Fig. 2-23.

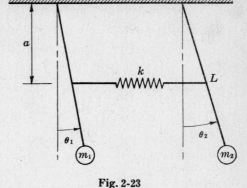

This problem is solved in Problem 10 by employing Newton's law of motion. Here it will be evident that Lagrange's equation is a more direct approach for solving vibration problems, and is particularly useful for systems having simple energy expressions.

For conservative systems without excitations, Lagrange's equation can be written as

$$\frac{d}{dt}\frac{\partial(\text{K.E.})}{\partial \dot{q}_i} - \frac{\partial(\text{K.E.})}{\partial q_i} + \frac{\partial(\text{P.E.})}{\partial q_i} = 0$$

Fig. 2-23

For this system, let θ_1 and θ_2 be the generalized coordinates. Thus the energy expressions are

$$\text{K.E.} = \tfrac{1}{2}m_1 L^2 \dot{\theta}_1^2 + \tfrac{1}{2}m_2 L^2 \dot{\theta}_2^2$$
$$\text{P.E.} = m_1 g L(1 - \cos \theta_1) + m_2 g L(1 - \cos \theta_2) + \tfrac{1}{2}k(a\theta_2 - a\theta_1)^2$$

and $\quad \dfrac{d}{dt}\dfrac{\partial(\text{K.E.})}{\partial \dot{\theta}_1} = m_1 L^2 \ddot{\theta}_1, \quad \dfrac{\partial(\text{K.E.})}{\partial \theta_1} = 0, \quad \dfrac{\partial(\text{P.E.})}{\partial \theta_1} = m_1 g L \sin \theta_1 - ka(a\theta_2 - a\theta_1)$

Hence the first equation of motion is given by

$$m_1 L^2 \ddot{\theta}_1 + m_1 g L \sin \theta_1 - a^2 k(\theta_2 - \theta_1) = 0$$

Similarly, $\dfrac{d}{dt}\dfrac{\partial(\text{K.E.})}{\partial \dot{\theta}_2} = m_2 L^2 \ddot{\theta}_2, \quad \dfrac{\partial(\text{K.E.})}{\partial \theta_2} = 0, \quad \dfrac{\partial(\text{P.E.})}{\partial \theta_2} = m_2 g L \sin \theta_2 + a^2 k(\theta_2 - \theta_1)$

and the second equation of motion is

$$m_2 L^2 \ddot{\theta}_2 + m_2 g L \sin \theta_2 + ka^2(\theta_2 - \theta_1) = 0$$

20. Use Lagrange's equation to find the equations of motion for the two-degree-of-freedom spring-mass system as shown in Fig. 2-24.

The two generalized coordinates for this system are x_1 and x_2 as shown. The energy of the system consists of kinetic energy due to the motion of the masses, and potential energy due to the action of the coupling spring k.

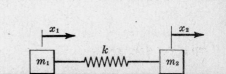

$$\text{K.E.} = \tfrac{1}{2}(m_1 \dot{x}_1^2 + m_2 \dot{x}_2^2)$$
$$\text{P.E.} = \tfrac{1}{2}k(x_2 - x_1)^2$$

Fig. 2-24

Lagrange's equation for a conservative system is

$$\frac{d}{dt}\frac{\partial(\text{K.E.})}{\partial \dot{q}_i} - \frac{\partial(\text{K.E.})}{\partial q_i} + \frac{\partial(\text{P.E.})}{\partial q_i} = 0$$

where $\quad \dfrac{d}{dt}\dfrac{\partial(\text{K.E.})}{\partial \dot{x}_1} = m_1 \ddot{x}_1, \quad \dfrac{\partial(\text{K.E.})}{\partial x_1} = 0, \quad \dfrac{\partial(\text{P.E.})}{\partial x_1} = -k(x_2 - x_1)$

Then the first equation of motion becomes

$$m_1 \ddot{x}_1 + k(x_1 - x_2) = 0$$

Similarly, $\quad \dfrac{d}{dt}\dfrac{\partial(\text{K.E.})}{\partial \dot{x}_2} = m_2 \ddot{x}_2, \quad \dfrac{\partial(\text{K.E.})}{\partial x_2} = 0, \quad \dfrac{\partial(\text{P.E.})}{\partial x_2} = k(x_2 - x_1)$

and the second equation of motion is

$$m_2 \ddot{x}_2 + k(x_2 - x_1) = 0$$

21. A double pendulum of lengths L_1 and L_2, masses m_1 and m_2 is shown in Fig. 2-25. Use Lagrange's equation to derive the equations of motion.

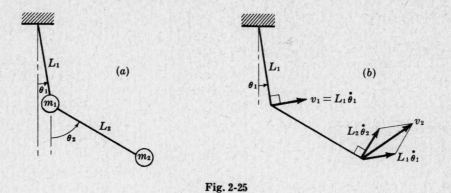

Fig. 2-25

The kinetic energy of the pendulum is given by

$$\text{K.E.} = \tfrac{1}{2}m_1 v_1^2 + \tfrac{1}{2}m_2 v_2^2$$

where $v_1^2 = (L_1\dot\theta_1)^2$ and $v_2^2 = (L_1\dot\theta_1)^2 + (L_2\dot\theta_2)^2 + 2L_1L_2\dot\theta_1\dot\theta_2 \cos(\theta_2-\theta_1)$ are the velocities of masses m_1 and m_2 respectively as shown in Fig. 2-25(a) and 2-25(b) above.

$$\text{P.E.} = m_1 g L_1(1-\cos\theta_1) + m_2 g[L_1(1-\cos\theta_1) + L_2(1-\cos\theta_2)]$$

Lagrange's equation is $\dfrac{d}{dt}\dfrac{\partial(\text{K.E.})}{\partial\dot q_i} + \dfrac{\partial(\text{P.E.})}{\partial q_i} - \dfrac{\partial(\text{K.E.})}{\partial q_i} = 0.$ Now

$$\frac{d}{dt}\frac{\partial(\text{K.E.})}{\partial\dot\theta_1} = \frac{d}{dt}\{m_1 L_1^2\dot\theta_1 + m_2[L_1^2\dot\theta_1 + L_1 L_2\dot\theta_2\cos(\theta_2-\theta_1)]\}$$

$$= m_1 L_1^2\ddot\theta_1 + m_2[L_1^2\ddot\theta_1 + L_1 L_2\ddot\theta_2\cos(\theta_2-\theta_1)] + L_1 L_2\dot\theta_2\frac{d}{dt}[m_2\cos(\theta_2-\theta_1)]$$

$$= m_1 L_1^2\ddot\theta_1 + m_2 L_1^2\ddot\theta_1 + m_2 L_1 L_2\ddot\theta_2$$

where $\sin\theta \doteq \theta$, $\cos(\theta_2-\theta_1) = 1$ and $\dfrac{d}{dt}[\cos(\theta_2-\theta_1)] = 0$ have been substituted because of the assumption of small angles of oscillation. Also,

$$\frac{\partial(\text{K.E.})}{\partial\theta_1} = 0, \qquad \frac{\partial(\text{P.E.})}{\partial\theta_1} = m_1 g L_1\sin\theta_1 + m_2 g L_1\sin\theta_1$$

Thus the first equation of motion is given by

$$(m_1+m_2)L_1\ddot\theta_1 + m_2 L_2\ddot\theta_2 + (m_1+m_2)g\theta_1 = 0$$

Similarly, $\dfrac{d}{dt}\dfrac{\partial(\text{K.E.})}{\partial\dot\theta_2} = \dfrac{d}{dt}[m_2 L_2^2\dot\theta_2 + m_2 L_1 L_2\dot\theta_1\cos(\theta_2-\theta_1)]$

$$= m_2 L_2^2\ddot\theta_2 + m_2 L_1 L_2\ddot\theta_1$$

$$\frac{\partial(\text{K.E.})}{\partial\theta_2} = 0, \qquad \frac{\partial(\text{P.E.})}{\partial\theta_2} = m_2 g L_2\sin\theta_2$$

and the second equation of motion is

$$L_2\ddot\theta_2 + g\theta_2 + L_1\ddot\theta_1 = 0$$

22. A spring connecting two equal rotors mounted on two identical circular shafts is shown in Fig. 2-26 below. If the values of the constants are

$$k = 5, \quad K = 90, \quad J = 1, \quad a = 2$$

determine the equations of motion and the natural frequencies of the system.

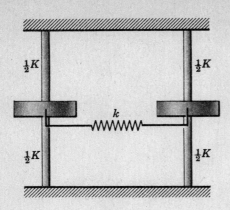 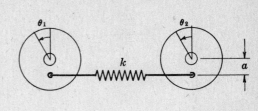

Fig. 2-26

Let θ_1 and θ_2 represent the angular displacements of the rotors. The energy expressions of the system can then be written as

$$\text{K.E.} = \tfrac{1}{2}J\dot{\theta}_1^2 + \tfrac{1}{2}J\dot{\theta}_2^2$$
$$\text{P.E.} = \tfrac{1}{2}K(\theta_1^2 + \theta_2^2) + \tfrac{1}{2}ka^2(\theta_1 - \theta_2)^2$$

where J is the moment of inertia of the rotor. Now

$$\frac{d}{dt}\frac{\partial(\text{K.E.})}{\partial\dot{\theta}_1} = J\ddot{\theta}_1, \qquad \frac{\partial(\text{K.E.})}{\partial\theta_1} = 0, \qquad \frac{\partial(\text{P.E.})}{\partial\theta_1} = K\theta_1 + ka^2(\theta_1 - \theta_2)$$

and hence Lagrange's equation yields

$$J\ddot{\theta}_1 + K\theta_1 + ka^2(\theta_1 - \theta_2) = 0$$

Similarly, the second equation of motion is given by

$$J\ddot{\theta}_2 + K\theta_2 + ka^2(\theta_2 - \theta_1) = 0$$

Substituting the given values for the constants into the equations of motion, we obtain

$$\ddot{\theta}_1 + 110\theta_1 - 20\theta_2 = 0$$
$$\ddot{\theta}_2 + 110\theta_2 - 20\theta_1 = 0$$

Assume that the motion is periodic and is composed of harmonic motion of various amplitudes and frequencies. Let

$$\theta_1 = A\sin(\omega t + \psi) \qquad \text{and} \qquad \theta_2 = B\sin(\omega t + \psi)$$

Substituting and simplifying,

$$(110 - \omega^2)A - 20B = 0$$
$$-20A + (110 - \omega^2)B = 0$$

and the frequency equation is

$$(110 - \omega^2)(110 - \omega^2) - 20^2 = 0$$

which gives $\omega_1 = 9.43$ and $\omega_2 = 11.3$ rad/sec.

23. A simple pendulum of length L and weight mg is pivoted to the mass M which slides without friction on a horizontal plane as shown in Fig. 2-27. Use Lagrange's equation to determine the equations of motion of the system.

Let $x(t)$ denote the displacement of the mass M and $\theta(t)$ denote the angular swing of the pendulum. The kinetic energy of the system is due to the motion of the mass M and the swing of the pendulum bob having a

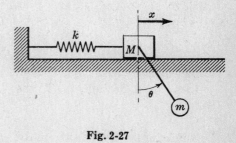

Fig. 2-27

mass m. The potential energy is derived from the spring (stretch or compress) and the position of the bob, as shown in Fig. 2-28.

$$\text{K.E.} = \tfrac{1}{2}M\dot{x}^2 + \tfrac{1}{2}m(\dot{x}^2 + L^2\dot{\theta}^2 + 2L\dot{x}\dot{\theta}\cos\theta)$$

$$\text{P.E.} = \tfrac{1}{2}kx^2 + mgL(1-\cos\theta)$$

and hence

$$\frac{d}{dt}\frac{\partial(\text{K.E.})}{\partial\dot{x}} = (M+m)\ddot{x} + mL\ddot{\theta}\cos\theta - mL\dot{\theta}^2\sin\theta$$

$$\frac{\partial(\text{K.E.})}{\partial x} = 0, \qquad \frac{\partial(\text{P.E.})}{\partial x} = kx$$

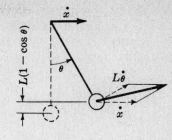

Fig. 2-28

Employing Lagrange's equation, we obtain

$$(M+m)\ddot{x} + mL\ddot{\theta}\cos\theta - mL\dot{\theta}^2\sin\theta + kx = 0$$

For small angles of oscillation, $\sin\theta \doteq \theta$, $\cos\theta = 1$. Then, neglecting higher order terms, the equation of motion becomes

$$(M+m)\ddot{x} + mL\ddot{\theta} + kx = 0$$

Similarly
$$\frac{d}{dt}\frac{\partial(\text{K.E.})}{\partial\dot{\theta}} = mL^2\ddot{\theta} + mL\ddot{x}, \qquad \frac{\partial(\text{K.E.})}{\partial\theta} = 0, \qquad \frac{\partial(\text{P.E.})}{\partial\theta} = mLg\theta$$

and the second equation of motion is

$$L\ddot{\theta} + g\theta + \ddot{x} = 0$$

24. Solve the physical pendulum as discussed in **Problem 13** by Lagrange's equation.

The kinetic energy of the system consists of two parts: (a) translational and (b) rotational.

$$(\text{K.E.})_a = \tfrac{1}{2}m_1(a_1\dot{\theta}_1)^2 + \tfrac{1}{2}m_2[(L^2\dot{\theta}_1^2 + a_2^2\dot{\theta}_2^2 + 2La_2\dot{\theta}_1\dot{\theta}_2\cos(\theta_1-\theta_2)]$$

$$(\text{K.E.})_b = \tfrac{1}{2}J_1\dot{\theta}_1^2 + \tfrac{1}{2}J_2\dot{\theta}_2^2$$

where J_1 and J_2 are the moments of inertia of pendulums 1 and 2 with respect to their pivoted points. Then

$$(\text{K.E.})_{\text{system}} = (\text{K.E.})_a + (\text{K.E.})_b$$

$$(\text{P.E.})_{\text{system}} = m_1ga_1(1-\cos\theta_1) + m_2g[L(1-\cos\theta_1) + a_2(1-\cos\theta_2)]$$

Lagrange's equation is $\dfrac{d}{dt}\dfrac{\partial(\text{K.E.})}{\partial\dot{q}_i} - \dfrac{\partial(\text{K.E.})}{\partial q_i} + \dfrac{\partial(\text{P.E.})}{\partial q_i} = 0$. Now

$$\frac{d}{dt}\frac{\partial(\text{K.E.})}{\partial\dot{\theta}_1} = \ddot{\theta}_1(m_1a_1^2 + m_2L^2 + J_1) + m_2La_2\ddot{\theta}_2\cos(\theta_1-\theta_2)$$

$$\frac{\partial(\text{K.E.})}{\partial\theta_1} = 0, \qquad \frac{\partial(\text{P.E.})}{\partial\theta_1} = m_1ga_1\theta_1 + m_2gL\theta_1$$

where $\sin\theta_1 \doteq \theta_1$, $\cos\theta_1 = 1$, and $m_2La_2\dot{\theta}_2\sin(\theta_1-\theta_2)(\dot{\theta}_1-\dot{\theta}_2) = 0$ because of the assumption of small angles of oscillation. Hence the first equation of motion is

$$(J_1 + m_1a_1^2 + m_2L^2)\ddot{\theta}_1 + (m_1a_1g + m_2gL)\theta_1 + m_2a_2L\ddot{\theta}_2 = 0$$

Similarly
$$\frac{d}{dt}\frac{\partial(\text{K.E.})}{\partial\dot{\theta}_2} = (m_2a_2^2 + J_2)\ddot{\theta}_2 + m_2La_2\ddot{\theta}_1\cos(\theta_1-\theta_2)$$

$$\frac{\partial(\text{K.E.})}{\partial\theta_2} = 0, \qquad \frac{\partial(\text{P.E.})}{\partial\theta_2} = m_2ga_2\theta_2$$

and thus
$$(J_2 + m_2a_2^2)\ddot{\theta}_2 + m_2ga_2\theta_2 + m_2a_2L\ddot{\theta}_1 = 0$$

25. A solid homogeneous cylinder of mass M and radius r rolls without slipping on a cart of mass m as shown in Fig. 2-29. The cart, connected by springs of constants k_1 and k_2, is free to slide on a horizontal surface. By the use of Lagrange's equation, find the equations of motion of the system.

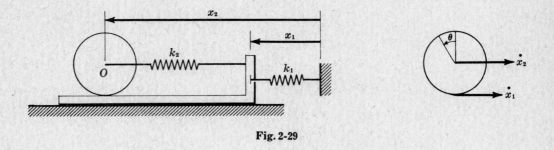

Fig. 2-29

$$\text{K.E.} = \tfrac{1}{2}m\dot{x}_1^2 + \tfrac{1}{2}M\dot{x}_2^2 + \tfrac{1}{2}J_0\dot{\theta}^2$$

where $J_0 = \tfrac{1}{2}Mr^2$ is the moment of inertia of the cylinder with respect to its center. Because the cylinder rolls without slipping, $\theta = (x_2 - x_1)/r$.

$$\text{K.E.} = \tfrac{1}{2}m\dot{x}_1^2 + \tfrac{1}{2}M\dot{x}_2^2 + \tfrac{1}{4}M(\dot{x}_2 - \dot{x}_1)^2$$
$$\text{P.E.} = \tfrac{1}{2}k_1x_1^2 + \tfrac{1}{2}k_2(x_2 - x_1)^2$$

Now $\dfrac{d}{dt}\dfrac{\partial(\text{K.E.})}{\partial \dot{x}_1} = m\ddot{x}_1 - \tfrac{1}{2}M(\ddot{x}_2 - \ddot{x}_1), \qquad \dfrac{\partial(\text{K.E.})}{\partial x_1} = 0, \qquad \dfrac{\partial(\text{P.E.})}{\partial x_1} = k_1x_1 - k_2(x_2 - x_1)$

Therefore the first equation of motion is

$$(m + M/2)\ddot{x}_1 + (k_1 + k_2)x_1 - (M/2)\ddot{x}_2 - k_2x_2 = 0$$

Similarly, $\dfrac{d}{dt}\dfrac{\partial(\text{K.E.})}{\partial \dot{x}_2} = M\ddot{x}_2 + \tfrac{1}{2}M(\ddot{x}_2 - \ddot{x}_1), \qquad \dfrac{\partial(\text{K.E.})}{\partial x_2} = 0, \qquad \dfrac{\partial(\text{P.E.})}{\partial x_2} = k_2(x_2 - x_1)$

and $\qquad\qquad\qquad (3M/2)\ddot{x}_2 + k_2x_2 - (M/2)\ddot{x}_1 - k_2x_1 = 0$

26. A circular cylinder of radius r and mass m rolls without slipping inside a semi-circular groove of radius R. Block M is supported by a spring of constant k and constrained to move without friction in the vertical guide as shown in Fig. 2-30. By the use of Lagrange's equation, find the equations of motion of the system.

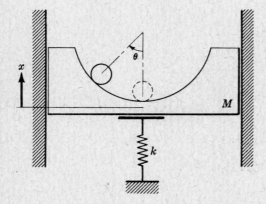

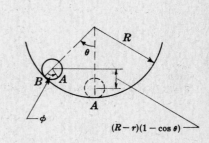

Fig. 2-30

The kinetic and potential energy expressions can be written as

$$\text{P.E.} = \tfrac{1}{2}kx^2 + mg(R-r)(1-\cos\theta)$$
$$\text{K.E.} = \tfrac{1}{2}M\dot{x}^2 + \tfrac{1}{2}mv^2 + \tfrac{1}{2}J_0(\dot{\phi}-\dot{\theta})^2$$

where the moment of inertia of the cylinder is $J_0 = \tfrac{1}{2}mr^2$. Also, arc length $AB = R\theta = r\phi$; hence $\phi = R\theta/r$. The velocity of the center of the cylinder is

$$v = \sqrt{[(R-r)\dot{\theta}]^2 + \dot{x}^2 - 2[(R-r)\dot{\theta}\,\dot{x}\cos(90°+\theta)]}$$

and so

$$v^2 = [(R-r)\dot{\theta}]^2 + \dot{x}^2 + 2(R-r)\dot{\theta}\,\dot{x}\sin\theta$$

Hence

$$\text{K.E.} = \tfrac{1}{2}M\dot{x}^2 + \tfrac{1}{4}m\dot{\theta}^2(R-r)^2 + \tfrac{1}{2}m[\dot{x}^2 + (R-r)^2\dot{\theta}^2 + 2\dot{x}\,\dot{\theta}(R-r)\sin\theta]$$

Assuming small angles of oscillation ($\sin\theta \doteq \theta$, $\cos\theta \doteq 1$) and neglect higher order terms, we have

$$\frac{d}{dt}\frac{\partial(\text{K.E.})}{\partial\dot{x}} = M\ddot{x} + m\ddot{x} + m\ddot{\theta}(R-r)\sin\theta + m\dot{\theta}^2(R-r)\cos\theta$$
$$= (M+m)\ddot{x} + m(R-r)\ddot{\theta}\sin\theta$$

$$\frac{\partial(\text{K.E.})}{\partial x} = 0, \qquad \frac{\partial(\text{P.E.})}{\partial x} = kx$$

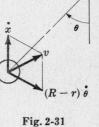

Thus the first equation of motion is obtained by Lagrange's equation

$$\frac{d}{dt}\frac{\partial(\text{K.E.})}{\partial\dot{x}} - \frac{\partial(\text{K.E.})}{\partial x} + \frac{\partial(\text{P.E.})}{\partial x} = 0$$

or

$$(M+m)\ddot{x} + kx + m(R-r)\ddot{\theta}\sin\theta = 0$$

Fig. 2-31

Similarly,

$$\frac{d}{dt}\frac{\partial(\text{K.E.})}{\partial\dot{\theta}} = \frac{3}{2}m(R-r)^2\ddot{\theta} + m\ddot{x}(R-r)\sin\theta, \qquad \frac{\partial(\text{K.E.})}{\partial\theta} = 0, \qquad \frac{\partial(\text{P.E.})}{\partial\theta} = mg(R-r)\sin\theta$$

and so

$$\tfrac{3}{2}(R-r)\ddot{\theta} + (\ddot{x}+g)\sin\theta = 0$$

27. Fig. 2-32 shows a two-degree-of-freedom spring-mass system with damping. Determine the equations of motion of the system by the use of Lagrange's equation.

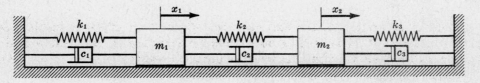

Fig. 2-32

Lagrange's equation for systems with damping is

$$\frac{d}{dt}\frac{\partial(\text{K.E.})}{\partial\dot{q}_i} - \frac{\partial(\text{K.E.})}{\partial q_i} + \frac{\partial(\text{P.E.})}{\partial q_i} + \frac{\partial(\text{D.E.})}{\partial\dot{q}_i} = 0$$

where

$$\text{D.E.} = \tfrac{1}{2}c_1\dot{x}_1^2 + \tfrac{1}{2}c_2(\dot{x}_1-\dot{x}_2)^2 + \tfrac{1}{2}c_3\dot{x}_2^2$$

is known as the dissipation energy of the system due to the presence of damping, represented by dashpots in this problem.

$$\text{K.E.} = \tfrac{1}{2}m_1\dot{x}_1^2 + \tfrac{1}{2}m_2\dot{x}_2^2$$
$$\text{P.E.} = \tfrac{1}{2}k_1x_1^2 + \tfrac{1}{2}k_2(x_1-x_2)^2 + \tfrac{1}{2}k_3x_2^2$$

Now

$$\frac{d}{dt}\frac{\partial(\text{K.E.})}{\partial\dot{x}_1} = m_1\ddot{x}_1, \qquad \frac{\partial(\text{K.E.})}{\partial x_1} = 0, \qquad \frac{\partial(\text{P.E.})}{\partial x_1} = k_1x_1 + k_2(x_1-x_2)$$

$$\frac{\partial(\text{D.E.})}{\partial\dot{x}_1} = c_1\dot{x}_1 + c_2(\dot{x}_1-\dot{x}_2)$$

and so the first equation of motion is

$$m_1\ddot{x}_1 + (c_1 + c_2)\dot{x}_1 + (k_1 + k_2)x_1 - c_2\dot{x}_2 - k_2x_2 = 0$$

Similarly, $\dfrac{d}{dt}\dfrac{\partial(\text{K.E.})}{\partial\dot{x}_2} = m_2\ddot{x}_2,$ $\dfrac{\partial(\text{K.E.})}{\partial x_2} = 0,$ $\dfrac{\partial(\text{P.E.})}{\partial x_2} = -k_2(x_1 - x_2) + k_3x_2$

$$\frac{\partial(\text{D.E.})}{\partial\dot{x}_2} = -c_2(\dot{x}_1 - \dot{x}_2) + c_3\dot{x}_2$$

and $m_2\ddot{x}_2 + (c_2 + c_3)\dot{x}_2 + (k_2 + k_3)x_2 - c_2\dot{x}_1 - k_2x_1 = 0$

28. Fig. 2-33 shows a damped spring-mass system having forced vibration. Find the equations of motion by the use of Lagrange's equation.

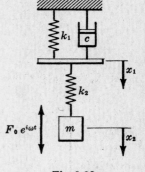

  Lagrange's equation for systems with damping and excitation can be written as

$$\frac{d}{dt}\frac{\partial(\text{K.E.})}{\partial\dot{q}_i} - \frac{\partial(\text{K.E.})}{\partial q_i} + \frac{\partial(\text{P.E.})}{\partial q_i} + \frac{\partial(\text{D.E.})}{\partial\dot{q}_i} = Q_i$$

where Q_i is the excitation. And for this system,

$$Q_1 = 0 \text{ and } Q_2 = F_0 e^{i\omega t}$$
$$\text{K.E.} = \tfrac{1}{2}m\dot{x}_2^2$$
$$\text{P.E.} = \tfrac{1}{2}k_1x_1^2 + \tfrac{1}{2}k_2(x_2 - x_1)^2$$
$$\text{D.E.} = \tfrac{1}{2}c\dot{x}_1^2$$

Fig. 2-33

Now $\dfrac{d}{dt}\dfrac{\partial(\text{K.E.})}{\partial\dot{x}_1} = 0,$ $\dfrac{\partial(\text{K.E.})}{\partial x_1} = 0,$ $\dfrac{\partial(\text{P.E.})}{\partial x_1} = k_1x_1 - k_2(x_2 - x_1),$ $\dfrac{\partial(\text{D.E.})}{\partial\dot{x}_1} = c\dot{x}_1$

and so the first equation of motion is

$$c\dot{x}_1 + (k_1 + k_2)x_1 - k_2x_2 = 0$$

Similarly, $\dfrac{d}{dt}\dfrac{\partial(\text{K.E.})}{\partial\dot{x}_2} = m\ddot{x}_2,$ $\dfrac{\partial(\text{K.E.})}{\partial x_2} = 0,$ $\dfrac{\partial(\text{P.E.})}{\partial x_2} = k_2(x_2 - x_1),$ $\dfrac{\partial(\text{D.E.})}{\partial\dot{x}_2} = 0$

and $m\ddot{x}_2 + k_2x_2 - k_2x_1 = F_0 e^{i\omega t}$

29. Two masses m_1 and m_2 are attached to a rigid weightless bar which is supported by springs k_1 and k_2 and dashpot c as shown in Fig. 2-34. If the motion of the bar is restricted to the plane of the paper, determine the equations of motion of the system by the use of Lagrange's equation.

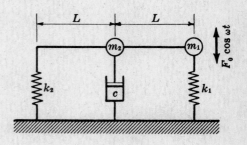

$$\text{K.E.} = \tfrac{1}{2}m_1\dot{x}_1^2 + \tfrac{1}{2}m_2\dot{x}_2^2$$
$$\text{P.E.} = \tfrac{1}{2}k_1x_1^2 + \tfrac{1}{2}k_2(2x_2 - x_1)^2$$
$$\text{D.E.} = \tfrac{1}{2}c\dot{x}_2^2$$
$$Q_1 = F_0\cos\omega t \text{ and } Q_2 = 0$$

  Lagrange's equation for this system can be expressed as

$$\frac{d}{dt}\frac{\partial(\text{K.E.})}{\partial\dot{x}_1} - \frac{\partial(\text{K.E.})}{\partial x_1} + \frac{\partial(\text{P.E.})}{\partial x_1} + \frac{\partial(\text{D.E.})}{\partial\dot{x}_1} = Q_1$$

$$\frac{d}{dt}\frac{\partial(\text{K.E.})}{\partial\dot{x}_2} - \frac{\partial(\text{K.E.})}{\partial x_2} + \frac{\partial(\text{P.E.})}{\partial x_2} + \frac{\partial(\text{D.E.})}{\partial\dot{x}_2} = Q_2$$

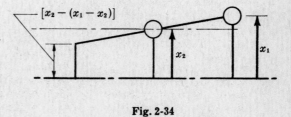

Fig. 2-34

Now $\quad \dfrac{d}{dt}\dfrac{\partial(\text{K.E.})}{\partial \dot{x}_1} = m\ddot{x}_1, \quad \dfrac{\partial(\text{K.E.})}{\partial x_1} = 0, \quad \dfrac{\partial(\text{P.E.})}{\partial x_1} = k_1 x_1 - k_2(2x_2 - x_1), \quad \dfrac{\partial(\text{D.E.})}{\partial \dot{x}_1} = 0$

$\quad\quad\quad \dfrac{d}{dt}\dfrac{\partial(\text{K.E.})}{\partial \dot{x}_2} = m\ddot{x}_2, \quad \dfrac{\partial(\text{K.E.})}{\partial x_2} = 0, \quad \dfrac{\partial(\text{P.E.})}{\partial x_2} = 2k_2(2x_2 - x_1), \quad \dfrac{\partial(\text{D.E.})}{\partial \dot{x}_2} = c\dot{x}_2$

and so the equations of motion are

$$m\ddot{x}_1 + (k_1 + k_2)x_1 - 2k_2 x_2 = F_0 \cos \omega t$$
$$m\ddot{x}_2 + c\dot{x}_2 + 4k_2 x_2 - 2k_2 x_1 = 0$$

EQUIVALENT SYSTEM

30. A rigid steel frame of mass M is connected by a taut wire of length $2L$ as shown in Fig. 2-35. A small mass m is attached to the inside of the frame by two springs of constant k. Determine the equivalent spring-mass system.

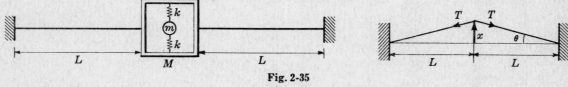

Fig. 2-35

For small angles of oscillation, $\sin \theta = \tan \theta = x/L$. Thus the restoring force on the steel frame due to the high tension in the wire is given by

$$F_x = 2Tx/L$$

where T is the tension in the wire, and remains constant for small angles of oscillation. The equivalent spring constant is therefore

$$k_{\text{eq}} = 2T/L$$

Thus the elastic action of the wire can be replaced by a spring of constant $k_{\text{eq}} = 2T/L$, and the equivalent system is then as shown in Fig. 2-36. The equations of motion of the two systems are identical and are

$$M\ddot{x}_1 + (2T/L + 2k)x_1 - 2kx_2 = 0$$
$$m\ddot{x}_2 + 2k(x_2 - x_1) = 0$$

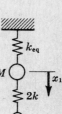

Fig. 2-36

31. A weightless stiff rod of length $2L$ is pivoted at its center and is restrained to move in the vertical plane by springs and masses at each end as shown in Fig. 2-37. Determine the equivalent system.

The equations of motion of the given system are

$$2mL^2\ddot{\theta} + 2kL^2\theta - kLx = T_0 \sin \omega t$$
$$m\ddot{x} + 2kx - kL\theta = 0$$

where θ denotes the angular displacement of the rod and x the rectilinear displacement of the mass m as shown.

For the two-degree-of-freedom spring-mass system as shown in Fig. 2-38 below, the equations of motion are

$$2m\ddot{x}_1 + 2kx_1 - kx_2 = F_0 \sin \omega t$$
$$m\ddot{x}_2 + 2kx_2 - kx_1 = 0$$

The two sets of equations of motion are not exactly the same but of the same form. Since equivalent systems must have the same natural frequencies, a comparison of the natural frequencies or the frequency equations of the two systems is required.

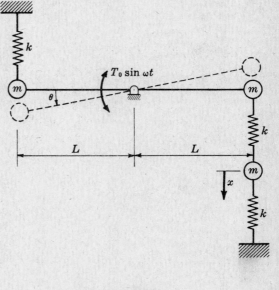

Fig. 2-37

Assume that the motion is periodic and is composed of harmonic motion of various amplitudes and frequencies. Let

$$x = A \cos(\omega t + \psi), \qquad \ddot{x} = -\omega^2 A \cos(\omega t + \psi)$$
$$\theta = B \cos(\omega t + \psi), \qquad \ddot{\theta} = -\omega^2 B \cos(\omega t + \psi)$$
$$x_1 = C \sin(\omega t + \psi), \qquad \ddot{x}_1 = -\omega^2 C \sin(\omega t + \psi)$$
$$x_2 = D \sin(\omega t + \psi), \qquad \ddot{x}_2 = -\omega^2 D \sin(\omega t + \psi)$$

Substituting these expressions into the two sets of equations of motion and simplifying, we obtain

$$(2k - m\omega^2)A - kLB = 0 \qquad \qquad (2k - 2m\omega^2)C - kD = 0$$
$$-kA + (2kL - 2mL\omega^2)B = 0 \quad \text{and} \quad -kC + (2k - m\omega^2)D = 0$$

The frequency equations are found by equating to zero the determinants of the coefficients of A and B, and of C and D. In each case we obtain

$$2m^2\omega^4 - 6mk\omega^2 + 3k^2 = 0$$

Hence the two systems are equivalent to each other.

The equivalence of the two systems can also be established from the equations of motion, considering that

$$x_1 = L\theta, \quad F_0 = T_0/L, \quad x_2 = x$$

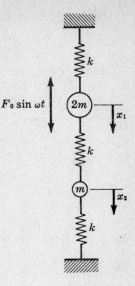

Fig. 2-38

FORCED VIBRATION

32. Use the Mechanical Impedance method to solve for the steady state responses of the general two-degree-of-freedom spring-mass system as shown in Fig. 2-39.

For forced vibration without damping, the equations of motion are given by

$$m_1 \ddot{x}_1 + (k_1 + k_2)x_1 - k_2 x_2 = F_0 \sin \omega t$$
$$m_2 \ddot{x}_2 + k_2 x_2 - k_2 x_1 = 0$$

Substituting $F_0 e^{i\omega t}$ for $F_0 \sin \omega t$, $X_1 e^{i\omega t}$ for x_1, and $X_2 e^{i\omega t}$ for x_2, the equations of motion become

$$m_1 i^2 \omega^2 X_1 e^{i\omega t} + (k_1 + k_2)X_1 e^{i\omega t} - k_2 X_2 e^{i\omega t} = F_0 e^{i\omega t}$$
$$m_2 i^2 \omega^2 X_2 e^{i\omega t} + k_2 X_2 e^{i\omega t} - k_2 X_1 e^{i\omega t} = 0$$

where $i = \sqrt{-1}$. Dividing through by $e^{i\omega t}$ and rearranging,

$$(k_1 + k_2 - m_1\omega^2)X_1 - k_2 X_2 = F_0$$
$$-k_2 X_1 + (k_2 - m_2\omega^2)X_2 = 0$$

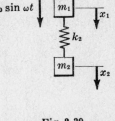

Fig. 2-39

and solving by Cramer's rule,

$$X_1 = \frac{\begin{vmatrix} F_0 & -k_2 \\ 0 & (k_2 - m_2\omega^2) \end{vmatrix}}{(k_1 + k_2 - m_1\omega^2)(k_2 - m_2\omega^2) - k_2^2}$$

$$X_2 = \frac{\begin{vmatrix} (k_1 + k_2 - m_1\omega^2) & F_0 \\ -k_2 & 0 \end{vmatrix}}{(k_1 + k_2 - m_1\omega^2)(k_2 - m_2\omega^2) - k_2^2}$$

or

$$X_1 = \frac{F_0(k_2 - m_2\omega^2)}{m_1 m_2 \omega^4 - (m_1 k_2 + m_2 k_2 + m_2 k_1)\omega^2 + k_1 k_2}$$

$$X_2 = \frac{F_0 k_2}{m_1 m_2 \omega^4 - (m_1 k_2 + m_2 k_2 + m_2 k_1)\omega^2 + k_1 k_2}$$

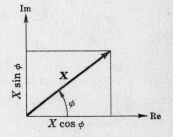

Fig. 2-40

Since the forcing function is $F_0 \sin \omega t = \text{Im}(F_0 e^{i\omega t})$, $x_1 = \text{Im}(\mathbf{X}_1 e^{i\omega t})$ or

$$x_1 = \text{Im}(X_1 e^{i\phi})(e^{i\omega t}) = \text{Im}(X_1 e^{i(\omega t + \phi)})$$
$$= \text{Im}[X_1 \cos(\omega t + \phi) + iX_1 \sin(\omega t + \phi)] = X_1 \sin(\omega t + \phi)$$

Similarly,
$$x_2 = X_2 \sin(\omega t + \phi)$$

But $\mathbf{X}_1 = X_1(\cos\phi + i\sin\phi)$; and the expression for $\mathbf{X}_1$ contains real quantities only. Thus $\phi = 0°$ or $180°$, i.e. the motion of the mass is either in phase or $180°$ out of phase with the excitation. Therefore $\mathbf{X}_1 = X_1$ and $\mathbf{X}_2 = X_2$.

The steady state responses are

$$x_1(t) = \frac{F_0(k_2 - m_2\omega^2)}{m_1 m_2 \omega^4 - (m_1 k_2 + m_2 k_2 + m_2 k_1)\omega^2 + k_1 k_2} \sin \omega t$$

$$x_2(t) = \frac{F_0 k_2}{m_1 m_2 \omega^4 - (m_1 k_2 + m_2 k_2 + m_2 k_1)\omega^2 + k_1 k_2} \sin \omega t$$

33. Determine the steady state vibration of the two masses m_1 and m_2 as shown in Fig. 2-41.

From the free-body diagrams, two force equations can be written:

$$m_1 \ddot{x}_1 = -k_1 x_1 - k_2(x_1 - x_2) + F_0 \cos \omega t$$
$$m_2 \ddot{x}_2 = -k_2(x_2 - x_1) - k_3 x_2$$

or

$$m_1 \ddot{x}_1 + (k_1 + k_2)x_1 - k_2 x_2 = F_0 \cos \omega t$$
$$m_2 \ddot{x}_2 + (k_2 + k_3)x_2 - k_2 x_1 = 0$$

Assume that the motion is periodic and is composed of harmonic motions of various amplitudes and frequencies. Let one of these harmonic components be

$$x_1 = A \cos(\omega t + \phi), \quad \ddot{x}_1 = -A\omega^2 \cos(\omega t + \phi)$$
$$x_2 = B \cos(\omega t + \phi), \quad \ddot{x}_2 = -B\omega^2 \cos(\omega t + \phi)$$

Substituting these values into the equations of motion gives

$$(k_1 + k_2 - m_1\omega^2)A - k_2 B = F_0$$
$$-k_2 A + (k_2 + k_3 - m_2\omega^2)B = 0$$

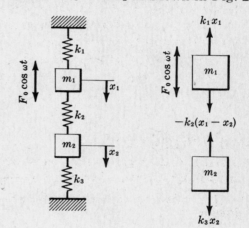

Fig. 2-41

By Cramer's rule, A and B are found to be

$$A = \frac{\begin{vmatrix} F_0 & -k_2 \\ 0 & (k_2 + k_3 - m_2\omega^2) \end{vmatrix}}{(k_1 + k_2 - m_1\omega^2)(k_2 + k_3 - m_2\omega^2) - k_2^2}$$

$$= \frac{F_0(k_2 + k_3 - m_2\omega^2)}{(k_1 + k_2 - m_1\omega^2)(k_2 + k_3 - m_2\omega^2) - k_2^2}$$

$$B = \frac{\begin{vmatrix} (k_1 + k_2 - m_1\omega^2) & F_0 \\ -k_2 & 0 \end{vmatrix}}{(k_1 + k_1 - m_1\omega^2)(k_2 + k_3 - m_2\omega^2) - k_2^2}$$

$$= \frac{F_0 k_2}{(k_1 + k_2 - m_1\omega^2)(k_2 + k_3 - m_2\omega^2) - k_2^2}$$

After the free vibrations have died away, the remaining vibration will be simple harmonic motion with a frequency equal to the forcing frequency. The phase angle is either $0°$ or $180°$, because the motion of the mass is either in phase or $180°$ out of phase with the excitation. Thus the steady state vibration is given by

$$x_1(t) = \frac{F_0(k_2 + k_3 - m_2\omega^2)}{(k_1 + k_2 - m_1\omega^2)(k_2 + k_3 - m_2\omega^2) - k_2^2} \cos \omega t$$

$$x_2(t) = \frac{F_0 k_2}{(k_1 + k_2 - m_1\omega^2)(k_2 + k_3 - m_2\omega^2) - k_2^2} \cos \omega t$$

34. The support of the spring-mass system as shown in Fig. 2-42 is given a forced sinusoidal displacement $x(t) = F_0 \cos \omega t$. Find the steady state vibrations of the masses.

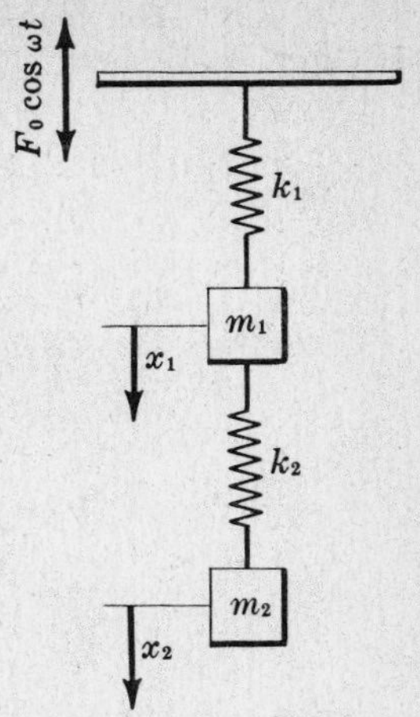

The equations of motion are given by

$$m_1 \ddot{x}_1 + k_1 x_1 + k_2(x_1 - x_2) = k_1 F_0 \cos \omega t$$
$$m_2 \ddot{x}_2 + k_2(x_2 - x_1) = 0$$

Since there is no damping present, the masses vibrate either in phase or $180°$ out of phase with the forced motion. Therefore, let

$$x_1 = A \cos \omega t, \qquad \ddot{x}_1 = -\omega^2 A \cos \omega t$$
$$x_2 = B \cos \omega t, \qquad \ddot{x}_2 = -\omega^2 B \cos \omega t$$

Substituting these values into equations of motion, we obtain

$$(k_1 + k_2 - m_1 \omega^2)A - k_2 B = k_1 F_0$$
$$-k_2 A + (k_2 - m_2 \omega^2)B = 0$$

Fig. 2-42

which when solved by Cramer's rule,

$$A = \frac{\begin{vmatrix} F_0 k_1 & -k_2 \\ 0 & (k_2 - m_2 \omega^2) \end{vmatrix}}{\begin{vmatrix} (k_1 + k_2 - m_1 \omega^2) & -k_2 \\ -k_2 & (k_2 - m_2 \omega^2) \end{vmatrix}}, \qquad B = \frac{\begin{vmatrix} (k_1 + k_2 - m_1 \omega^2) & F_0 k_1 \\ -k_2 & 0 \end{vmatrix}}{\begin{vmatrix} (k_1 + k_2 - m_1 \omega^2) & -k_2 \\ -k_2 & (k_2 - m_2 \omega^2) \end{vmatrix}}$$

Thus the steady state forced vibrations are

$$x_1(t) = \frac{F_0 k_1 (k_2 - m_2 \omega^2)}{m_1 m_2 \omega^4 - [k_1 m_2 + k_2 m_1 + k_2 m_2]\omega^2 + k_1 k_2} \cos \omega t$$

$$x_2(t) = \frac{k_1 k_2 F_0}{m_1 m_2 \omega^4 - [k_1 m_2 + k_2 m_1 + k_2 m_2]\omega^2 + k_1 k_2} \cos \omega t$$

35. A block of mass m, resting on a frictionless horizontal plane, is connected through a spring of constant k to a homogeneous uniform rod of mass M and length L as shown in Fig. 2-43. Determine the steady state response of the block.

Applying the force equation $\Sigma F = ma$ to the block, and the moment equation $\Sigma M = J\ddot{\theta}$ to the pivoted point O, we have

$$m\ddot{x} + kx - kL\theta = F_0 \sin \omega t$$
$$J\ddot{\theta} + (kL^2 + Mga)\theta - kLx = 0$$

where the moment of inertia of the rod with respect to point O is $J = ML^2/3$.

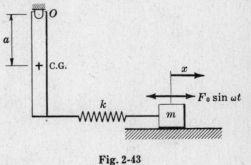

Fig. 2-43

Let $x = A \sin \omega t$, $\theta = B \sin \omega t$; then $\ddot{x} = -\omega^2 A \sin \omega t$, $\ddot{\theta} = -\omega^2 B \sin \omega t$. Substituting these values into the equations of motion,

$$(k - m\omega^2)A - kLB = F_0$$
$$-kLA + (kL^2 + Mga - J\omega^2)B = 0$$

Solving, $\qquad A = \dfrac{\begin{vmatrix} F_0 & -kL \\ 0 & (kL^2 + Mga - J\omega^2) \end{vmatrix}}{\begin{vmatrix} (k - m\omega^2) & -kL \\ -kL & (kL^2 + Mga - J\omega^2) \end{vmatrix}}$

and so $\quad x(t) = \dfrac{F_0(kL^2 + Mga - J\omega^2)}{mJ\omega^4 - (kJ + mkL^2 + mMga)\omega^2 + kMga} \sin \omega t$

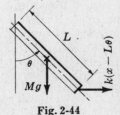

Fig. 2-44

DYNAMIC VIBRATION ABSORBER

36. Fig. 2-45(a) shows that mass M is having forced vibration. In order to cut down as much as possible the amplitude of vibration of mass M, an auxiliary spring-mass is added to mass M as shown in Fig. 2-45(b). Specify the auxiliary system.

In Fig. 2-45(b), the system is of two degrees of freedom with a forcing function acting on mass M. The equations of motion are

$$M\ddot{x}_1 + k_1 x_1 + k_2(x_1 - x_2) = F_0 \sin \omega t$$
$$m\ddot{x}_2 + k_2(x_2 - x_1) = 0$$

Let $x_1 = A \sin \omega t$ and $x_2 = B \sin \omega t$, and thus $\ddot{x}_1 = -\omega^2 A \sin \omega t$, $\ddot{x}_2 = -\omega^2 B \sin \omega t$. Substituting these values into the equations of motion,

$$(k_1 + k_2 - M\omega^2)A - k_2 B = F_0$$
$$-k_2 A + (k_2 - m\omega^2)B = 0$$

Solving, $\quad A = \dfrac{F_0(k_2 - m\omega^2)}{(k_1 + k_2 - M\omega^2)(k_2 - m\omega^2) - k_2^2}$

In order to cut down the amplitude of vibration of mass M, i.e. $A = 0$, $(k_2 - m\omega^2)$ must equal zero. Hence $k_2 = m\omega^2$ and $\omega^2 = k_2/m$.

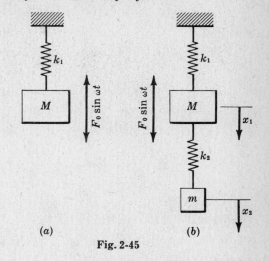

(a)   (b)

Fig. 2-45

The absorber must be, therefore, so designed that its natural frequency is equal to the impressed frequency. When this happens, the amplitude of vibration of mass M is practically zero. In general, an absorber is used only when the natural frequency of the original system is close to the forcing frequency. Hence $k_1/M = k_2/m$ is approximately true for the entire system.

37. A small reciprocating machine weighs 50 lb and runs at a constant speed of 6000 rpm. After it was installed, it was found that the forcing frequency is too close to the natural frequency of the system. What dynamic vibration absorber should be added if the nearest natural frequency of the system should be at least 20% from the impressed frequency?

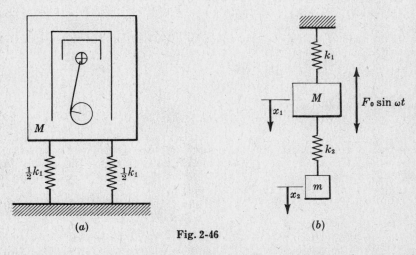

(a)   (b)

Fig. 2-46

After the absorber is added to the reciprocating machine, the entire system becomes a two-degree-of-freedom system, which is simplified and represented by Fig. 2-46(b).

The amplitudes of the steady state vibration of the two masses M and m can be found from earlier discussions in Problem 36, and are given by

$$A = \frac{F_0(k_2 - m\omega^2)}{(k_1 + k_2 - M\omega^2)(k_2 - m\omega^2) - k_2^2}, \qquad B = \frac{F_0 k_2}{(k_1 + k_2 - M\omega^2)(k_2 - m\omega^2) - k_2^2}$$

The natural frequencies of this entire system can be found from

$$(k_1 + k_2 - M\omega^2)(k_2 - m\omega^2) - k_2^2 = 0$$

and these frequencies should be at least 20% from the impressed frequency in order to avoid resonance.

Dividing the last expression by $k_1 k_2$ yields

$$(1 + k_2/k_1 - M\omega^2/k_1)(1 - m\omega^2/k_2) - k_2/k_1 = 0$$

But $k_1/M = \omega_n^2$, $k_2/m = \omega_n^2$, $\omega^2/\omega_n^2 = r^2$. Substituting these values into the above equation, we obtain

$$r^4 - (2 + k_2/k_1)r^2 + 1 = 0$$

and since $r = 0.8$, $k_2/k_1 = 0.21$.

As discussed in Problem 36, $k_1/M = k_2/m$. Hence $k_2/k_1 = m/M = 0.21$ and $m = 0.21M = 10.5$.

Thus the required absorber should weigh 10.5 lb and have a spring of constant $0.21k_1$.

DAMPED FREE VIBRATION

38. Investigate the motion of a two-degree-of-freedom spring-mass system having damped free vibration as shown in Fig. 2-47.

The equations of motion are given by

$$m_1 \ddot{x}_1 + (c_1 + c_2)\dot{x}_1 + (k_1 + k_2)x_1 - c_2\dot{x}_2 - k_2 x_2 = 0$$
$$m_2 \ddot{x}_2 + c_2\dot{x}_2 + k_2 x_2 - c_2\dot{x}_1 - k_2 x_1 = 0$$

Since the components of vibration for a damped system are non-periodic, or oscillatory with diminishing amplitudes, let

$$x_1 = Ae^{st}, \quad x_2 = Be^{st}$$

be the general form of solution. Substituting these expressions into the differential equations and dividing through by e^{st}, we have

$$[m_1 s^2 + (c_1 + c_2)s + (k_1 + k_2)]A - (c_2 s + k_2)B = 0$$
$$[m_2 s^2 + c_2 s + k_2]B - (c_2 s + k_2)A = 0$$

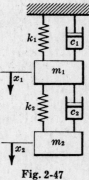

Fig. 2-47

As in the case of free undamped vibration, the solution for this set of algebraic equations has meaning only if

$$\begin{vmatrix} [m_1 s^2 + (c_1 + c_2)s + (k_1 + k_2)] & -(c_2 s + k_2) \\ -(c_2 s + k_2) & (m_2 s^2 + c_2 s + k_2) \end{vmatrix} = 0$$

Expanding the determinant,

$$[m_1 s^2 + (c_1 + c_2)s + k_1 + k_2][m_2 s^2 + c_2 s + k_2] - (c_2 s + k_2)^2 = 0$$

which is known as the characteristic equation of the system. In comparison with the frequency equation of an undamped system, they have similar forms. The solution of this characteristic equation will yield four values for s. Therefore the complete general motion of the system can be expressed as

$$x_1(t) = A_1 e^{s_1 t} + A_2 e^{s_2 t} + A_3 e^{s_3 t} + A_4 e^{s_4 t}$$
$$x_2(t) = B_1 e^{s_1 t} + B_2 e^{s_2 t} + B_3 e^{s_3 t} + B_4 e^{s_4 t}$$

where the four unknown coefficients A_1, A_2, A_3, and A_4 (because $B_1 = \lambda_1 A_1$, $B_2 = \lambda_2 A_2$, $B_3 = \lambda_3 A_3$, and $B_4 = \lambda_4 A_4$) are to be found from the four initial conditions of the system, namely $x_1(0)$, $x_2(0)$, $\dot{x}_1(0)$, and $\dot{x}_2(0)$.

The amplitude ratios are found from the algebraic equations with coefficients A and B:

$$\frac{A_i}{B_i} = \frac{c_2 s_i + k_2}{m_1 s_i^2 + (c_1 + c_2)s_i + k_1 + k_2} = \frac{m_2 s_i^2 + c_2 s_i + k_2}{c_2 s_i + k_2} = \frac{1}{\lambda_i}$$

where $i = 1, 2, 3, 4$.

Suppose the roots of the characteristic equation are complex; then there must be the corresponding complex conjugate roots, i.e.,

$$s_1 = -(r + id) \quad \text{and} \quad s_2 = -(r - id)$$

and so
$$
\begin{aligned}
A_1 e^{-(r+id)t} + A_2 e^{-(r-id)t} &= e^{-rt}[A_1 e^{-idt} + A_2 e^{idt}] \\
&= e^{-rt}[A_1 \cos dt - iA_1 \sin dt + A_2 \cos dt + iA_2 \sin dt] \\
&= Ce^{-rt} \sin(dt + \psi)
\end{aligned}
$$

Similarly
$$B_1 e^{s_1 t} + B_2 e^{s_2 t} = De^{-rt} \sin(dt + \psi)$$

Therefore the general motion is

$$
\begin{aligned}
x_1(t) &= Ce^{-rt} \sin(dt + \psi) + A_3 e^{s_3 t} + A_4 e^{s_4 t} \\
x_2(t) &= De^{-rt} \sin(dt + \psi) + B_3 e^{s_3 t} + B_4 e^{s_4 t}
\end{aligned}
$$

which are oscillatory with diminishing amplitude and aperiodic.

FORCED VIBRATION WITH DAMPING

39. The system shown in Fig. 2-48 is having forced vibration. Determine the general motion of the system.

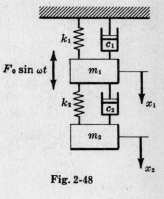

Fig. 2-48

The equations of motion for forced damped vibration are

$$
\begin{aligned}
m_1 \ddot{x}_1 + (c_1 + c_2)\dot{x}_1 + (k_1 + k_2)x_1 - c_2 \dot{x}_2 - k_2 x_2 &= F_0 \sin \omega t \\
m_2 \ddot{x}_2 + c_2 \dot{x}_2 + k_2 x_2 - c_2 \dot{x}_1 - k_2 x_1 &= 0
\end{aligned}
$$

Using the Mechanical Impedance method, substitute $F_0 e^{i\omega t}$ for $F_0 \sin \omega t$, $X_1 e^{i\omega t}$ for x_1, and $X_2 e^{i\omega t}$ for x_2. Rearranging and dividing through by $e^{i\omega t}$, the equations of motion become

$$
\begin{aligned}
[(k_1 + k_2) - m_1 \omega^2 + i(c_1 + c_2)\omega]\mathbf{X}_1 - (k_2 + ic_2\omega)\mathbf{X}_2 &= F_0 \\
-(k_2 + ic_2\omega)\mathbf{X}_1 + [k_2 - m_2 \omega^2 + ic_2\omega]\mathbf{X}_2 &= 0
\end{aligned}
$$

Solving by Cramer's rule,

$$
\mathbf{X}_1 = \frac{\begin{vmatrix} F_0 & -(k_2 + ic_2\omega) \\ 0 & (k_2 - m_2\omega^2 + ic_2\omega) \end{vmatrix}}{[(k_1 + k_2) - m_1\omega^2 + i(c_1 + c_2)\omega][k_2 - m_2\omega^2 + ic_2\omega] - (k_2 + ic_2\omega)^2}
$$

which is of the form $(A + iB)/(C + iD)$ or $(G + iH)$, and $X_1 = \sqrt{\dfrac{A^2 + B^2}{C^2 + D^2}}$ or

$$
X_1 = \sqrt{\frac{F_0^2(k_2 - m_2\omega^2)^2 + c_2^2\omega^2}{(m_1 m_2 \omega^4 - m_1 k_2 \omega^2 - m_2 k_1 \omega^2 - c_1 c_2 \omega^2 + k_1 k_2)^2 + (k_1 c_2 \omega + k_2 c_1 \omega - m_1 c_2 \omega^3 - m_2 c_1 \omega^3 + m_2 c_2 \omega^3)^2}}
$$

Similarly,

$$
X_2 = \sqrt{\frac{F_0^2(k_2^2 + c_2^2\omega^2)}{(m_1 m_2 \omega^4 - m_1 k_2 \omega^2 - m_2 k_1 \omega^2 - c_1 c_2 \omega^2 + k_1 k_2)^2 + (k_1 c_2 \omega + k_2 c_1 \omega - m_1 c_2 \omega^3 - m_2 c_1 \omega^3 + m_2 c_2 \omega^3)^2}}
$$

From the theory of complex variables,

$$
\begin{aligned}
\mathbf{X}_1 &= F_0(G + iH) = X_1 e^{i\phi_1} = X_1(\cos \phi_1 + i \sin \phi_1) \\
\mathbf{X}_2 &= F_0(K + iL) = X_2 e^{i\phi_2} = X_2(\cos \phi_2 + i \sin \phi_2)
\end{aligned}
$$

where $\phi_1 = \tan^{-1}\dfrac{H}{G}$ and $\phi_2 = \tan^{-1}\dfrac{L}{K}$.

But the forcing function is $F_0 \sin \omega t = \text{Im}(F_0 e^{i\omega t})$, and so

$$
\begin{aligned}
x_1 &= \text{Im}(\mathbf{X}_1 e^{i\omega t}) = \text{Im}(X_1 e^{i(\omega t + \phi_1)}) = X_1 \sin(\omega t + \phi_1) \\
x_2 &= \text{Im}(\mathbf{X}_2 e^{i\omega t}) = \text{Im}(X_2 e^{i(\omega t + \phi_2)}) = X_2 \sin(\omega t + \phi_2)
\end{aligned}
$$

where the expressions for X_1, X_2, ϕ_1, and ϕ_2 are given above.

If the forcing function is $F_0 \cos \omega t$, then

$$x_1 = \mathrm{Re}\,(X_1^{i(\omega t + \phi_1)}) = X_1 \cos(\omega t + \phi_1)$$
$$x_2 = \mathrm{Re}\,(X_2^{i(\omega t + \phi_2)}) = X_2 \cos(\omega t + \phi_2)$$

or if the forcing function is $F_0 e^{i\omega t}$, then

$$x_1 = X_1 \sin(\omega t + \phi_1) + iX_1 \cos(\omega t + \phi_1)$$
$$x_2 = X_2 \sin(\omega t + \phi_2) + iX_2 \cos(\omega t + \phi_2)$$

40. A two-degree-of-freedom spring-mass system is having forced vibration as shown in Fig. 2-49. Assume the value of air damping is $c = 1$ lb-in/sec and the spring constants $k_1 = k_2 = 1$ lb/in. Determine the steady state damped vibrations of the masses if $W_1 = W_2 = 1$ lb, $\omega = 1$.

The equations of motion are given by

$$m_1 \ddot{x}_1 + c\dot{x}_1 + k_1 x_1 + k_2(x_1 - x_2) = F_0 \cos t$$
$$m_2 \ddot{x}_2 + c\dot{x}_2 + k_2(x_2 - x_1) = 0$$

where c represents air resistance.

Using the Mechanical Impedance method, and replacing $F_0 \cos t$ by $F_0 e^{it}$, x_1 by $X_1 e^{it}$, and x_2 by $X_2 e^{it}$, the equations of motion become

$$[(k_1 + k_2) - m + ic]X_1 - k_2 X_2 = F_0$$
$$(k_2 - m_2 + ic)X_2 - k_2 X_1 = 0$$

Substituting the values $k_1 = k_2 = 1$, $m_1 = m_2 = 1$, and $c = 1$ into the equations of motion, we have

$$(1 + i)X_1 - X_2 = F_0, \qquad -X_1 + iX_2 = 0$$

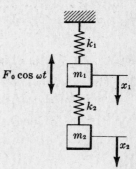

Fig. 2-49

Solving by Cramer's rule,

$$\mathbf{X}_1 = \frac{\begin{vmatrix} F_0 & -1 \\ 0 & i \end{vmatrix}}{\begin{vmatrix} (1+i) & -1 \\ -1 & i \end{vmatrix}} = \frac{iF_0}{i-2}, \qquad \mathbf{X}_2 = \frac{\begin{vmatrix} (1+i) & F_0 \\ -1 & 0 \end{vmatrix}}{\begin{vmatrix} (1+i) & -1 \\ -1 & i \end{vmatrix}} = \frac{F_0}{i-2}$$

Simplifying,

$$\mathbf{X}_1 = F_0(0.2 - 0.4i), \qquad \mathbf{X}_2 = F_0(-0.4 - 0.2i)$$

and hence

$$X_1 = F_0\sqrt{(0.2^2 + 0.4^2)} = 0.45F_0, \qquad X_2 = F_0\sqrt{(0.2^2 + 0.4^2)} = 0.45F_0$$
$$\phi_1 = \tan^{-1}(-0.4/0.2) = -63.4°, \qquad \phi_2 = \tan^{-1}(-0.2/-0.4) = 26.6°$$

But the forcing function is $F_0 \cos t = \mathrm{Re}\,(F_0 e^{it})$ and $\mathbf{X}_1 = X_1 e^{i\phi_1}$. Then

$$x_1 = \mathrm{Re}\,(\mathbf{X}_1 e^{it}) = \mathrm{Re}\,(X_1 e^{i\phi_1} e^{it}) = X_1 \cos(t + \phi_1)$$

Similarly

$$x_2 = X_2 \cos(t + \phi_2)$$

Therefore the steady state damped vibrations are given by

$$x_1 = 0.45F_0 \cos(t - 63.4°)$$
$$x_2 = 0.45F_0 \cos(t + 26.6°)$$

ORTHOGONALITY PRINCIPLE

41. Prove the orthogonality principle for a general two-degree-of-freedom vibrating system by energy considerations.

In general, the motion of the two-degree-of-freedom vibrating system is given by

$$x_1(t) = A_1 \sin(\omega_1 t + \psi_1) + A_2 \sin(\omega_2 t + \psi_2)$$
$$x_2(t) = B_1 \sin(\omega_1 t + \psi_1) + B_2 \sin(\omega_2 t + \psi_2)$$

where A_1, A_2, B_1, B_2 are the amplitudes of vibration of the two masses; ψ_1, ψ_2 the phase angles; and ω_1, ω_2 the natural frequencies of the system.

The kinetic energy of the system is

$$\text{K.E.} = \tfrac{1}{2}m_1 v_1^2 + \tfrac{1}{2}m_2 v_2^2$$

and so

$$
\begin{aligned}
(\text{K.E.})_{\max} &= \tfrac{1}{2}m_1(\dot{x}_1)^2_{\max} + \tfrac{1}{2}m_2(\dot{x}_2)^2_{\max} \\
&= \tfrac{1}{2}m_1(A_1\omega_1 + A_2\omega_2)^2 + \tfrac{1}{2}m_2(B_1\omega_1 + B_2\omega_2)^2 \\
&= \tfrac{1}{2}m_1[(A_1\omega_1)^2 + (A_2\omega_2)^2 + 2(A_1\omega_1)(A_2\omega_2)] + \tfrac{1}{2}m_2[(B_1\omega_1)^2 + (B_2\omega_2)^2 + 2(B_1\omega_1)(B_2\omega_2)]
\end{aligned}
$$

When the system is having principal modes of vibration, the energy expression becomes

$$(\text{K.E.})_{\max} = \tfrac{1}{2}[m_1(A_1\omega_1)^2 + m_2(B_1\omega_1)^2] + \tfrac{1}{2}[m_1(A_2\omega_2)^2 + m_2(B_2\omega_2)^2]$$

Since the two maximum kinetic energy expressions must be equal,

$$m_1(A_1\omega_1)(A_2\omega_2) + m_2(B_1\omega_1)(B_2\omega_2) = 0$$

where ω_1 and ω_2 are the natural frequencies of a two-degree-of-freedom system, and thus cannot always be zero; and hence

$$m_1(A_1 A_2) + m_2(B_1 B_2) = 0$$

42. The mass m, connected by springs $k_1, k_2, \ldots, k_n$, as shown in Fig. 2-50, is constrained to move on the frictionless x-y plane. If the mass m is displaced from its static equilibrium position O and released, show that its principal modes of vibration are orthogonal.

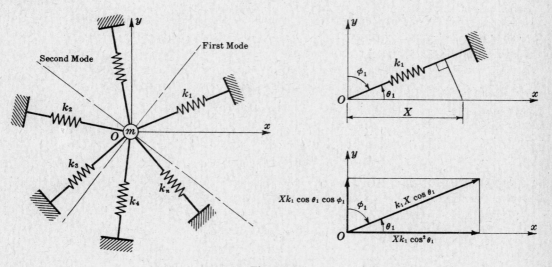

Fig. 2-50

Since the mass m is constrained to move on the x-y plane, this is a two-degree-of-freedom system. When the appropriate initial conditions are applied to the mass, it will either vibrate at the first principal mode with natural frequency ω_1 or at the second principal mode with natural frequency ω_2, i.e.,

$$x(t) = A_1 \sin(\omega_1 t + \psi_1) \quad \text{and} \quad y(t) = \lambda_1 A_1 \sin(\omega_1 t + \psi_1)$$
$$x(t) = A_2 \sin(\omega_2 t + \psi_2) \quad \text{and} \quad y(t) = \lambda_2 A_2 \sin(\omega_2 t + \psi_2)$$

where $\omega_1 \neq \omega_2$.

Let the mass m be displaced X from its static equilibrium position. Spring k_1 is then deformed $X \cos\theta_1$ as shown. The x-component of this spring force is $k_1(X\cos\theta_1)(\cos\theta_1) = Xk_1\cos^2\theta_1$, while the y-component of this same spring force is $Xk_1\cos\theta_1\cos\phi_1$. Similarly, the x-components of the spring forces for springs $k_2, \ldots, k_n$ are $Xk_2\cos^2\theta_2, \ldots, Xk_n\cos^2\theta_n$; and the y-components of the same spring forces are $Xk_2\cos\theta_2\cos\phi_2, \ldots, Xk_n\cos\theta_n\cos\phi_n$. Therefore the x-component of all the spring forces is given by

$$X(k_1\cos^2\theta_1 + k_2\cos^2\theta_2 + \cdots + k_n\cos^2\theta_n) = (k_{xx})X$$

and the y-component of all the spring forces is

$$X(k_1\cos\theta_1\cos\phi_1 + k_2\cos\theta_2\cos\phi_2 + \cdots + k_n\cos\theta_n\cos\phi_n) = (k_{yx})X$$

In a similar manner, spring k_1 is deformed by $Y \cos \phi_1$ when the mass m is displaced Y from the static equilibrium position. The x-component of this spring force is $Yk_1 \cos \phi_1 \cos \theta_1$, while the y-component of this same spring force is $Yk_1 \cos^2 \phi_1$ as shown. Therefore the x-component of all the spring forces due to a displacement Y of the mass m in the y-direction is given by

$$Y(k_1 \cos \phi_1 \cos \theta_1 + k_2 \cos \phi_2 \cos \theta_2 + \cdots + k_n \cos \phi_n \cos \theta_n) = (k_{xy})Y$$

and the y-component is

$$Y(k_1 \cos^2 \phi_1 + k_2 \cos^2 \phi_2 + \cdots + k_n \cos^2 \phi_n) = (k_{yy})Y$$

The equations of motion of the mass m for a general x-y displacement can be expressed as

$$m\ddot{x} + (k_{xx})x + (k_{xy})y = 0$$
$$m\ddot{y} + (k_{yy})y + (k_{yx})x = 0$$

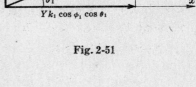

As indicated in the early part of this problem, the amplitudes x and y of the mass vibrating in the principal modes are in the same ratios λ_1 or λ_2. The motion of the mass is along a straight line through the origin O for principal modes of vibration.

Fig. 2-51

Assume that the amplitudes of vibration are unity; then

$$A_1 = \cos \theta_1', \quad A_2 = \cos \theta_2', \quad B_1 = \cos \phi_1', \quad B_2 = \cos \phi_2'$$

which are the direction cosines of the lines of vibration at the principal modes. If $\cos \theta_1' \cos \theta_2' + \cos \phi_1' \cos \phi_2' = 0$, the two lines of vibration will be perpendicular to each other.

When the mass is vibrating at the first principal mode, the equations of motion become

$$-m\omega_1^2 A_1 + (k_{xx})A_1 + (k_{xy})B_1 = 0$$
$$-m\omega_1^2 B_1 + (k_{yy})B_1 + (k_{yx})A_1 = 0$$

and when the mass is vibrating at the second principal mode, the equations of motion are

$$-m\omega_2^2 A_2 + (k_{xx})A_2 + (k_{xy})B_2 = 0$$
$$-m\omega_2^2 B_2 + (k_{yy})B_2 + (k_{yx})A_2 = 0$$

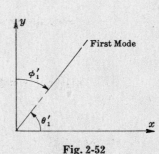

Fig. 2-52

Multiplying the first equation of motion shown above by $-A_2$, the second by $-B_2$, the third by A_1, the fourth by B_1, and adding them together, we obtain

$$m\omega_1^2 A_1 A_2 - (k_{xx})A_1 A_2 - (k_{xy})A_2 B_1 = 0$$
$$m\omega_1^2 B_1 B_2 - (k_{yy})B_1 B_2 - (k_{yx})A_1 B_2 = 0$$
$$-m\omega_2^2 A_1 A_2 + (k_{xx})A_1 A_2 + (k_{xy})A_1 B_2 = 0$$
$$-m\omega_2^2 B_1 B_2 + (k_{yy})B_1 B_2 + (k_{yx})A_2 B_1 = 0$$
$$mA_1 A_2 (\omega_2^2 - \omega_1^2) + mB_1 B_2 (\omega_2^2 - \omega_1^2) = 0$$

or

$$m(A_1 A_2 + B_1 B_2)(\omega_2^2 - \omega_1^2) = 0$$

which indicates that

$$A_1 A_2 + B_1 B_2 = 0$$

and hence

$$\cos \theta_1' \cos \theta_2' + \cos \phi_1' \cos \phi_2' = 0$$

i.e. the two straight lines along which the two principal modes vibrate are perpendicular to each other.

43. If $k_1 = k_2 = k$, and $m_1 = m_2 = m$, show that the principal modes of vibration of the system as given in Problem 1 are orthogonal.

Substituting these given values into the frequency equation, we obtain

$$\omega_1 = 0.62 \sqrt{k/m} \quad \text{and} \quad \omega_2 = 1.62 \sqrt{k/m} \text{ rad/sec}$$

From the amplitude ratio equations, the general motions of the masses are given by

$$x_1(t) = A_1 \sin(0.62\sqrt{k/m}\, t + \psi_1) + A_2 \sin(1.62\sqrt{k/m}\, t + \psi_2)$$
$$x_2(t) = 1.62A_1 \sin(0.62\sqrt{k/m}\, t + \psi_1) - 0.63A_2 \sin(1.62\sqrt{k/m}\, t + \psi_2)$$

from which $A_1A_2 + B_1B_2 = A_1A_2 + (1.62A_1)(-0.63A_2) = 0$

Therefore the principal modes of vibration of the system are orthogonal.

SEMI-DEFINITE SYSTEM

44. Two blocks of mass m_1 and m_2 connected together by a spring of constant k, are resting on a smooth horizontal surface as shown in Fig. 2-53. Obtain an expression for the natural frequencies of the system.

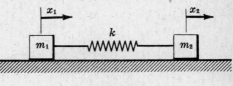

Fig. 2-53

Applying $\Sigma F = ma$,

$$m_1 \ddot{x}_1 + k(x_1 - x_2) = 0$$
$$m_2 \ddot{x}_2 + k(x_2 - x_1) = 0$$

Assume that the motion is periodic and is composed of harmonic motions of various amplitudes and frequencies. Let one of these harmonic components be

$$x_1 = A \sin(\omega t + \psi), \qquad \ddot{x}_1 = -\omega^2 A \sin(\omega t + \psi)$$
$$x_2 = B \sin(\omega t + \psi), \qquad \ddot{x}_2 = -\omega^2 B \sin(\omega t + \psi)$$

Substituting these values into the equations of motion yields

$$(k - m_1\omega^2)A - kB = 0$$
$$-kA + (k - m_2\omega^2)B = 0$$

The frequency equation is obtained by equating to zero the determinant of the coefficients A and B, i.e.,

$$\begin{vmatrix} (k - m_1\omega^2) & -k \\ -k & (k - m_2\omega^2) \end{vmatrix} = 0$$

Expanding,

$$m_1 m_2 \omega^4 - k(m_1 + m_2)\omega^2 = 0 \qquad \text{or} \qquad \omega^2[m_1 m_2 \omega^2 - k(m_1 + m_2)] = 0$$

and the natural frequencies of the system are then equal to

$$\omega_1 = 0 \quad \text{and} \quad \omega_2 = \sqrt{\frac{k(m_1 + m_2)}{m_1 m_2}} \text{ rad/sec}$$

Since one of the natural frequencies of the system is equal to zero, the system is not oscillating. In other words, the two masses move as one whole unit and have no relative motion between them. This is known as a semi-definite system.

45. A simple pendulum is hinged at the center of mass M as shown in Fig. 2-54. Mass M is sliding on a frictionless horizontal surface. Find the natural frequencies of the system.

Take x and θ as the coordinates necessary to specify the configuration of the system. The energy expressions of the system are given by

$$\text{K.E.} = \tfrac{1}{2}M\dot{x}^2 + \tfrac{1}{2}m[\dot{x}^2 + (L\dot{\theta})^2 + 2\dot{x}L\dot{\theta}\cos\theta]$$
$$\text{P.E.} = mgL(1 - \cos\theta)$$

For small angles of oscillation, $\sin\theta \doteq \theta$, $\cos\theta = 1 - \theta^2/2$; hence

$$\text{K.E.} = \tfrac{1}{2}m(\dot{x} + L\dot{\theta})^2 + \tfrac{1}{2}M\dot{x}^2, \quad \text{P.E.} = \tfrac{1}{2}mgL\theta^2$$

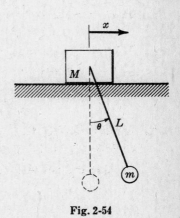

Fig. 2-54

Lagrange's equation is

$$\frac{d}{dt}\frac{\partial(\text{K.E.})}{\partial \dot{q}_i} - \frac{\partial(\text{K.E.})}{\partial q_i} + \frac{\partial(\text{P.E.})}{\partial q_i} = 0$$

Letting $q_i = x$, we have

$$\frac{d}{dt}\frac{\partial(\text{K.E.})}{\partial \dot{x}} = M\ddot{x} + m(\ddot{x} + L\ddot{\theta}), \qquad \frac{\partial(\text{K.E.})}{\partial x} = 0, \qquad \frac{\partial(\text{P.E.})}{\partial x} = 0$$

Thus the first equation of motion is given by

$$(M + m)\ddot{x} + mL\ddot{\theta} = 0$$

For the second equation of motion, let $q_i = \theta$; then

$$\frac{d}{dt}\frac{\partial(\text{K.E.})}{\partial \dot{\theta}} = mL(\ddot{x} + L\ddot{\theta}), \qquad \frac{\partial(\text{K.E.})}{\partial \theta} = 0, \qquad \frac{\partial(\text{P.E.})}{\partial \theta} = mgL\theta$$

and so

$$\ddot{x} + L\ddot{\theta} + g\theta = 0$$

Assume that the motion is periodic and is composed of harmonic motion of various amplitudes and frequencies. Let one of these components be

$$x = A\cos(\omega t + \psi), \qquad \ddot{x} = -\omega^2 A\cos(\omega t + \psi)$$
$$\theta = B\cos(\omega t + \psi), \qquad \ddot{\theta} = -\omega^2 B\cos(\omega t + \psi)$$

Substituting these values into the equations of motion and dividing through by $\cos(\omega t + \psi)$,

$$-\omega^2(M + m)A - mL\omega^2 B = 0$$
$$-\omega^2 A - (\omega^2 L - g)B = 0$$

The frequency equation is obtained by equating to zero the determinant of the coefficients of A and B, i.e.,

$$\begin{vmatrix} -(M + m)\omega^2 & -mL\omega^2 \\ -\omega^2 & -(\omega^2 L - g) \end{vmatrix} = 0$$

or

$$\omega^2[(M + m)(\omega^2 L - g) - mL\omega^2] = 0$$

and hence

$$\omega_1 = 0 \quad \text{and} \quad \omega_2 = \sqrt{\frac{g(M + m)}{LM}} \text{ rad/sec}$$

Since one of the natural frequencies of the system is equal to zero, this is a semi-definite system. At this zero frequency the system is having translational motion and no oscillation.

Supplementary Problems

EQUATION OF MOTION

46. Derive the equations of motion of the system shown in Fig. 2-55. The circular cylinder has a mass m and radius r, and rolls without slipping inside the circular groove of radius R.

   Ans. $(M + m)\ddot{x}_1 + 2kx_1 + m(R - r)\ddot{\theta} = 0$

   $$\frac{3(R - r)}{2}\ddot{\theta} + g\theta + \ddot{x}_1 = 0$$

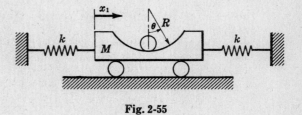

Fig. 2-55

47. Two identical pendulums are rigidly fastened to each end of a steel shaft of torsional stiffness K, as shown in Fig. 2-56 below. The mass of the pendulum bob is m, and the weightless stiff rod of the pendulum is of length L. Assuming that the shaft is resting on frictionless bearing, derive the equations of motion of the system.

Ans. $mL^2\ddot{\theta}_1 + (mgL + K)\theta_1 - K\theta_2 = 0$

$mL^2\ddot{\theta}_2 + (mgL + K)\theta_2 - K\theta_1 = 0$

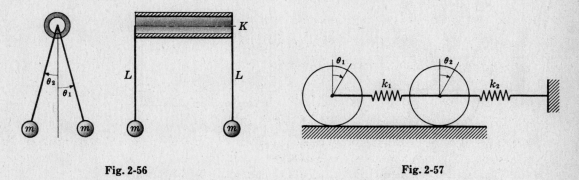

Fig. 2-56   Fig. 2-57

48. Fig. 2-57 above shows that two identical circular cylinders of mass m and radius r are linked together by spring k_1. If the cylinders are free to roll on the horizontal surface, derive the equations of motion of the system.

Ans. $(3m/2)\ddot{\theta}_1 + k_1\theta_1 - k_1\theta_2 = 0$

$(3m/2)\ddot{\theta}_2 + (k_1 + k_2)\theta_2 - k_1\theta_1 = 0$

FREQUENCY EQUATION

49. Derive the frequency equation of the system shown in Fig. 2-58 below. The pulleys are weightless.

Ans. $(m_1 + m_2)\omega^4 - (k_1 m_2 + k_2 m_1 + 4k_2 m_2)\omega^2 + k_1 k_2 = 0$

50. The rectangular block of mass m is supported by four springs at the corners as shown in Fig. 2-59 below. If motion in the vertical plane only is permitted, determine the frequency equation.

Ans. $J_0 m\omega^4 - (2k_x J_0 + 2k_x mh^2 + 2k_y mb^2)\omega^2 + 4k_x k_y b^2 = 0$, where J_0 is the moment of inertia of the block with respect to its center of mass.

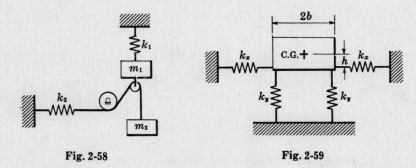

Fig. 2-58   Fig. 2-59   Fig. 2-60

51. For the coupled-pendulum as shown in Fig. 2-60 above, find the frequency equation.

Ans. $4mL^2\omega^4 - (6mLg + 5kL^2)\omega^2 + (2mg^2 + 3Lgk) = 0$

52. A weightless rigid rod with two equal masses m attached to its ends is connected to two springs as shown in Fig. 2-61. Derive an expression for the frequency equation of the system.

Ans. $J_0 m\omega^4 - (k_1 + k_2)J_0 + (k_1 L_1 + k_2 L_2)m\omega^2 + k_1 k_2 (L_1 + L_2)^2 = 0$, where J_0 is the moment of inertia of the rod with respect to its center of gravity O.

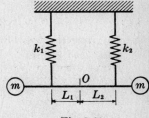

Fig. 2-61

NATURAL FREQUENCIES

53. In Fig. 2-62 below assume the tension in the wire remains constant for small angles of oscillation. Derive expressions for the natural frequencies.

  Ans. $\omega_1 = \sqrt{T/mL}$, $\omega_2 = 1.42\sqrt{T/Lm}$ rad/sec

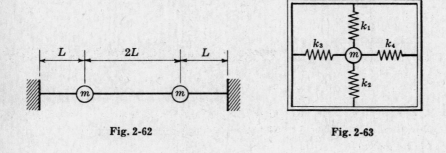

Fig. 2-62   Fig. 2-63

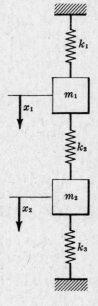

54. Mass m is suspended inside a rigid frame by four springs as shown in Fig. 2-63 above. Determine the natural frequencies of vibration.

  Ans. $\omega_1 = \sqrt{(k_1 + k_2)/m}$, $\omega_2 = \sqrt{(k_3 + k_4)/m}$ rad/sec

55. Calculate the natural frequencies of the general two-degree-of-freedom spring-mass system as shown in Fig. 2-64. $k_1 = 3$, $k_2 = 5$, and $k_3 = 2$ lb/in, and the weights of m_1 and m_2 are 6 and 4 lb respectively.

  Ans. $\omega_2 = 2.59$, $\omega_1 = 0.49$ rad/sec

56. Two identical circular cylinders are linked together as shown in Fig. 2-65 below. Determine the natural frequencies of the system.

  Ans. $\omega_1 = 0$, $\omega_2 = \sqrt{4k/3m}$ rad/sec

Fig. 2-64

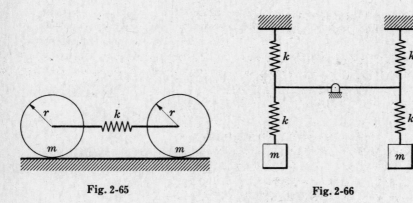

Fig. 2-65   Fig. 2-66

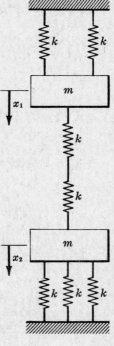

57. Assuming the connecting rod is weightless, determine the frequencies of oscillation of the system shown in Fig. 2-66 above.

  Ans. $\omega_1 = \sqrt{k/m}$, $\omega_2 = \sqrt{k/2m}$ rad/sec

58. Calculate the natural frequencies of the system shown in Fig. 2-67.

  Ans. $\omega_1 = 1.96\sqrt{k/m}$, $\omega_2 = 2.16\sqrt{k/m}$ rad/sec

59. A bar of mass M is supported by two springs of modulus k and $2k$ as shown in Fig. 2-68 below. If the motion of the bar is restrained to the plane of the paper, find the natural frequencies.

  Ans. $\omega_1 = 1.64\sqrt{k/M}$, $\omega_2 = 3.08\sqrt{k/M}$ rad/sec

Fig. 2-67

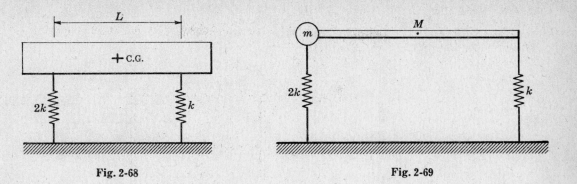

Fig. 2-68   Fig. 2-69

60. For the system shown in Fig. 2-64 above, the weight of each of the two masses $m_1 = m_2$ is 10 lb, and each spring stiffness is equal to $k = 1000$ lb/in. Calculate the frequencies.

Ans. $\omega_1 = 10.00$, $\omega_2 = 17.3$ rad/sec

61. Restraining motion of the rod in the vertical plane, find the frequencies of the system as shown in Fig. 2-69 above.

Ans. $\omega_1 = 0.85\sqrt{2k/(m+M)}$ and $\omega_2 = 1.81\sqrt{2k/(m+M)}$ rad/sec, where M is the mass of the rod.

62. If $m_1 = m_2 = 1$ lb-sec²/in, $k_1 = 200$ and $k_2 = 400$ lb/in, find the frequencies of the system shown in Fig. 2-70.

Ans. $\omega_1 = 9.37$, $\omega_2 = 30.2$ rad/sec

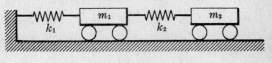

Fig. 2-70

PRINCIPAL MODES OF VIBRATION

63. If $m_1 = m_2 = 10/386$ lb-sec²/in and $k_1 = k_2 = k_3 = 1000$ lb/in, determine the general motion of the system shown in Fig. 2-64 above.

Ans. $x_1(t) = A_1 \sin(10t + \psi_1) + A_2 \sin(17t + \psi_2)$
$x_2(t) = A_1 \sin(10t + \psi_1) - A_2 \sin(17t + \psi_2)$

64. For the system shown in Fig. 2-64 above, $m_1 = 1$ and $m_2 = 2$ lb-sec²/ft, $k_1 = 20$, $k_2 = 10$ and $k_3 = 30$ lb/ft, and an initial velocity of 10 ft/sec is imparted to mass m_1. What is the resulting motion of the masses?

Ans. $x_1(t) = 1.97 \sin(4.05t) + 0.35 \sin(5.81t)$
$x_2(t) = 1.42 \sin(4.05t) - 0.97 \sin(5.81t)$

65. For Problem 62, calculate $x_1(t)$ and $x_2(t)$ for the initial conditions:

(a) $x_1(0) = 0.3$, $\dot{x}_1(0) = 0$, $x_2(0) = 0$, $\dot{x}_2(0) = 0$
(b) $x_1(0) = 0.3$, $\dot{x}_1(0) = 0$, $x_2(0) = 0$, $\dot{x}_2(0) = 5.0$

Ans. (a) $x_1(t) = 0.114 \cos(9.37t) + 0.186 \cos(30.2t)$
$x_2(t) = 0.145 \cos(9.37t) - 0.145 \cos(30.2t)$

(b) $x_1(t) = -0.117 \cos(9.37t + 167°) - 0.186 \cos(30.2t - 177°)$
$x_2(t) = -0.149 \cos(9.37t + 167°) + 0.145 \cos(30.2t - 177°)$

66. Determine the appropriate initial conditions for the first and second modes of vibration of the system discussed in Problem 65.

Ans.   First mode: (1) $\dot{x}_1(0) = \dot{x}_2(0) = 0$, $x_2(0) = 1.28x_1(0)$
(2) $x_1(0) = x_2(0) = 0$, $\dot{x}_2(0) = 1.28\dot{x}_1(0)$
(3) $x_2(0) = 1.28x_1(0)$, $\dot{x}_2(0) = 1.28\dot{x}_1(0)$

Second mode: (1) $\dot{x}_1(0) = \dot{x}_2(0) = 0$, $x_2(0) = -0.78x_1(0)$
(2) $x_1(0) = x_2(0) = 0$, $\dot{x}_2(0) = -0.78\dot{x}_1(0)$
(3) $x_2(0) = -0.78x_1(0)$, $\dot{x}_2(0) = -0.78\dot{x}_1(0)$

67. A stiff uniform bar of length $4L$ and mass M is suspended by two springs of equal stiffness k as shown in Fig. 2-71. Determine the appropriate initial conditions for the first and second principal modes of vibration.

Ans. First mode: (1) $\dot{x}_1(0) = \dot{x}_2(0) = 0, \;\; x_1(0) = -2.81Lx_2(0)$

                     (2) $x_1(0) = x_2(0) = 0, \;\; \dot{x}_1(0) = -2.81L\dot{x}_2(0)$

                     (3) $x_1(0) = -2.81Lx_2(0), \;\; \dot{x}_1(0) = -2.81L\dot{x}_2(0)$

     Second mode: (1) $\dot{x}_1(0) = \dot{x}_2(0) = 0, \;\; x_1(0) = 0.47Lx_2(0)$

                     (2) $x_1(0) = x_2(0) = 0, \;\; \dot{x}_1(0) = 0.47L\dot{x}_2(0)$

                     (3) $x_1(0) = 0.47Lx_2(0), \;\; \dot{x}_1(0) = 0.47L\dot{x}_2(0)$

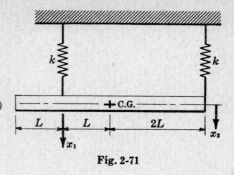

Fig. 2-71

COORDINATE COUPLING

68. Two equal masses m_1 and m_2 are attached to a wire with high initial tension as shown in Fig. 2-72 below. Determine the principal coordinates of the system.

     Ans. $p_1 = y_1 + y_2, \;\; p_2 = y_1 - y_2$

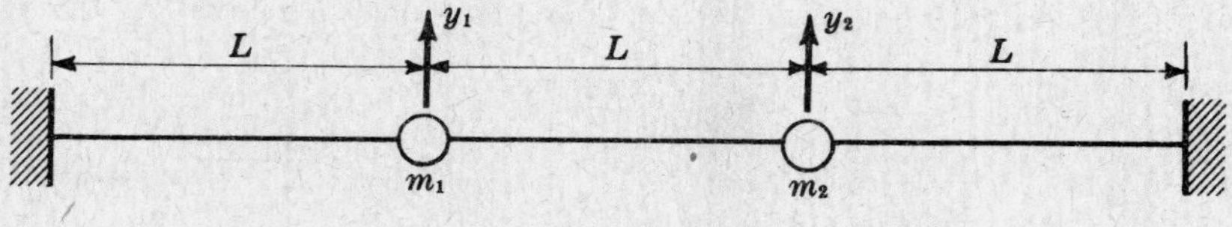

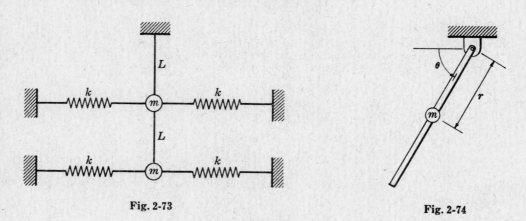

Fig. 2-72

69. Determine the coupling terms in Problem 19.

     Ans. $a^2k(\theta_2 - \theta_1)$, static coupling term.

70. Determine the coupling terms in Problem 24.

     Ans. $m_2a_2L\ddot{\theta}_1, \;\; m_2a_2L\ddot{\theta}_2$; both are dynamic coupling terms.

LAGRANGE'S EQUATION

71. A double pendulum is connected by four springs of equal stiffness as shown in Fig. 2-73 below. For small angles of oscillation, find its frequencies by the use of Lagrange's equation.

     Ans. $\omega_2 = \sqrt{2k/m + 3.12g/L}, \;\; \omega_1 = \sqrt{2k/m + 0.58g/L}$ rad/sec

Fig. 2-73

Fig. 2-74

72. A small mass m is free to slide on a homogeneous uniform rod of mass M and length L which is pivoted at one end as shown in Fig. 2-74 above. Derive the equations of motion by the use of Lagrange's equation.

     Ans. $(ML^2 + mr^2)\ddot{\theta} + 2mr\dot{r}\dot{\theta} - (mr + ML)g\cos\theta = 0$

            $m\ddot{r} - m\dot{\theta}^2r + mg(1 - \sin\theta) = 0$

73. A circular homogeneous cylinder of mass M and radius R rolls without slipping inside a circular surface of radius $3R$. A small mass m, connected by two equal springs of modulus k, is initially at the center of the cylinder at the equilibrium position as shown in Fig. 2-75 below. Derive expressions for the equations of motion of the system by the use of Lagrange's equation.

Ans. $4(MR^2 + J_0 + mR^2)\ddot{\theta} + 2(M + m)gR\theta + 2mR\ddot{r} - 2mgr = 0$

$m\ddot{r} + 2kr + 2mR\ddot{\theta} - 2mg\theta = 0$

where J_0 is the moment of inertia of the cylinder.

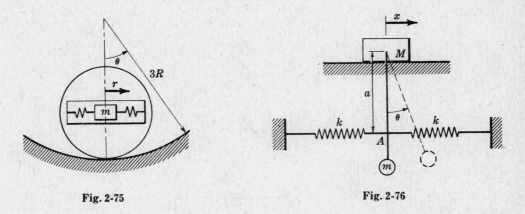

Fig. 2-75   **Fig. 2-76**

74. A particle of mass m is moving on a horizontal plane under the action of an attractive force which is a function of the displacement, i.e. $F(t) = f(1/r^2)$. Determine the equations of motion of the particle by the use of Lagrange's equation.

Ans. $r\ddot{\theta} + 2\dot{r}\dot{\theta} = 0$

$m\ddot{r} + k/r^2 - mr\dot{\theta}^2 = 0$

75. The block of mass M moves along a smooth horizontal plane, and carries a simple pendulum of length L and mass m as shown in Fig. 2-76 above. Two equal springs of modulus k are connected to the pendulum at point A. Determine the equations of motion describing small oscillation of the system about the equilibrium position by the use of the Lagrange's equation.

Ans. $(M + m)\ddot{x} + 2kx + mL\ddot{\theta} + 2ak\theta = 0$

$mL^2\ddot{\theta} + (mgL + 2ka^2)\theta + mL\ddot{x} + 2akx = 0$

EQUIVALENT SYSTEM

76. A simple pendulum of length L and mass m is suspended by two equal springs of constant k as shown in Fig. 2-77 below. Restraining the motion of the pendulum in the plane of the paper, show that the system is equivalent to a mathematical pendulum of length $(L + mg/k)$.

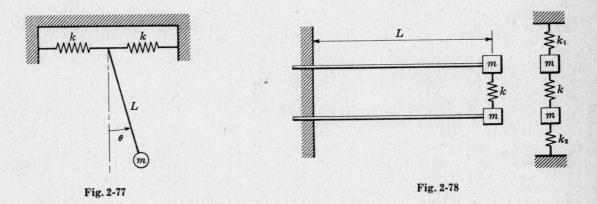

Fig. 2-77   **Fig. 2-78**

77. Assuming the cantilevers are weightless, find the equivalent spring-mass system for the system shown in Fig. 2-78 above.   *Ans.* $k_1 = k_2 = 3EI/L^3$

78. For the system shown in Fig. 2-79 below, determine a dynamically equivalent system and its frequency. Let J_0 be the moment of inertia of the pulley.

Ans. Let $x = ax_e$; then $k_e = a^2(r/R)^2 k$, $M_e = (M + J_0/R^2)(\dot{x}^2/\dot{x}_e^2)$, and $\omega_n = \sqrt{kr^2/(MR^2 + J_0^2)}$ rad/sec.

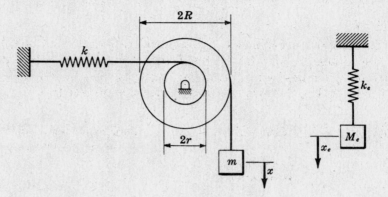

Fig. 2-79

FORCED VIBRATION

79. Determine the steady state vibration of the system shown in Fig. 2-80.

Ans.
$$x_1 = \frac{F_1[(k_2+k_3)/m_1m_2 - \omega^2/m_1] + F_2k_2/m_1m_2}{m_1m_2\omega^4 - [m_2(k_1+k_2) + m_1(k_2+k_3)]\omega^2 + k_1k_2 + k_2k_3 + k_3k_1} \cos\omega t$$

$$x_2 = \frac{F_2[(k_1+k_2)/m_1m_2 - \omega^2/m_2] + F_1k_2/m_1m_2}{m_1m_2\omega^4 - [m_2(k_1+k_2) + m_1(k_2+k_3)]\omega^2 + k_1k_2 + k_2k_3 + k_3k_1} \cos\omega t$$

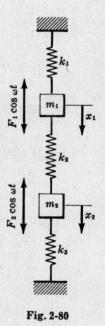

80. A simple spring-mass system is having forced vibration. The modulus of the spring is 1000 lb/in, and the weight of the mass is 10 lb. The forcing function is $10\cos 250t$. Specify the spring constant and the weight of an absorber such that the natural frequencies of the entire system will be 25% from the impressed frequency.   *Ans.* $k = 560$ lb/in, $W = 3.48$ lb

81. An air compressor, running at a constant speed of 1200 rpm, is having large amplitude of vibration. A vibration absorber is added. The weight of the compressor is 500 lb, and the compressor has an unbalance of 1.00 in-lb. Calculate the weight and spring constant of the absorber if the natural frequencies of the system should be at least 10% from the impressed frequency. *Ans.* $W = 22.2$ lb, $k = 824$ lb/in

82. Show that the orthogonality principle is satisfied for Problem 3 and Problem 5.

Fig. 2-80

Chapter 3

Several Degrees of Freedom

INTRODUCTION

When n independent coordinates are required to specify the positions of the masses of a system, the system is of *n degrees of freedom*. A block supported by springs, for example, could have six degrees of freedom, three in translation and three in rotation. The systems shown in Fig. 3-1 are of multiple degrees of freedom.

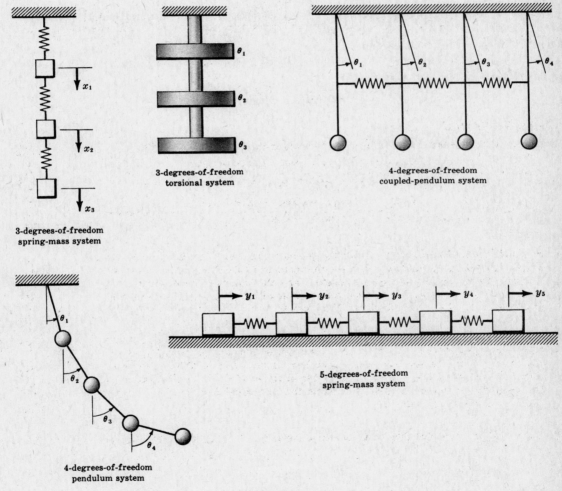

3-degrees-of-freedom
spring-mass system

3-degrees-of-freedom
torsional system

4-degrees-of-freedom
coupled-pendulum system

4-degrees-of-freedom
pendulum system

5-degrees-of-freedom
spring-mass system

Fig. 3-1

In principle, the vibration analysis of a system of n degrees of freedom is not too much different from that of two degrees of freedom; similar approaches and methods can be employed for their solution and analysis. Due to the increased number of possibilities, however, the task of finding the natural frequencies and of evaluating the responses usually requires much more mathematical work.

71

EQUATION OF MOTION

The motion of a vibratory system of n degrees of freedom is represented by n differential equations of motion which can be obtained by Newton's second law of motion, by Lagrange's equation, or by the *Influence Coefficients method*. Since the equations of motion are not entirely independent, a simultaneous solution of these equations requires the complete evaluation of determinants of nth order. This will yield all the natural frequencies of the system.

Other methods of solution are the *Stodola method, Holzer's method,* and the *Matrix Iteration*. These are more direct numerical approaches, and have been frequently employed for the analysis of multiple-degree-of-freedom vibratory systems.

INFLUENCE COEFFICIENTS

An *influence coefficient,* denoted by α_{12}, is defined as the static deflection of the system at position 1 due to a unit force applied at position 2 when the unit force is the only force acting. The influence coefficient is therefore a convenient method to keep account of all the induced deflections due to various applied forces, and to set up the differential equations of motion for the system.

It can be shown that the following expression is true:

$$\alpha_{ij} = \alpha_{ji}$$

where $\quad \alpha_{ij} =$ deflection at position i due to a unit force applied at position j,
$\qquad \alpha_{ji} =$ deflection at position j due to a unit force applied at position i.

This is *Maxwell's reciprocal theorem.*

An influence coefficient α_{ij}, on the other hand, can be interpreted to mean the angular displacement at coordinate i due to a unit torque applied at coordinate j for rotational motions.

MATRICES

The use of matrices in vibration analysis not only simplifies the work involved, but also helps us to understand the procedure of solution. This is particularly true for systems of several degrees of freedom.

The differential equations of motion of a system of n masses can be expressed in general as

$$m_{11}\ddot{q}_1 + m_{12}\ddot{q}_2 + \cdots + m_{1n}\ddot{q}_n + k_{11}q_1 + k_{12}q_2 + \cdots + k_{1n}q_n = 0$$
$$m_{21}\ddot{q}_1 + m_{22}\ddot{q}_2 + \cdots + m_{2n}\ddot{q}_n + k_{21}q_1 + k_{22}q_2 + \cdots + k_{2n}q_n = 0$$
$$\cdots\cdots\cdots\cdots\cdots\cdots\cdots\cdots\cdots\cdots\cdots\cdots\cdots\cdots\cdots\cdots\cdots$$
$$m_{n1}\ddot{q}_1 + m_{n2}\ddot{q}_2 + \cdots + m_{nn}\ddot{q}_n + k_{n1}q_1 + k_{n2}q_2 + \cdots + k_{nn}q_n = 0$$

In matrix notation, these differential equations are written as

$$\begin{bmatrix} m_{11} & m_{12} & \dots & m_{1n} \\ m_{21} & m_{22} & \dots & m_{2n} \\ \dots & \dots & \dots & \dots \\ m_{n1} & m_{n2} & \cdots & m_{nn} \end{bmatrix} \begin{Bmatrix} \ddot{q}_1 \\ \ddot{q}_2 \\ . \\ \ddot{q}_n \end{Bmatrix} + \begin{bmatrix} k_{11} & k_{12} & \dots & k_{1n} \\ k_{21} & k_{22} & \dots & k_{2n} \\ \dots & \dots & \dots & \dots \\ k_{n1} & k_{n2} & \dots & k_{nn} \end{bmatrix} \begin{Bmatrix} q_1 \\ q_2 \\ . \\ q_n \end{Bmatrix} = \begin{Bmatrix} 0 \\ 0 \\ . \\ 0 \end{Bmatrix}$$

or may be expressed simply as

$$[M]\{\ddot{q}\} + [K]\{q\} = 0$$

where $[M]$ is called the *inertia matrix* and $[K]$ the *stiffness matrix.*

The simplified matrix equation can also be expressed as

$$\{\ddot{q}\} + [M]^{-1}[K]\{q\} = 0$$

or

$$\{\ddot{q}\} + [C]\{q\} = 0$$

where $[C]$ is called the *dynamic matrix*.

The natural frequencies are obtained from the characteristic equation $|I - C| = 0$ derived from λ-matrix theory: $[I - C] = 0$ where I is the unit diagonal matrix and C the dynamic matrix.

Let p_i be the principal coordinates, then

$$
\begin{aligned}
q_1 &= \alpha_{11}p_1 + \alpha_{12}p_2 + \cdots + \alpha_{1n}p_n \\
q_2 &= \alpha_{21}p_1 + \alpha_{22}p_2 + \cdots + \alpha_{2n}p_n \\
&\cdots\cdots\cdots\cdots\cdots\cdots\cdots\cdots\cdots\cdots\cdots\cdots\cdots \\
q_n &= \alpha_{n1}p_1 + \alpha_{n2}p_2 + \cdots + \alpha_{nn}p_n
\end{aligned}
$$

In matrix notation, this becomes

$$[q] = [\alpha][p] \quad \text{and} \quad [p] = [\alpha]^{-1}[q]$$

where $[\alpha]$ is called the *transformation matrix*.

MATRIX ITERATION

This is an iterative procedure that leads to the principal modes of vibration of a system and its natural frequencies.

Displacements of the masses are estimated, from which the matrix equation of the system is written. The influence coefficients of the system are substituted into the matrix equation which is then expanded. Normalization of the displacement and expansion of the matrix is repeated. This process is continued until the first mode repeats itself to any desired degree of accuracy.

For the next higher modes and natural frequencies, the orthogonality principle is used to obtain a new matrix equation that is free from any lower modes. The iterative procedure is repeated.

THE STODOLA METHOD

The Stodola method is an *iterative process* used for the calculation of the principal modes and natural frequencies of free undamped vibrating systems. It is a *physical approach* and there is no need to derive the differential equations of motion.

In general, the inertia force is a maximum at the same time the deflection is a maximum, and is in the direction opposite to that of the deflection. In other words, the inertia force is interpreted as the dynamic loading. When the system is vibrating at one of its principal modes with natural frequency ω, it is acted upon by inertia forces $-m_i\ddot{x}_i$, where $x_i = A_i \sin \omega t$, and hence

$$-m_i\ddot{x}_i = \omega^2 m_i x_i$$

Initially, the configuration of a principal mode is assumed, and the corresponding inertia forces and spring forces are calculated. Then the spring deflections are determined from these spring forces. These deflections are employed to start the next iteration. If the assumed configuration happens to be a principal mode, the computed deflections will be the same as those to start with. If the assumed deflections are different from those of a principal mode, the iteration process is repeated until it converges.

THE HOLZER METHOD

The Holzer method is a *tabular method* used for the determination of natural frequency for free or forced vibration with or without damping. It is based on successive assumptions of the natural frequency of the system, each followed by the calculation of the configuration governed by that assumed frequency. It can be used to compute all the natural frequencies of a system, and each calculation is completely independent of the others. The Holzer method is particularly useful for calculating the frequencies of torsional vibration in shafts.

For both ends free systems,

$$x_i = x_{i-1} - (\omega^2/k_i) \sum_1^{i-1} m_j x_j$$

where x, ω, m, k are the displacement, natural frequency, mass, and spring constant of the system respectively.

For one end fixed and one end free systems,

$$x_i = x_{i-1} - (\omega^2/k_i) \sum_1^{i-1} m_j x_j$$

For both ends fixed systems,

$$x_i = x_{i-1} + (1/k_i) \left[k_1 x_1 - \omega^2 \sum_1^{i-1} m_j x_j \right]$$

For an assumed value of ω, begin the process by assuming unit amplitude of vibration for the first mass. The amplitudes and inertia forces for all the remaining masses are then calculated. For the last mass of the system, its amplitude of vibration should be zero for fixed ends; and for free ends, the total inertia force is zero. The remaining values (amplitude or inertia force) for each of the assumed frequencies are then plotted against the assumed values of the natural frequency to give the true frequencies of the system.

THE MECHANICAL IMPEDANCE METHOD

Certain vibration problems are conveniently solved by the Mechanical Impedance method which makes use of the fact that the impedances of the spring, dashpot, and mass are respectively k, $ic\omega$, and $-m\omega^2$. This method will yield the steady state responses for forced vibration, and will lead to the frequency equation of the system for free vibration. The application of this method can be simplified to the following *four steps* for multiple-degree-of-freedom systems:

(1) Multiply the amplitude of each connecting point or junction of the system by the impedances of the elements connected to it.

(2) Subtract from this quantity the "slippage terms", which can be defined as the products of the impedances of the elements attached to the junction and the amplitudes of their opposite ends.

(3) Set this quantity equal to zero for free vibration, and equal to the maximum value of the sinusoidal force for forced vibration. If more than one force is applied to the junction, proper account must be taken of their phase relations.

(4) Solve the equations for the amplitudes of vibration. The expression for the amplitude of each junction can be put into the form $F/(A + iB)$. The numerical value of the amplitude is $F/\sqrt{A^2 + B^2}$, and the motion lags behind the force by the angle whose tangent is B/A.

THE ORTHOGONALITY PRINCIPLE

For three-degree-of-freedom systems, this is

$$m_1A_1A_2 + m_2B_1B_2 + m_3C_1C_2 = 0$$
$$m_1A_1A_3 + m_2B_1B_3 + m_3C_1C_3 = 0$$
$$m_1A_2A_3 + m_2B_2B_3 + m_3C_2C_3 = 0$$

where the m's are the masses of the system; A's, B's, C's are the amplitudes of vibration at different modes.

For n-degree-of-freedom systems, the orthogonality principle is

$$\sum_{i=1}^{n} m_i A_i^r A_i^s = 0$$

where $r \neq s$ are the principal modes of vibration of the system.

Solved Problems

THREE-DEGREE-OF-FREEDOM SYSTEM

1. Calculate the natural frequencies of the three-degree-of-freedom spring-mass system as shown in Fig. 3-2.

The differential equations of motion can be obtained by applying $\Sigma F = ma$ to each mass.

$$4m\ddot{x}_1 = -3kx_1 - k(x_1 - x_2)$$
$$2m\ddot{x}_2 = k(x_1 - x_2) - k(x_2 - x_3)$$
$$m\ddot{x}_3 = k(x_2 - x_3)$$

Rearranging,

$$4m\ddot{x}_1 + 4kx_1 - kx_2 = 0$$
$$2m\ddot{x}_2 + 2kx_2 - kx_3 - kx_1 = 0$$
$$m\ddot{x}_3 + kx_3 - kx_2 = 0$$

Assume that the motion is periodic, and is composed of harmonic components of various amplitudes and frequencies. Let

$$x_1 = A \cos(\omega t + \psi), \quad \ddot{x}_1 = -\omega^2 A \cos(\omega t + \psi)$$
$$x_2 = B \cos(\omega t + \psi), \quad \ddot{x}_2 = -\omega^2 B \cos(\omega t + \psi)$$
$$x_3 = C \cos(\omega t + \psi), \quad \ddot{x}_3 = -\omega^2 C \cos(\omega t + \psi).$$

Substituting these values into the equations of motion yields

$$(4k - 4m\omega^2)A - kB = 0$$
$$-kA + (2k - 2m\omega^2)B - kC = 0$$
$$-kB + (k - m\omega^2)C = 0$$

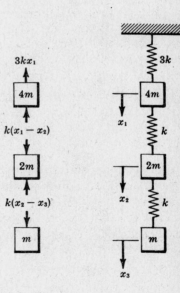

Fig. 3-2

The frequency equation is obtained by equating to zero the determinant of the coefficients of $A, B,$ and C:

$$\begin{vmatrix} (4k - 4m\omega^2) & -k & 0 \\ -k & (2k - 2m\omega^2) & -k \\ 0 & -k & (k - m\omega^2) \end{vmatrix} = 0$$

Expand the determinant to get

$$(k - m\omega^2)(8m^2\omega^4 - 16km\omega^2 + 3k^2) = 0$$

and hence the natural frequencies are $\omega_1 = 0.46\sqrt{k/m}$, $\omega_2 = \sqrt{k/m}$, $\omega_3 = 1.34\sqrt{k/m}$ rad/sec.

2. Determine the frequency equation for a general three-degree-of-freedom spring-mass system as shown in Fig. 3-3.

The differential equations of motion can be obtained by applying Newton's second law to each of the masses.

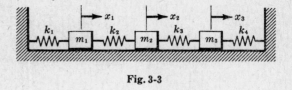

$$m_1 \ddot{x}_1 = -k_1 x_1 - k_2 (x_1 - x_2)$$
$$m_2 \ddot{x}_2 = -k_2 (x_2 - x_1) - k_3 (x_2 - x_3)$$
$$m_3 \ddot{x}_3 = -k_3 (x_3 - x_2) - k_4 x_3$$

Fig. 3-3

Rearranging,

$$m_1 \ddot{x}_1 + (k_1 + k_2) x_1 - k_2 x_2 = 0$$
$$m_2 \ddot{x}_2 + (k_2 + k_3) x_2 - k_3 x_3 - k_2 x_1 = 0$$
$$m_3 \ddot{x}_3 + (k_3 + k_4) x_3 - k_3 x_2 = 0$$

Assume that the motion is periodic and is composed of harmonic components of various amplitudes and frequencies. Let

$$x_1 = A \sin(\omega t + \psi), \qquad \ddot{x}_1 = -\omega^2 A \sin(\omega t + \psi)$$
$$x_2 = B \sin(\omega t + \psi), \qquad \ddot{x}_2 = -\omega^2 B \sin(\omega t + \psi)$$
$$x_3 = C \sin(\omega t + \psi), \qquad \ddot{x}_3 = -\omega^2 C \sin(\omega t + \psi)$$

Substituting these values into the equations of motion yields

$$(k_1 + k_2 - m_1 \omega^2) A - k_2 B = 0$$
$$-k_2 A + (k_2 + k_3 - m_2 \omega^2) B - k_3 C = 0$$
$$-k_3 B + (k_3 + k_4 - m_3 \omega^2) C = 0$$

The frequency equation of the system is found by equating to zero the determinant of the coefficients of $A, B,$ and C:

$$\begin{vmatrix} (k_1 + k_2 - m_1 \omega^2) & -k_2 & 0 \\ -k_2 & (k_2 + k_3 - m_2 \omega^2) & -k_3 \\ 0 & -k_3 & (k_3 + k_4 - m_3 \omega^2) \end{vmatrix} = 0$$

Expand the determinant to obtain

$$\omega^6 - [(k_1 + k_2)/m_1 + (k_2 + k_3)/m_2 + (k_3 + k_4)/m_3] \omega^4$$
$$+ [(k_1 k_2 + k_2 k_3 + k_3 k_1)/m_1 m_2 + (k_2 k_3 + k_3 k_4 + k_4 k_2)/m_2 m_3 + (k_1 + k_2)(k_3 + k_4)/m_3 m_1] \omega^2$$
$$- (k_1 k_2 k_3 + k_2 k_3 k_4 + k_3 k_4 k_1 + k_4 k_1 k_2)/m_1 m_2 m_3 = 0$$

3. Calculate the natural frequencies of the three unequal masses attached at the quarter points of a high tension string as shown in Fig. 3-4 below.

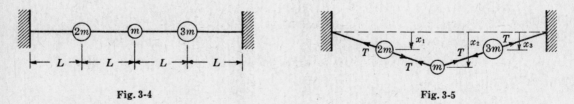

Fig. 3-4   **Fig. 3-5**

Assume the tension in the string is T and remains unchanged for small angles of oscillation. Consider the mass $2m$ as shown in Fig. 3-5 above. Apply $\Sigma F = ma$.

$$2m \ddot{x}_1 = T \left[\frac{(x_2 - x_1)}{L} \right] - T \left[\frac{x_1}{L} \right]$$

$$m \ddot{x}_2 = -T \left[\frac{(x_2 - x_1)}{L} \right] - T \left[\frac{(x_2 - x_3)}{L} \right]$$

$$3m \ddot{x}_3 = T \left[\frac{(x_2 - x_3)}{L} \right] - T \left[\frac{x_3}{L} \right]$$

Assume the motion of oscillation is periodic and is composed of harmonic components of various amplitudes and frequencies. Let

$$x_1 = A \sin(\omega t + \psi), \qquad \ddot{x}_1 = -\omega^2 A \sin(\omega t + \psi)$$
$$x_2 = B \sin(\omega t + \psi), \qquad \ddot{x}_2 = -\omega^2 B \sin(\omega t + \psi)$$
$$x_3 = C \sin(\omega t + \psi), \qquad \ddot{x}_3 = -\omega^2 C \sin(\omega t + \psi)$$

Substituting these values into the equations of motion yields

$$(2T/L - 2m\omega^2)A - (T/L)B = 0$$
$$-(T/L)A + (2T/L - m\omega^2)B - (T/L)C = 0$$
$$-(T/L)B + (2T/L - 3m\omega^2)C = 0$$

The frequency of the system is then obtained by equating to zero the determinant of the coefficients of $A, B,$ and C:

$$\begin{vmatrix} (2T/L - 2m\omega^2) & -T/L & 0 \\ -T/L & (2T/L - m\omega^2) & -T/L \\ 0 & -T/L & (2T/L - 3m\omega^2) \end{vmatrix} = 0$$

Expand the determinant to obtain

$$6m^3\omega^6 - (22Tm^2/L)\omega^4 + (19T^2m/L^2)\omega^2 - 4T^3/L^3 = 0$$

from which $\omega_1 = 0.56\sqrt{T/Lm}, \quad \omega_2 = 0.83\sqrt{T/Lm}, \quad \omega_3 = 1.59\sqrt{T/Lm}$ rad/sec.

4. Calculate the natural frequencies of the system shown in Fig. 3-6. The coupling springs are unstressed when the pendulums are in the vertical positions.

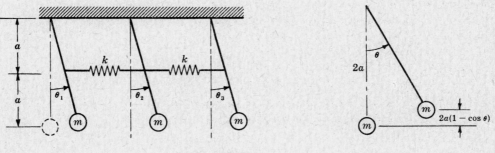

Fig. 3-6

The kinetic energy of the system is

$$\text{K.E.} = \tfrac{1}{2}m(2a\dot{\theta}_1)^2 + \tfrac{1}{2}m(2a\dot{\theta}_2)^2 + \tfrac{1}{2}m(2a\dot{\theta}_3)^2$$

where $2a$ is the length of all the pendulums and the θ's are the angular displacements.

The potential energy of the system is

$$\text{P.E.} = 2mga[(1 - \cos\theta_1) + (1 - \cos\theta_2) + (1 - \cos\theta_3)] + \tfrac{1}{2}k(a\theta_2 - a\theta_1)^2 + \tfrac{1}{2}k(a\theta_3 - a\theta_2)^2$$

Lagrange's equation is

$$\frac{d}{dt}\frac{\partial(\text{K.E.})}{\partial\dot{q}_i} - \frac{\partial(\text{K.E.})}{\partial q_i} + \frac{\partial(\text{P.E.})}{\partial q_i} = 0$$

where q_i, the generalized coordinates, are $\theta_1, \theta_2,$ and θ_3 for this system. Now

$$\frac{d}{dt}\frac{\partial(\text{K.E.})}{\partial\dot{\theta}_1} = 4ma^2\ddot{\theta}_1, \qquad \frac{\partial(\text{K.E.})}{\partial\theta_1} = 0, \qquad \frac{\partial(\text{P.E.})}{\partial\theta_1} = 2mga\sin\theta_1 - ka(a\theta_2 - a\theta_1)$$

and hence the first equation of motion is

$$4ma^2\ddot{\theta}_1 + (2mag + ka^2)\theta_1 - ka^2\theta_2 = 0$$

in which $\sin\theta_1 \doteq \theta$ has been assumed for small oscillations. Similarly, the remaining equations of motion are

$$4ma^2\ddot{\theta}_2 + (2mag + 2ka^2)\theta_2 - ka^2\theta_1 - ka^2\theta_3 = 0$$
$$4ma^2\ddot{\theta}_3 + (2mag + ka^2)\theta_3 - ka^2\theta_2 = 0$$

Assume the motion is periodic and is composed of harmonic components of various amplitudes and frequencies. Let

$$\theta_1 = A \sin(\omega t + \psi), \qquad \ddot{\theta}_1 = -\omega^2 A \sin(\omega t + \psi)$$
$$\theta_2 = B \sin(\omega t + \psi), \qquad \ddot{\theta}_2 = -\omega^2 B \sin(\omega t + \psi)$$
$$\theta_3 = C \sin(\omega t + \psi), \qquad \ddot{\theta}_3 = -\omega^2 C \sin(\omega t + \psi)$$

Substituting these values into the equations of motion and simplifying,

$$(2mag + ka^2 - 4ma^2\omega^2)A - (ka^2)B = 0$$
$$-(ka^2)A + (2mag + 2ka^2 - 4ma^2\omega^2)B - (ka^2)C = 0$$
$$-(ka^2)B + (2mag + ka^2 - 4ma^2\omega^2)C = 0$$

and the frequency equation is

$$\begin{vmatrix} (2mag + ka^2 - 4ma^2\omega^2) & -ka^2 & 0 \\ -ka^2 & (2mag + 2ka^2 - 4ma^2\omega^2) & -ka^2 \\ 0 & -ka^2 & (2mag + ka^2 - 4ma^2\omega^2) \end{vmatrix} = 0$$

Expand the determinant to obtain

$$(2mag + ka^2 - 4ma^2\omega^2)[(2mag + ka^2 - 4ma^2\omega^2)(2mag + 2ka^2 - 4ma^2\omega^2) - 2k^2a^4] = 0$$

Solving, $\omega_1 = \sqrt{g/2a}$, $\omega_2 = \sqrt{g/2a + k/4m}$, $\omega_3 = \sqrt{g/2a + 3k/4m}$ rad/sec.

5. Determine the natural frequencies of the three-degree-of-freedom spring-mass system shown in Fig. 3-7.

Employing $\Sigma F = ma$, we obtain

$$m\ddot{x}_1 = -kx_1 - k(x_1 - x_3) - k(x_1 - x_2)$$
$$m\ddot{x}_2 = -kx_2 - k(x_2 - x_1) - k(x_2 - x_3)$$
$$m\ddot{x}_3 = -kx_3 - k(x_3 - x_1) - k(x_3 - x_2)$$

or

$$m\ddot{x}_1 + 3kx_1 - kx_2 - kx_3 = 0$$
$$m\ddot{x}_2 + 3kx_2 - kx_3 - kx_1 = 0$$
$$m\ddot{x}_3 + 3kx_3 - kx_1 - kx_2 = 0$$

Assume the motion is periodic and is composed of harmonic components of various amplitudes and frequencies. Let

$$x_1 = A \cos(\omega t + \psi), \qquad \ddot{x}_1 = -\omega^2 A \cos(\omega t + \psi)$$
$$x_2 = B \cos(\omega t + \psi), \qquad \ddot{x}_2 = -\omega^2 B \cos(\omega t + \psi)$$
$$x_3 = C \cos(\omega t + \psi), \qquad \ddot{x}_3 = -\omega^2 C \cos(\omega t + \psi)$$

Substituting these values into the equations of motion and simplifying, we obtain

$$(3k - m\omega^2)A - kB - kC = 0$$
$$-kA + (3k - m\omega^2)B - kC = 0$$
$$-kA - kB + (3k - m\omega^2)C = 0$$

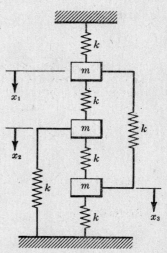

Fig. 3-7

The frequency equation is obtained from

$$\begin{vmatrix} (3k - m\omega^2) & -k & -k \\ -k & (3k - m\omega^2) & -k \\ -k & -k & (3k - m\omega^2) \end{vmatrix} = 0$$

Expand to obtain

$$\omega^6 - (9k/m)\omega^4 + (24k^2/m^2)\omega^2 - (16k^3/m^3) = 0$$

from which $\omega_1 = \sqrt{k/m}$, $\omega_2 = \omega_3 = 2\sqrt{k/m}$ rad/sec. Since two of the natural frequencies of the system are equal, the system is said to be *degenerate*.

6. Derive the equations of motion of the triple pendulum as shown in Fig. 3-8 below by the use of Lagrange's equation. What are the frequencies of oscillation? $L_1 = L_2 = L_3 = L$, and $m_1 = m_2 = m_3 = m$.

The kinetic energy of the triple pendulum is

$$\text{K.E.} = \tfrac{1}{2}mv^2 = \tfrac{1}{2}m(v_1^2 + v_2^2 + v_3^2)$$

where $v_1^2 = (L\dot{\theta}_1)^2$

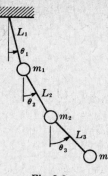

$$v_2^2 = (L\dot{\theta}_1)^2 + (L\dot{\theta}_2)^2 + 2L^2\dot{\theta}_1\dot{\theta}_2 \cos(\theta_2 - \theta_1)$$
$$= L^2(\dot{\theta}_1^2 + \dot{\theta}_2^2 + 2\dot{\theta}_1\dot{\theta}_2) = L^2(\dot{\theta}_1 + \dot{\theta}_2)^2$$

$$v_3^2 = v_2^2 + (L\dot{\theta}_3)^2 + 2v_2(L\dot{\theta}_3)\cos(\theta_3 - \theta_2 + \theta_1)$$
$$= L^2(\dot{\theta}_1^2 + \dot{\theta}_2^2 + \dot{\theta}_3^2 + 2\dot{\theta}_1\dot{\theta}_3 + 2\dot{\theta}_2\dot{\theta}_3 + 2\dot{\theta}_1\dot{\theta}_2)$$

are the velocities of the bobs as shown in Fig. 3-9.   Then

$$\text{K.E.} = (mL^2/2)(3\dot{\theta}_1^2 + 2\dot{\theta}_2^2 + \dot{\theta}_3^2 + 4\dot{\theta}_1\dot{\theta}_2 + 2\dot{\theta}_1\dot{\theta}_3 + 2\dot{\theta}_2\dot{\theta}_3)$$

Fig. 3-8

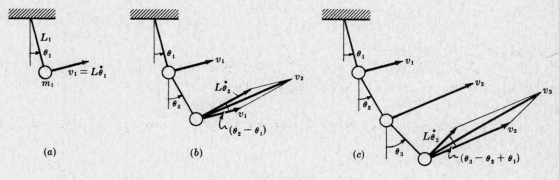

Fig. 3-9

The potential energy due to mass m_1 is $mgL(1 - \cos\theta_1)$; due to mass m_2 is $mgL[(1 - \cos\theta_1) + (1 - \cos\theta_2)]$; and due to mass m_3 is $mgL[(1 - \cos\theta_1) + (1 - \cos\theta_2) + (1 - \cos\theta_3)]$.   Thus the potential energy of the system is

$$\text{P.E.} = 3mgL(1 - \cos\theta_1) + 2mgL(1 - \cos\theta_2) + mgL(1 - \cos\theta_3)$$

Lagrange's equation is

$$\frac{d}{dt}\frac{\partial(\text{K.E.})}{\partial \dot{q}_i} - \frac{\partial(\text{K.E.})}{\partial q_i} + \frac{\partial(\text{P.E.})}{\partial q_i} = 0$$

When $q_1 = \theta_1$, Lagrange's equation yields

$$3L\ddot{\theta}_1 + 3g\theta_1 + 2L\ddot{\theta}_2 + L\ddot{\theta}_3 = 0$$

Similarly for $q_2 = \theta_2$,

$$2L\ddot{\theta}_1 + 2L\ddot{\theta}_2 + 2g\theta_2 + L\ddot{\theta}_3 = 0$$

and for $q_3 = \theta_3$,

$$L\ddot{\theta}_1 + L\ddot{\theta}_2 + L\ddot{\theta}_3 + g\theta_3 = 0$$

Assume the motions are periodic and are composed of harmonic components of various frequencies and amplitudes.   Let

$$\theta_1 = A\sin(\omega t + \psi), \qquad \ddot{\theta}_1 = -\omega^2 A\sin(\omega t + \psi)$$
$$\theta_2 = B\sin(\omega t + \psi), \qquad \ddot{\theta}_2 = -\omega^2 B\sin(\omega t + \psi)$$
$$\theta_3 = C\sin(\omega t + \psi), \qquad \ddot{\theta}_3 = -\omega^2 C\sin(\omega t + \psi)$$

Substituting these values into the equations of motion, we obtain

$$(3g - 3L\omega^2)A - 2L\omega^2 B - L\omega^2 C = 0$$
$$-2L\omega^2 A + (2g - 2L\omega^2)B - L\omega^2 C = 0$$
$$-L\omega^2 A - L\omega^2 B + (g - L\omega^2)C = 0$$

which give the frequency equation of the system in the determinant form

$$\begin{vmatrix} (3g - 3L\omega^2) & -2L\omega^2 & -L\omega^2 \\ -2L\omega^2 & (2g - 2L\omega^2) & -L\omega^2 \\ -L\omega^2 & -L\omega^2 & (g - L\omega^2) \end{vmatrix} = 0$$

Expand the determinant and simplify to obtain

$$L^3\omega^6 - 9gL^2\omega^4 + 18g^2L\omega^2 - 6g^3 = 0$$

from which $\omega_1 = 0.65\sqrt{g/L}$, $\omega_2 = 1.52\sqrt{g/L}$,   and $\omega_3 = 2.5\sqrt{g/L}$ rad/sec.

7. A taut string with three equal masses attached is shown in Fig. 3-10. The tension in the string can be assumed to remain constant for small angles of oscillation. If an excitation $F_0 \sin \omega t$ is applied to the center mass, determine the steady-state motion.

The equations of motion can be obtained from $\Sigma F = ma$.

$$m_1 \ddot{x}_1 = -(T/L)x_1 - (T/L)(x_1 - x_2)$$
$$m_2 \ddot{x}_2 = -(T/L)(x_2 - x_1) - (T/L)(x_2 - x_3) + F_0 \sin \omega t$$
$$m_3 \ddot{x}_3 = -(T/L)(x_3 - x_2) - (T/L)x_3$$

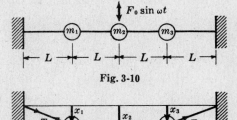

Fig. 3-10

Rearranging,

$$m_1 \ddot{x}_1 + (2T/L)x_1 - (T/L)x_2 = 0$$
$$m_2 \ddot{x}_2 + (2T/L)x_2 - (T/L)x_3 - (T/L)x_1 = F_0 \sin \omega t$$
$$m_3 \ddot{x}_3 + (2T/L)x_3 - (T/L)x_2 = 0$$

Fig. 3-11

Assume the motion is periodic and is composed of harmonic components of various amplitudes and frequencies. Let

$$x_1 = A \sin(\omega t + \psi), \qquad \ddot{x}_1 = -\omega^2 A \sin(\omega t + \psi)$$
$$x_2 = B \sin(\omega t + \psi), \qquad \ddot{x}_2 = -\omega^2 B \sin(\omega t + \psi)$$
$$x_3 = C \sin(\omega t + \psi), \qquad \ddot{x}_3 = -\omega^2 C \sin(\omega t + \psi)$$

Substituting these values into the equations of motion, we obtain

$$(2T/L - \omega^2 m)A - (T/L)B = 0$$
$$-(T/L)A + (2T/L - \omega^2 m)B - (T/L)C = F_0$$
$$-(T/L)B + (2T/L - \omega^2 m)C = 0$$

Solving by Cramer's rule

$$A = \frac{(F_0 T/L)(2T/L - \omega^2 m)}{(Lm/T)^3\omega^6 - 6(Lm/T)^2\omega^4 + (10Lm/T)\omega^2 - 4}$$

$$B = \frac{(F_0 T/L)(2T/L - \omega^2 m)^2}{(Lm/T)^3\omega^6 - 6(Lm/T)^2\omega^4 + 10(Lm/T)\omega^2 - 4}$$

$$C = \frac{(F_0 T/L)(2T/L - \omega^2 m)}{(Lm/T)^3\omega^6 - 6(Lm/T)^2\omega^4 + 10(Lm/T)\omega^2 - 4}$$

Therefore the steady state motion of the masses are given by

$$x_1(t) = A \sin \omega t$$
$$x_2(t) = B \sin \omega t$$
$$x_3(t) = C \sin \omega t$$

where the values for the coefficients A, B, and C are as given above.

8. Three wooden blocks of equal unit mass, connected by springs of stiffness 1 lb/in as shown in Fig. 3-12, are resting on a frictionless surface. If block m_3 is given an initial displacement of 1 inch, determine the resulting motion of the system.

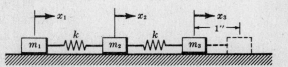

Fig. 3-12

The differential equations of motion of the system are given by

$$m_1\ddot{x}_1 + k(x_1 - x_2) = 0$$
$$m_2\ddot{x}_2 + k(x_2 - x_1) + k(x_2 - x_3) = 0 \qquad (1)$$
$$m_3\ddot{x}_3 + k(x_3 - x_2) = 0$$

With the given values, equations (1) can be simplified to

$$\ddot{x}_1 + x_1 - x_2 = 0$$
$$\ddot{x}_2 + 2x_2 - x_3 - x_1 = 0 \qquad (2)$$
$$\ddot{x}_3 + x_3 - x_2 = 0$$

Assume the motion is periodic and is composed of harmonic components of various amplitudes and frequencies. Let

$$x_1 = A \sin(\omega t + \psi), \qquad \ddot{x}_1 = -\omega^2 A \sin(\omega t + \psi)$$
$$x_2 = B \sin(\omega t + \psi), \qquad \ddot{x}_2 = -\omega^2 B \sin(\omega t + \psi)$$
$$x_3 = C \sin(\omega t + \psi), \qquad \ddot{x}_3 = -\omega^2 C \sin(\omega t + \psi)$$

Substituting these values into the equations of motion, we obtain

$$(1 - \omega^2)A - B = 0$$
$$-A + (2 - \omega^2)B - C = 0 \qquad (3)$$
$$-B + (1 - \omega^2)C = 0$$

The solution to this set of algebraic equations is obtained from

$$\begin{vmatrix} (1 - \omega^2) & -1 & 0 \\ -1 & (2 - \omega^2) & -1 \\ 0 & -1 & (1 - \omega^2) \end{vmatrix} = 0$$

Expand the determinant to obtain the frequency equation of the system:

$$(1 - \omega^2)(\omega^2 - 3)\omega^2 = 0$$

from which $\omega_1 = 0$, $\omega_2 = 1$, $\omega_3 = \sqrt{3}$ rad/sec.

But the general motion of a three-degree-of-freedom system can be written as

$$x_1 = A_1 \sin(\omega_1 t + \psi_1) + A_2 \sin(\omega_2 t + \psi_2) + A_3 \sin(\omega_3 t + \psi_3)$$
$$x_2 = B_1 \sin(\omega_1 t + \psi_1) + B_2 \sin(\omega_2 t + \psi_2) + B_3 \sin(\omega_3 t + \psi_3)$$
$$x_3 = C_1 \sin(\omega_1 t + \psi_1) + C_2 \sin(\omega_2 t + \psi_2) + C_3 \sin(\omega_3 t + \psi_3)$$

where the amplitudes B's and C's can be expressed in terms of the A's from the amplitude ratios obtained from the algebraic equations.

From the first equation of (3),

$$\frac{B}{A} = (1 - \omega^2), \qquad \text{and} \qquad \frac{B_1}{A_1} = 1, \quad \frac{B_2}{A_2} = 0, \quad \frac{B_3}{A_3} = -2$$

where the three amplitude ratios are obtained from substituting ω_1, ω_2, ω_3, respectively into the amplitude equation for ω. Using the first and the third equations of (3), we obtain

$$\frac{C}{A} = \frac{(1 - \omega^2)}{(1 - \omega^2)}, \qquad \text{and} \qquad \frac{C_1}{A_1} = 1, \quad \frac{C_2}{A_2} = \frac{0}{0}, \quad \frac{C_3}{A_3} = 1$$

Since the amplitude ratio C_2/A_2 is indeterminate from the above equation, use the second equation of (3) to obtain

$$-A_2 + (2 - \omega_2^2)B_2 - C_2 = 0$$
or
$$-A_2 + B_2 - C_2 = 0$$

where ω is replaced by ω_2. This can be written as

$$-1 + (B_2/A_2) - (C_2/A_2) = 0$$

But $(B_2/A_2) = 0$, so $C_2/A_2 = -1$.

In terms of A_1, A_2, A_3, the general motion becomes

$$x_1 = A_1 \sin(\omega_1 t + \psi_1) + A_2 \sin(\omega_2 t + \psi_2) + A_3 \sin(\omega_3 t + \psi_3)$$
$$x_2 = A_1 \sin(\omega_1 t + \psi_1) - 2A_3 \sin(\omega_3 t + \psi_3)$$
$$x_3 = A_1 \sin(\omega_1 t + \psi_1) - A_2 \sin(\omega_2 t + \psi_2) + A_3 \sin(\omega_3 t + \psi_3)$$

The six initial conditions are $x_1(0) = 0$, $x_2(0) = 0$, $x_3(0) = 1$, $\dot{x}_1(0) = 0$, $\dot{x}_2(0) = 0$, $\dot{x}_3(0) = 0$; then

$$A_1 \sin \psi_1 + A_2 \sin \psi_2 + A_3 \sin \psi_3 = 0 \tag{4}$$

$$A_1 \sin \psi_1 - 2A_3 \sin \psi_3 = 0 \tag{5}$$

$$A_1 \sin \psi_1 - A_2 \sin \psi_2 + A_3 \sin \psi_3 = 1 \tag{6}$$

$$A_1 \cos \psi_1 + A_2 \cos \psi_2 + \sqrt{3}\, A_3 \cos \psi_3 = 0 \tag{7}$$

$$A_1 \cos \psi_1 - 2\sqrt{3}\, A_3 \cos \psi_3 = 0 \tag{8}$$

$$A_1 \cos \psi_1 - A_2 \cos \psi_2 + \sqrt{3}\, A_3 \cos \psi_3 = 0 \tag{9}$$

Adding equations $(7), (8), (9)$ gives $3A_1 \cos \psi_1 = 0$, or $\psi_1 = \pi/2$; and from $(7), (8)$ we obtain $\psi_2 = \psi_3 = \pi/2$. From equations $(4), (5), (6)$ we get $A_1 = 1/3$, $A_2 = -1/2$ and $A_3 = 1/6$. Thus the general motion is

$$x_1(t) = \tfrac{1}{3} - \tfrac{1}{2} \sin(t + \pi/2) + (\tfrac{1}{6}) \sin(\sqrt{3}\, t + \pi/2)$$

$$x_2(t) = \tfrac{1}{3} - (\tfrac{1}{3}) \sin(\sqrt{3}\, t + \pi/2)$$

$$x_3(t) = \tfrac{1}{3} + \tfrac{1}{2} \sin(t + \pi/2) + (\tfrac{1}{6}) \sin(\sqrt{3}\, t + \pi/2)$$

9.   The tension T in the string with three equal masses attached to it can be assumed to remain constant for small angles of oscillation. If the masses are placed at equal distances apart as shown in Fig. 3-13, determine the natural frequencies and the principal modes of vibration of the system.

The equations of motion are given by

$$m_1 \ddot{x}_1 + (2T/L)x_1 - (T/L)x_2 = 0$$

$$m_2 \ddot{x}_2 + (2T/L)x_2 - (T/L)x_3 - (T/L)x_1 = 0 \tag{1}$$

$$m_3 \ddot{x}_3 + (2T/L)x_3 - (T/L)x_2 = 0$$

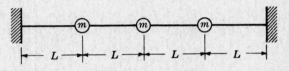

Fig. 3-13

Assume the motion is periodic and is composed of harmonic components of various amplitudes and frequencies. Let

$$\begin{aligned} x_1 &= A \sin(\omega t + \psi), & \ddot{x}_1 &= -\omega^2 A \sin(\omega t + \psi) \\ x_2 &= B \sin(\omega t + \psi), & \ddot{x}_2 &= -\omega^2 B \sin(\omega t + \psi) \\ x_3 &= C \sin(\omega t + \psi), & \ddot{x}_3 &= -\omega^2 C \sin(\omega t + \psi) \end{aligned} \tag{2}$$

Substituting these values into the equations of motion, we obtain

$$(2T/L - \omega^2 m)A - (T/L)B = 0$$

$$-(T/L)A + (2T/L - \omega^2 m)B - (T/L)C = 0 \tag{3}$$

$$-(T/L)B + (2T/L - \omega^2 m)C = 0$$

and the frequency equation is obtained by equating to zero the determinant of the coefficients of A, B, C:

$$(Lm/T)^3 \omega^6 - 6(Lm/T)^2 \omega^4 + (10Lm/T)\omega^2 - 4 = 0$$

Solving, $\omega_1 = \sqrt{0.6T/Lm}$, $\omega_2 = \sqrt{2T/Lm}$, $\omega_3 = \sqrt{3.4T/Lm}$ rad/sec.

From the first and the third equations in (3),

$$B/A = B/C = 2 - (mL\omega^2)/T$$

For the first mode of vibration, let $\omega^2 = \omega_1^2 = 0.6T/Lm$. Then $B_1/A_1 = B_1/C_1 = 1.4$ or $B_1 = 1.4A_1 = 1.4C_1$. See Fig. 3-14(a).

For the second mode of vibration, let $\omega^2 = \omega_2^2 = 2T/Lm$. Then $B_2/A_2 = B_2/C_2 = 0$; and from the second equation in (3), $A_2 = -C_2$. See Fig. 3-14(b).

For the third mode of vibration, let $\omega^2 = \omega_3^2 = 3.4T/Lm$. Then $B_3/A_3 = B_3/C_3 = -1.4$. See Fig. 3-14(c).

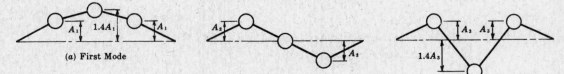

(a) First Mode

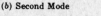

(b) Second Mode

(c) Third Mode

Fig. 3-14

10. A mass m at the end of a rubber band of length L is whirling about point O as shown in Fig. 3-15. Use Lagrange's equation to derive expressions for the generalized forces acting on m during free oscillations.

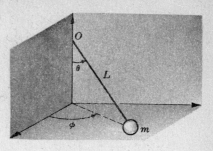

Fig. 3-15

Lagrange's equation for free vibrations may be written in the form

$$\frac{d}{dt}\frac{\partial(\text{K.E.})}{\partial \dot{q}_i} - \frac{\partial(\text{K.E.})}{\partial q_i} = Q_i$$

without formally including P.E., the potential energy. The generalized forces Q_i are already those derivable from a potential.

The kinetic energy for the mass is given by

$$\text{K.E.} = \tfrac{1}{2}m(\dot{L}^2 + L^2\dot{\theta}^2 + L^2\dot{\phi}^2\sin^2\theta)$$

and

$$\frac{d}{dt}\frac{\partial(\text{K.E.})}{\partial \dot{L}} = m\ddot{L}, \qquad \frac{\partial(\text{K.E.})}{\partial L} = m(L\dot{\theta}^2 + L\dot{\phi}^2\sin^2\theta)$$

$$\frac{d}{dt}\frac{\partial(\text{K.E.})}{\partial \dot{\theta}} = mL^2\ddot{\theta} + 2mL\dot{L}\dot{\theta}, \qquad \frac{\partial(\text{K.E.})}{\partial \theta} = mL^2\dot{\phi}^2\sin\theta\cos\theta$$

$$\frac{d}{dt}\frac{\partial(\text{K.E.})}{\partial \dot{\phi}} = 2mL\dot{L}\dot{\phi}\sin^2\theta + mL^2\ddot{\phi}\sin^2\theta + 2mL^2\dot{\phi}\sin\theta\cos\theta, \qquad \frac{\partial(\text{K.E.})}{\partial \phi} = 0$$

Therefore the expressions for the generalized forces are

$$m(\ddot{L} - L\dot{\theta}^2 - L\dot{\phi}^2\sin^2\theta) = Q_L$$
$$m(L^2\ddot{\theta} + 2L\dot{L}\dot{\theta} - L^2\dot{\phi}^2\sin\theta\cos\theta) = Q_\theta$$
$$m(2L\dot{L}\dot{\phi}\sin^2\theta + L^2\ddot{\phi}\sin^2\theta + 2mL^2\dot{\phi}\sin\theta\cos\theta) = Q_\phi$$

11. A mass m is supported on leaf springs having structural damping g and stiffness k. It is connected through an element of stiffness k_1 and inertial damping G, and restrained by a spring of stiffness k_2 and viscous damping c as shown in Fig. 3-16. Determine the differential equations of motion.

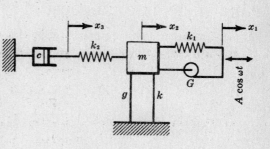

Fig. 3-16

Inertial damping can be expressed as an energy-dissipating force whose magnitude is proportional to the amplitude of the acceleration change across the element with constant of proportionality G, while *structural damping* is an energy-dissipating force whose magnitude is proportional to the amplitude of the displacement change across the leaf springs with constant of proportionality gk.

Drawing the rotating force vector diagram shown in Fig. 3-17, it can be seen that structural damping leads viscous damping by 90° whereas inertial damping lags viscous damping, i.e.,

$c\dot{x}$ for viscous damping

$igkx$ for structural damping

$-iG\ddot{x}$ for inertial damping.

But for forced vibration,

$$\dot{x} = i\omega x, \quad \text{or} \quad x = \dot{x}/i\omega, \quad \ddot{x} = i\omega\dot{x}$$

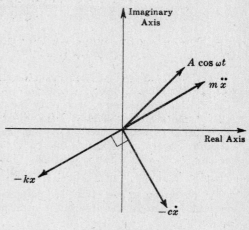

Fig. 3-17

So structural damping becomes

$$igkx \;=\; igk(\dot{x}/i\omega) \;=\; (gk/\omega)\dot{x}$$

and inertial damping becomes

$$-iG\ddot{x} \;=\; -iG(i\omega\dot{x}) \;=\; (G\omega)\dot{x}$$

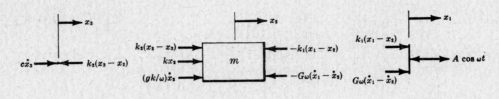

Fig. 3-18

From the free body diagrams, Fig. 3-18, the differential equations of motion are

$$G\omega(\dot{x}_1 - \dot{x}_2) + k_1(x_1 - x_2) \;=\; A\cos\omega t$$
$$m\,\ddot{x}_2 + (gk/\omega)\dot{x}_2 + kx_2 + k_2(x_2 - x_3) + G\omega(\dot{x}_2 - \dot{x}_1) + k_1(x_2 - x_1) \;=\; 0$$
$$c\dot{x}_3 + k_2(x_3 - x_2) \;=\; 0$$

Rearrange to obtain

$$G\omega\dot{x}_1 + k_1 x_1 - G\omega\dot{x}_2 - k_1 x_2 \;=\; A\cos\omega t$$
$$m\,\ddot{x}_2 + (gk/\omega + G\omega)\dot{x}_2 + (k + k_2 + k_1)x_2 - k_2 x_3 - G\omega\dot{x}_1 - k_1 x_1 \;=\; 0$$
$$c\dot{x}_3 + k_2 x_3 - k_2 x_2 \;=\; 0$$

FOUR-DEGREE-OF-FREEDOM SYSTEM

12. Assuming all the contacting surfaces are smooth, write the differential equations of motion for the system as shown in Fig. 3-19.

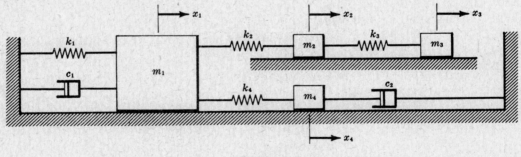

Fig. 3-19

Since there are four masses connected by springs and dashpots in the system, four independent coordinates are necessary to specify the configuration of the system at any time. With x_1, x_2, x_3, x_4 as shown, the positions of all the masses at any instant of time different from zero can be determined. So this is a four-degree-of-freedom system. The differential equations of motion can be easily obtained by first drawing the free-body diagrams for each mass separately and then applying Newton's second law to each mass successively to obtain as many equations of motion as the number of degrees of freedom. This procedure can be applied to simple spring-mass systems of many degrees of freedom.

Referring to Fig. 3-20, for mass m_1, assume x_1 is greater than x_2; the differential equation of motion is given by $\Sigma F = ma$:

$$m_1\,\ddot{x}_1 \;=\; -k_1 x_1 - k_2(x_1 - x_2) - k_4(x_1 - x_4) - c_1\dot{x}_1$$

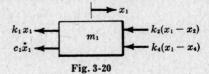

Fig. 3-20

For mass m_2, the equation of motion is

$$m_2 \ddot{x}_2 = k_2(x_1 - x_2) - k_3(x_2 - x_3)$$

For mass m_3, the equation of motion is similarly given by

$$m_3 \ddot{x}_3 = k_3(x_2 - x_3)$$

and for mass m_4,

$$m_4 \ddot{x}_4 = k_4(x_1 - x_4) - c_2 \dot{x}_4$$

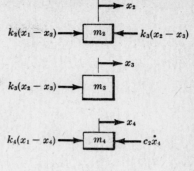

Therefore, the four differential equations of motion representing the given system can be rewritten as

$$m_1 \ddot{x}_1 + (k_1 + k_2 + k_4)x_1 - k_2 x_2 - k_4 x_4 + c_1 \dot{x}_1 = 0$$
$$m_2 \ddot{x}_2 + (k_3 + k_2)x_2 - k_3 x_3 - k_2 x_1 = 0$$
$$m_3 \ddot{x}_3 + k_3 x_3 - k_3 x_2 = 0$$
$$m_4 \ddot{x}_4 + c_2 \dot{x}_4 + k_4 x_4 - k_4 x_1 = 0$$

Fig. 3-20

13. Derive the frequency equation of the general four-degree-of-freedom spring-mass system as shown in Fig. 3-21.

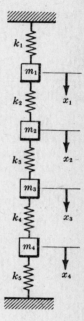

The equations of motion are given by $\Sigma F = ma$:

$$m_1 \ddot{x}_1 + (k_1 + k_2)x_1 - k_2 x_2 = 0$$
$$m_2 \ddot{x}_2 + (k_2 + k_3)x_2 - k_3 x_3 - k_2 x_1 = 0$$
$$m_3 \ddot{x}_3 + (k_3 + k_4)x_3 - k_4 x_4 - k_3 x_2 = 0$$
$$m_4 \ddot{x}_4 + (k_4 + k_5)x_4 - k_4 x_3 = 0$$

Assume that the motion is periodic and is composed of harmonic components of various amplitudes and frequencies. Let

$$x_1 = A \sin(\omega t + \psi), \qquad \ddot{x}_1 = -\omega^2 A \sin(\omega t + \psi)$$
$$x_2 = B \sin(\omega t + \psi), \qquad \ddot{x}_2 = -\omega^2 B \sin(\omega t + \psi)$$
$$x_3 = C \sin(\omega t + \psi), \qquad \ddot{x}_3 = -\omega^2 C \sin(\omega t + \psi)$$
$$x_4 = D \sin(\omega t + \psi), \qquad \ddot{x}_4 = -\omega^2 D \sin(\omega t + \psi)$$

Substituting these values into the equations of motion and simplifying,

$$(k_1 + k_2 - m_1\omega^2)A - k_2 B = 0$$
$$-k_2 A + (k_2 + k_3 - m_2\omega^2)B - k_3 C = 0$$
$$-k_3 B + (k_3 + k_4 - m_3\omega^2)C - k_4 D = 0$$
$$-k_4 C + (k_4 + k_5 - m_4\omega^2)D = 0$$

Fig. 3-21

The frequency equation is obtained by equating to zero the determinant of the coefficients of $A, B, C,$ and D:

$$\begin{vmatrix} (k_1 + k_2 - m_1\omega^2) & -k_2 & 0 & 0 \\ -k_2 & (k_2 + k_3 - m_2\omega^2) & -k_3 & 0 \\ 0 & -k_3 & (k_3 + k_4 - m_3\omega^2) & -k_4 \\ 0 & 0 & -k_4 & (k_4 + k_5 - m_4\omega^2) \end{vmatrix} = 0$$

Rewrite the determinant in this form:

$$\begin{vmatrix} a & -k_2 & 0 & 0 \\ -k_2 & b & -k_3 & 0 \\ 0 & -k_3 & c & -k_4 \\ 0 & 0 & -k_4 & d \end{vmatrix} = \frac{1}{a^2} \begin{vmatrix} ab - k_2^2 & -ak_3 & 0 \\ -ak_3 & ac & -ak_4 \\ 0 & -ak_4 & ad \end{vmatrix}$$

$$= abcd - (k_2^2 cd + k_3^2 da + k_4^2 ab) + k_2^2 k_4^2 = 0$$

or

$$(k_1 + k_2 - m_1\omega^2)(k_2 + k_3 - m_2\omega^2)(k_3 + k_4 - m_3\omega^2)(k_4 + k_5 - m_4\omega^2)$$
$$- \{[k_2^2(k_3 + k_4 - m_3\omega^2)(k_4 + k_5 - m_4\omega^2)] + [k_3^2(k_4 + k_5 - m_4\omega^2)(k_1 + k_2 - m_1\omega^2)]$$
$$+ [k_4^2(k_1 + k_2 - m_1\omega^2)(k_2 + k_3 - m_2\omega^2)]\} + k_2^2 k_4^2 = 0$$

Expand this expression to obtain

$$\omega^8 - \left[\frac{k_1+k_2}{m_1} + \frac{k_2+k_3}{m_2} + \frac{k_3+k_4}{m_3} + \frac{k_4+k_5}{m_4}\right]\omega^6$$

$$+ \left[\frac{k_1k_2 + k_2k_3 + k_3k_1}{m_1m_2} + \frac{k_2k_3 + k_3k_4 + k_4k_2}{m_2m_3} + \frac{k_3k_4 + k_4k_5 + k_5k_3}{m_3m_4}\right.$$

$$\left. + \frac{(k_1+k_2)(k_3+k_4)}{m_1m_3} + \frac{(k_2+k_3)(k_4+k_5)}{m_2m_4} + \frac{(k_4+k_5)(k_1+k_2)}{m_4m_1}\right]\omega^4$$

$$- \left[\frac{k_1k_2k_3 + k_2k_3k_4 + k_3k_4k_1 + k_4k_1k_2}{m_1m_2m_3} + \frac{k_2k_3k_4 + k_3k_4k_5 + k_4k_5k_2 + k_5k_2k_3}{m_2m_3m_4}\right.$$

$$\left. + \frac{(k_1+k_2)(k_3k_4 + k_4k_5 + k_5k_3)}{m_3m_4m_1} + \frac{(k_4+k_5)(k_1k_2 + k_2k_3 + k_3k_1)}{m_4m_1m_2}\right]\omega^2$$

$$+ \frac{k_1k_2k_3k_4 + k_2k_3k_4k_5 + k_3k_4k_5k_1 + k_4k_5k_1k_2 + k_5k_1k_2k_3}{m_1m_2m_3m_4} = 0$$

MATRICES

14. The motion of the two masses shown in Fig. 3-22 is restricted to the plane of the paper. For small angles of oscillation, the motion in the perpendicular directions can be taken independent of each other. Calculate the natural frequencies, the inertia matrix, the stiffness matrix, and the dynamic matrix.

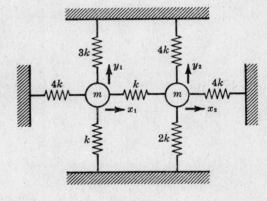

Fig. 3-22

The equations of motion are given by

$$m\ddot{x}_1 + 5kx_1 - kx_2 = 0$$
$$m\ddot{x}_2 + 5kx_2 - kx_1 = 0$$
$$m\ddot{y}_1 + 4ky_1 = 0$$
$$m\ddot{y}_2 + 6ky_2 = 0$$

Substituting $x_1 = q_1$, $x_2 = q_2$, $y_1 = q_3$, $y_2 = q_4$, the equations of motion become

$$\begin{bmatrix} m & 0 & 0 & 0 \\ 0 & m & 0 & 0 \\ 0 & 0 & m & 0 \\ 0 & 0 & 0 & m \end{bmatrix}\begin{bmatrix} \ddot{q}_1 \\ \ddot{q}_2 \\ \ddot{q}_3 \\ \ddot{q}_4 \end{bmatrix} + \begin{bmatrix} 5k & -k & 0 & 0 \\ -k & 5k & 0 & 0 \\ 0 & 0 & 4k & 0 \\ 0 & 0 & 0 & 6k \end{bmatrix}\begin{bmatrix} q_1 \\ q_2 \\ q_3 \\ q_4 \end{bmatrix} = 0$$

and so the inertia matrix $[M] = \begin{bmatrix} m & 0 & 0 & 0 \\ 0 & m & 0 & 0 \\ 0 & 0 & m & 0 \\ 0 & 0 & 0 & m \end{bmatrix}$ and the stiffness matrix $[K] = \begin{bmatrix} 5k & -k & 0 & 0 \\ -k & 5k & 0 & 0 \\ 0 & 0 & 4k & 0 \\ 0 & 0 & 0 & 6k \end{bmatrix}$

The dynamic matrix $[C] = [M]^{-1}[K]$ where $[M]^{-1} = \begin{bmatrix} 1/m & 0 & 0 & 0 \\ 0 & 1/m & 0 & 0 \\ 0 & 0 & 1/m & 0 \\ 0 & 0 & 0 & 1/m \end{bmatrix}$, and so

$$[C] = \begin{bmatrix} 1/m & 0 & 0 & 0 \\ 0 & 1/m & 0 & 0 \\ 0 & 0 & 1/m & 0 \\ 0 & 0 & 0 & 1/m \end{bmatrix}\begin{bmatrix} 5k & -k & 0 & 0 \\ -k & 5k & 0 & 0 \\ 0 & 0 & 4k & 0 \\ 0 & 0 & 0 & 6k \end{bmatrix} = \begin{bmatrix} 5k/m & -k/m & 0 & 0 \\ -k/m & 5k/m & 0 & 0 \\ 0 & 0 & 4k/m & 0 \\ 0 & 0 & 0 & 6k/m \end{bmatrix}$$

Use the λ-matrix theory to obtain

$$f(\lambda) = [\lambda I - C] = \begin{bmatrix} (\lambda - 5k/m) & (k/m) & 0 & 0 \\ (k/m) & (\lambda - 5k/m) & 0 & 0 \\ 0 & 0 & (\lambda - 4k/m) & 0 \\ 0 & 0 & 0 & (\lambda - 6k/m) \end{bmatrix}$$

and the characteristic equation is $\Delta(f) = |\lambda I - C| = 0$, i.e.,

$$\begin{vmatrix} (\lambda - 5k/m) & (k/m) & 0 & 0 \\ (k/m) & (\lambda - 5k/m) & 0 & 0 \\ 0 & 0 & (\lambda - 4k/m) & 0 \\ 0 & 0 & 0 & (\lambda - 6k/m) \end{vmatrix} = 0$$

and hence　$\omega_1 = \sqrt{4k/m}$,　$\omega_2 = \sqrt{6k/m}$,　$\omega_3 = \sqrt{4k/m}$,　$\omega_4 = \sqrt{6k/m}$　rad/sec.

15. Determine the principal coordinates of the spring-mass system as shown in Fig. 3-22, Page 86, by matrix method.

The principal coordinates are given by

$$\{p\} = [\alpha]^{-1}\{q\}$$

where $[\alpha]^{-1}$ is the inverse of the transformation matrix $[\alpha]$, which is formed by all the principal modes of vibration of the system.

The natural frequencies have been found to be

$$\omega_1 = \sqrt{4k/m},\quad \omega_2 = \sqrt{6k/m},\quad \omega_3 = \sqrt{4k/m},\quad \omega_4 = \sqrt{6k/m}$$

and from λ-matrix theory, it can be shown that the transformation matrix is given by

$$[\alpha] = \begin{bmatrix} 1 & 1 & 0 & 0 \\ 1 & -1 & 0 & 0 \\ 0 & 0 & 1 & 0 \\ 0 & 0 & 0 & 1 \end{bmatrix}$$

Now　$[\alpha]^{-1} = \dfrac{\{\text{Adjoint } [\alpha]\}^T}{|\alpha|}$　where　$\{\text{Adjoint } [\alpha]\}^T = \begin{bmatrix} 1 & -1 & 0 & 0 \\ 1 & 1 & 0 & 0 \\ 0 & 0 & 1 & 0 \\ 0 & 0 & 0 & 1 \end{bmatrix}$　and　$|\alpha| = 2$.　Hence

$$[\alpha]^{-1} = \begin{bmatrix} \frac{1}{2} & -\frac{1}{2} & 0 & 0 \\ \frac{1}{2} & \frac{1}{2} & 0 & 0 \\ 0 & 0 & \frac{1}{2} & 0 \\ 0 & 0 & 0 & \frac{1}{2} \end{bmatrix} \quad \text{and} \quad \{p\} = [\alpha]^{-1}\{q\} = \begin{bmatrix} \frac{1}{2} & -\frac{1}{2} & 0 & 0 \\ \frac{1}{2} & \frac{1}{2} & 0 & 0 \\ 0 & 0 & \frac{1}{2} & 0 \\ 0 & 0 & 0 & \frac{1}{2} \end{bmatrix} \begin{bmatrix} q_1 \\ q_2 \\ q_3 \\ q_4 \end{bmatrix}$$

where q_1, q_2, q_3, q_4 are the generalized coordinates representing x_1, x_2, y_1, y_2 respectively. Therefore,

$$p_1 = q_1/2 - q_2/2 = x_1/2 - x_2/2 \qquad p_3 = q_3/2 = y_1/2$$
$$p_2 = q_1/2 + q_2/2 = x_1/2 + x_2/2 \qquad p_4 = q_4/2 = y_2/2$$

INFLUENCE COEFFICIENTS

16. Determine the influence coefficients of the three-degree-of-freedom spring-mass system as shown in Fig. 3-23 below.

From definition, influence coefficient α_{ij} is the deflection at the coordinate i due to a unit force applied at coordinate j. For a three-degree-of-freedom system, there will be nine influence coefficients. They are $\alpha_{11}, \alpha_{12}, \alpha_{13}, \alpha_{21}, \alpha_{22}, \alpha_{23}, \alpha_{31}, \alpha_{32}$, and α_{33}.

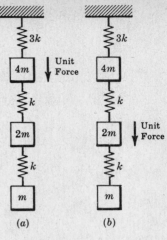

When a unit force is applied to mass $4m$ as shown in Fig. 3-23(a), the spring of stiffness $3k$ will stretch $1/3k$, equal to α_{11}; thus $\alpha_{11} = 1/3k$.

When mass $4m$ deflects $\alpha_{11} = 1/3k$ under the action of a unit force, then masses $2m$ and m will simply move downward by the same amount, i.e.,

$$\alpha_{21} = \alpha_{31} = \alpha_{11} = 1/3k$$

By Maxwell's reciprocal principle, $\alpha_{ij} = \alpha_{ji}$. Hence $\alpha_{31} = \alpha_{13}$, $\alpha_{12} = \alpha_{21}$, and so

$$\alpha_{11} = \alpha_{12} = \alpha_{13} = \alpha_{21} = \alpha_{31} = 1/3k$$

To find α_{22}, apply a unit force to mass $2m$ as shown in Fig. 3-23(b). The two springs $3k$ and k are in series, and their equivalent spring constant is given by

$$1/k_{\text{eq}} = 1/3k + 1/k \quad \text{or} \quad k_{\text{eq}} = 3k/4$$

The deflection is F/k_{eq}, or $1/(3k/4) = 4/3k = \alpha_{22}$; and as mass m hangs on mass $2m$, $\alpha_{32} = \alpha_{22}$. Now $\alpha_{32} = \alpha_{23}$ and hence

$$\alpha_{22} = \alpha_{23} = \alpha_{32}$$

To find α_{33}, apply a unit force to mass m. The three springs are in series and their equivalent spring stiffness is given by

$$1/k_{\text{eq}} = 1/3k + 1/k + 1/k = 7/3k \quad \text{or} \quad k_{\text{eq}} = 3k/7$$

and

$$\alpha_{33} = F/k_{\text{eq}} = 1/(3k/7) = 7/3k$$

The influence coefficients of the system are

$$\alpha_{11} = 1/3k, \quad \alpha_{12} = 1/3k, \quad \alpha_{13} = 1/3k$$
$$\alpha_{21} = 1/3k, \quad \alpha_{22} = 4/3k, \quad \alpha_{23} = 4/3k$$
$$\alpha_{31} = 1/3k, \quad \alpha_{32} = 4/3k, \quad \alpha_{33} = 7/3k$$

Fig. 3-23

17. Determine the influence coefficients of the triple pendulum of lengths L_1, L_2, L_3 and masses m_1, m_2, m_3 as shown in Fig. 3-24.

Apply a unit horizontal force to the mass m_1 of the pendulum as shown in Fig. 3-25 below and write force equations about mass m_1. Since m_1 is in equilibrium,

$$T \sin \theta = 1 \qquad (1)$$
$$T \cos \theta = g(m_1 + m_2 + m_3) \qquad (2)$$

Divide equation (1) by (2) to obtain

$$\tan \theta = 1/g(m_1 + m_2 + m_3)$$

Fig. 3-24

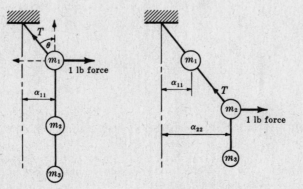

Fig. 3-25   Fig. 3-26   Fig. 3-27

For small angles of oscillation, $\tan \theta \doteq \sin \theta$; and from the configuration of the system, $\sin \theta = \alpha_{11}/L_1$. Hence

$$\alpha_{11} = L_1/g(m_3 + m_2 + m_1)$$

and $\alpha_{11} = \alpha_{21} = \alpha_{31}$ from the geometry of the system.

When a unit horizontal force is applied to m_2 as shown in Fig. 3-26 above, mass m_1 will be displaced a distance α_{11}, but m_2 and m_3 will each be displaced an additional distance equal to $L_2/g(m_2 + m_3)$. Therefore,

$$\alpha_{12} = \alpha_{11} \qquad \text{and} \qquad \alpha_{22} = \alpha_{32} = \alpha_{11} + L_2/g(m_2 + m_3)$$

Similarly, when a unit horizontal force is the only force acting on mass m_3 as shown in Fig. 3-27 above, mass m_1 will be simply displaced a distance α_{11}, and m_2 a distance $[\alpha_{11} + L_2/g(m_2 + m_3)]$, while m_3 will be displaced an additional distance equal to L_3/gm_3; then

$$\alpha_{13} = \alpha_{11}, \qquad \alpha_{23} = \alpha_{22}, \qquad \alpha_{33} = \alpha_{22} + L_3/gm_3$$

Thus the influence coefficients are given by

$$\alpha_{11} = \alpha_{12} = \alpha_{13} = \frac{L_1}{g(m_1 + m_2 + m_3)}$$

$$\alpha_{21} = \frac{L_1}{g(m_1 + m_2 + m_3)}, \qquad \alpha_{22} = \alpha_{23} = \frac{L_1}{g(m_1 + m_2 + m_3)} + \frac{L_2}{g(m_2 + m_3)}$$

$$\alpha_{31} = \frac{L_1}{g(m_1 + m_2 + m_3)}, \qquad \alpha_{32} = \frac{L_1}{g(m_1 + m_2 + m_3)} + \frac{L_2}{g(m_2 + m_3)},$$

$$\alpha_{33} = \frac{L_1}{g(m_1 + m_2 + m_3)} + \frac{L_2}{g(m_2 + m_3)} + \frac{L_3}{gm_3}$$

18. Calculate the influence coefficients of the three-degree-of-freedom spring-mass system as shown in Fig. 3-28, where all the masses are equal to m and all the springs equal to k.

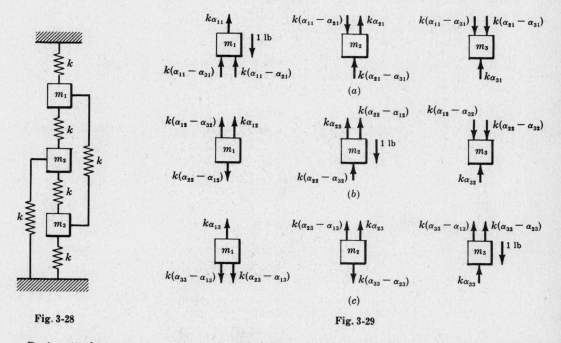

Fig. 3-28   Fig. 3-29

Designate the masses as $m_1, m_2,$ and m_3. Apply a one lb unit force to m_1. From the free-body force diagrams, Fig. 3-29(a),

$$k\alpha_{11} + k(\alpha_{11} - \alpha_{31}) + k(\alpha_{11} - \alpha_{21}) = 1 \qquad\qquad 3k\alpha_{11} - k\alpha_{21} - k\alpha_{31} = 1$$
$$k(\alpha_{11} - \alpha_{21}) = k(\alpha_{21} - \alpha_{31}) + k\alpha_{21} \qquad \text{or} \qquad 3\alpha_{21} - \alpha_{31} - \alpha_{11} = 0$$
$$k(\alpha_{21} - \alpha_{31}) + k(\alpha_{11} - \alpha_{31}) = k\alpha_{31} \qquad\qquad 3\alpha_{31} - \alpha_{11} - \alpha_{21} = 0$$

which gives

$$\alpha_{11} = 1/2k, \qquad \alpha_{21} = 1/4k, \qquad \alpha_{31} = 1/4k$$

Similarly, the following force equations will be obtained when a unit force is applied to mass m_2 as shown in Fig. 3-29(b) above:

$$k(\alpha_{12} - \alpha_{32}) + k\alpha_{12} = k(\alpha_{22} - \alpha_{12}) \qquad\qquad 3\alpha_{12} - \alpha_{22} - \alpha_{32} = 0$$
$$k(\alpha_{22} - \alpha_{12}) + k\alpha_{22} + k(\alpha_{22} - \alpha_{32}) = 1 \quad \text{or} \quad 3k\alpha_{22} - k\alpha_{32} - k\alpha_{12} = 1$$
$$k(\alpha_{22} - \alpha_{32}) + k(\alpha_{12} - \alpha_{32}) = k\alpha_{32} \qquad\qquad 3\alpha_{32} - \alpha_{12} - \alpha_{22} = 0$$

from which

$$\alpha_{12} = 1/4k, \quad \alpha_{22} = 1/2k, \quad \alpha_{32} = 1/4k$$

And when a unit force is applied to mass m_3 as shown in Fig. 3-29(c) above, we obtain

$$k(\alpha_{33} - \alpha_{13}) + k(\alpha_{23} - \alpha_{13}) = k\alpha_{13} \qquad\qquad 3\alpha_{13} - \alpha_{23} - \alpha_{33} = 0$$
$$k(\alpha_{23} - \alpha_{13}) + k\alpha_{23} = k(\alpha_{33} - \alpha_{23}) \quad \text{or} \quad 3\alpha_{23} - \alpha_{33} - \alpha_{13} = 0$$
$$k(\alpha_{33} - \alpha_{13}) + k(\alpha_{33} - \alpha_{23}) + k\alpha_{33} = 1 \qquad\qquad 3k\alpha_{33} - k\alpha_{13} - k\alpha_{23} = 1$$

from which

$$\alpha_{13} = 1/4k, \quad \alpha_{23} = 1/4k, \quad \alpha_{33} = 1/2k$$

The influence coefficients of the system are then

$$\alpha_{11} = 1/2k \qquad \alpha_{12} = 1/4k \qquad \alpha_{13} = 1/4k$$
$$\alpha_{21} = 1/4k \qquad \alpha_{22} = 1/2k \qquad \alpha_{23} = 1/4k$$
$$\alpha_{31} = 1/4k \qquad \alpha_{32} = 1/4k \qquad \alpha_{33} = 1/2k$$

19. Calculate the influence coefficients of a dynamic system consisting of three equal masses attached to a taut string as shown in Fig. 3-30.

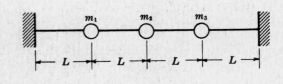

Fig. 3-30

The tension T in the string can be assumed to remain unchanged for small angles of oscillation. α_{11} is the deflection at position 1 due to a unit force applied to position 1.

At the position shown in Fig. 3-31, the unit force is balanced by the tension forces exerted by the string. For small angles of oscillation this can be written as

$$(\alpha_{11}/L)T + (\alpha_{11}/3L)T = 1$$

which gives $\alpha_{11} = 3L/4T$.

α_{21} and α_{31} are the deflections of the masses m_2 and m_3 due to a unit force applied to m_1. They are given by

$$\alpha_{21} = \tfrac{2}{3}(\alpha_{11}) = L/2T, \qquad \alpha_{31} = \tfrac{1}{3}(\alpha_{11}) = L/4T$$

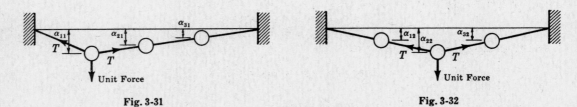

Fig. 3-31   Fig. 3-32

To determine α_{22}, apply a unit force to mass m_2 as shown in Fig. 3-32. The forces acting at mass m_2 are the applied unit force and the tension forces; then

$$(\alpha_{22}/2L)T + (\alpha_{22}/2L)T = 1$$

which gives $\alpha_{22} = L/T$, and $\alpha_{12} = \alpha_{32} = L/2T$.

By symmetry, $\alpha_{11} = \alpha_{33} = 3L/4T$; and by Maxwell's reciprocal theorem, $\alpha_{12} = \alpha_{21}$, $\alpha_{13} = \alpha_{31}$, $\alpha_{23} = \alpha_{32}$. Thus the influence coefficients of the system are

$$\alpha_{11} = 3L/4T, \qquad \alpha_{12} = L/2T, \qquad \alpha_{13} = L/4T$$
$$\alpha_{21} = L/2T, \qquad \alpha_{22} = L/T, \qquad \alpha_{23} = L/2T$$
$$\alpha_{31} = L/4T, \qquad \alpha_{32} = L/2T, \qquad \alpha_{33} = 3L/4T$$

20. Prove Maxwell's reciprocal theorem $\alpha_{ij} = \alpha_{ji}$ for the simply supported beam with two concentrated loads acting as shown in Fig. 3-33.

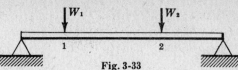
Fig. 3-33

The four influence coefficients of the system are $\alpha_{11}, \alpha_{12}, \alpha_{21}, \alpha_{22}$. It is necessary to show that $\alpha_{12} = \alpha_{21}$ in order to prove Maxwell's reciprocal theorem. This can be done by applying the loads in two cycles.

For the first cycle, apply W_1 first and then W_2. When W_1 is alone at position 1, the influence coefficients are α_{11}, α_{21} and

$$\text{P.E.} = \tfrac{1}{2}W_1^2\alpha_{11}$$

When W_2 is applied after W_1 is on, the additional energy of the system is $\tfrac{1}{2}W_2^2\alpha_{22} + W_1(W_2\alpha_{12})$ and the total energy is therefore $\tfrac{1}{2}W_1^2\alpha_{11} + \tfrac{1}{2}W_2^2\alpha_{22} + W_1(W_2\alpha_{12})$.

For the second cycle, apply W_2 first and then W_1. In a similar fashion, the total energy of the system is given by $\tfrac{1}{2}W_2^2\alpha_{22} + \tfrac{1}{2}W_1^2\alpha_{11} + W_2(W_1\alpha_{21})$.

Since at the ends of both cycles of application of loads the same state prevails, the two energy expressions must be the same. Thus by equating the two energy expressions, we obtain $\alpha_{12} = \alpha_{21}$.

It can be shown that the Maxwell's reciprocal theorem can be extended to systems with several loads acting.

21. In Fig. 3-34, assume the beam is weightless and has constant flexible rigidity EI. Use influence coefficients to determine the differential equations of motion.

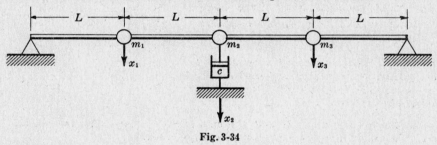
Fig. 3-34

From influence coefficient theory, the total deflections at positions 1, 2, and 3 are given by

$$x_1 = -m_1\ddot{x}_1\alpha_{11} - m_2\ddot{x}_2\alpha_{12} - m_3\ddot{x}_3\alpha_{13} - c\dot{x}_2\alpha_{12}$$
$$x_2 = -m_1\ddot{x}_1\alpha_{21} - m_2\ddot{x}_2\alpha_{22} - m_3\ddot{x}_3\alpha_{23} - c\dot{x}_2\alpha_{22}$$
$$x_3 = -m_1\ddot{x}_1\alpha_{31} - m_2\ddot{x}_2\alpha_{32} - m_3\ddot{x}_3\alpha_{33} - c\dot{x}_2\alpha_{32}$$

From Strength of Materials,

$$\alpha_{11} = \frac{9L^3}{12EI}, \qquad \alpha_{21} = \frac{11L^3}{12EI}, \qquad \alpha_{22} = \frac{16L^3}{12EI}$$

and from the symmetry of the system

$$\alpha_{33} = \alpha_{11} = \frac{9L^3}{12EI}, \qquad \alpha_{32} = \alpha_{12} = \frac{11L^3}{12EI}, \qquad \alpha_{13} = \alpha_{31} = \frac{7L^3}{12EI}$$

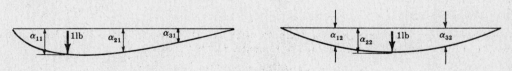
Fig. 3-35

Finally, by Maxwell's reciprocal theorem, $\alpha_{12} = \alpha_{21}$ and $\alpha_{23} = \alpha_{32}$. Thus the equations of motion take the following final form:

$$(9m_1\ddot{x}_1 + 11m_2\ddot{x}_2 + 11c\dot{x}_2 + 7m_3\ddot{x}_3)(L^3/12EI) + x_1 = 0$$
$$(16m_2\ddot{x}_2 + 16c\dot{x}_2 + 11m_3\ddot{x}_3 + 11m_1\ddot{x}_1)(L^3/12EI) + x_2 = 0$$
$$(9m_3\ddot{x}_3 + 7m_1\ddot{x}_1 + 11m_2\ddot{x}_2 + 11c\dot{x}_2)(L^3/12EI) + x_3 = 0$$

MATRIX ITERATION

22. Use matrix iteration to determine the natural frequencies of the system shown in Fig. 3-36.

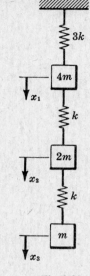

From influence coefficient theory, the equation of motion can be written as

$$-x_1 = \alpha_{11}4m\,\ddot{x}_1 + \alpha_{12}2m\,\ddot{x}_2 + \alpha_{13}m\,\ddot{x}_3$$
$$-x_2 = \alpha_{21}4m\,\ddot{x}_1 + \alpha_{22}2m\,\ddot{x}_2 + \alpha_{23}m\,\ddot{x}_3$$
$$-x_3 = \alpha_{31}4m\,\ddot{x}_1 + \alpha_{32}2m\,\ddot{x}_2 + \alpha_{33}m\,\ddot{x}_3$$

When $\ddot{x}_i$ is replaced by $-\omega^2 x_i$, the equations take the form

$$x_1 = 4\alpha_{11}mx_1\omega^2 + 2\alpha_{12}mx_2\omega^2 + \alpha_{13}mx_3\omega^2$$
$$x_2 = 4\alpha_{21}mx_1\omega^2 + 2\alpha_{22}mx_2\omega^2 + \alpha_{23}mx_3\omega^2$$
$$x_3 = 4\alpha_{31}mx_1\omega^2 + 2\alpha_{32}mx_2\omega^2 + \alpha_{33}mx_3\omega^2$$

In matrix notation, this becomes

$$\begin{bmatrix} x_1 \\ x_2 \\ x_3 \end{bmatrix} = \omega^2 m \begin{bmatrix} 4\alpha_{11} & 2\alpha_{12} & \alpha_{13} \\ 4\alpha_{21} & 2\alpha_{22} & \alpha_{23} \\ 4\alpha_{31} & 2\alpha_{32} & \alpha_{33} \end{bmatrix} \begin{bmatrix} x_1 \\ x_2 \\ x_3 \end{bmatrix}$$

The values for the influence coefficients were found in Problem 16, Page 88, to be

$$\alpha_{11} = \alpha_{12} = \alpha_{21} = \alpha_{13} = \alpha_{31} = 1/3k, \qquad \alpha_{22} = \alpha_{32} = \alpha_{23} = 4/3k, \qquad \alpha_{33} = 7/3k$$

When these values are substituted into the matrix equation, we obtain

$$\begin{bmatrix} x_1 \\ x_2 \\ x_3 \end{bmatrix} = \frac{\omega^2 m}{3k} \begin{bmatrix} 4 & 2 & 1 \\ 4 & 8 & 4 \\ 4 & 8 & 7 \end{bmatrix} \begin{bmatrix} x_1 \\ x_2 \\ x_3 \end{bmatrix}$$

Fig. 3-36

To start the iteration process, estimate the configuration of the first mode. Let $x_1 = 1$, $x_2 = 2$, $x_3 = 4$.

First iteration:

$$\begin{bmatrix} 1 \\ 2 \\ 4 \end{bmatrix} = \frac{\omega^2 m}{3k} \begin{bmatrix} 4 & 2 & 1 \\ 4 & 8 & 4 \\ 4 & 8 & 7 \end{bmatrix} \begin{bmatrix} 1 \\ 2 \\ 4 \end{bmatrix} = \frac{\omega^2 m}{3k} \begin{bmatrix} 12 \\ 36 \\ 48 \end{bmatrix} = \frac{\omega^2 m}{3k}(12) \begin{bmatrix} 1 \\ 3 \\ 4 \end{bmatrix}$$

Second iteration:

$$\begin{bmatrix} 1 \\ 3 \\ 4 \end{bmatrix} = \frac{\omega^2 m}{3k} \begin{bmatrix} 4 & 2 & 1 \\ 4 & 8 & 4 \\ 4 & 8 & 7 \end{bmatrix} \begin{bmatrix} 1 \\ 3 \\ 4 \end{bmatrix} = \frac{\omega^2 m}{3k} \begin{bmatrix} 14.0 \\ 44.0 \\ 56.0 \end{bmatrix} = \frac{\omega^2 m}{3k}(14) \begin{bmatrix} 1.0 \\ 3.2 \\ 4.0 \end{bmatrix}$$

Third iteration:

$$\begin{bmatrix} 1 \\ 3.2 \\ 4 \end{bmatrix} = \frac{\omega^2 m}{3k} \begin{bmatrix} 4 & 2 & 1 \\ 4 & 8 & 4 \\ 4 & 8 & 7 \end{bmatrix} \begin{bmatrix} 1 \\ 3.2 \\ 4 \end{bmatrix} = \frac{\omega^2 m}{3k} \begin{bmatrix} 14.4 \\ 45.6 \\ 57.6 \end{bmatrix} = \frac{\omega^2 m}{3k}(14.4) \begin{bmatrix} 1.00 \\ 3.18 \\ 4.00 \end{bmatrix}$$

Since the ratio obtained here is very close to the initial value,

$$\begin{bmatrix} 1.0 \\ 3.2 \\ 4.0 \end{bmatrix} = \frac{14.4 m\omega^2}{3k} \begin{bmatrix} 1.00 \\ 3.18 \\ 4.00 \end{bmatrix} \qquad \text{or} \qquad 1 = (14.4m\omega^2)/3k \qquad \text{and} \qquad \omega_1 = 0.46\sqrt{k/m} \text{ rad/sec}$$

To obtain the second principal mode, the orthogonality principle is used:

$$m_1 A_1 A_2 + m_2 B_1 B_2 + m_3 C_1 C_2 = 0$$

For the first and second mode, this becomes

$$4m(1)A_2 + 2m(3.2)B_2 + m(4)C_2 = 0$$

or

$$A_2 = -1.6B_2 - C_2, \qquad B_2 = B_2, \qquad C_2 = C_2$$

and in matrix form

$$\begin{bmatrix} A_2 \\ B_2 \\ C_2 \end{bmatrix} = \begin{bmatrix} 0 & -1.6 & -1.0 \\ 0 & 1 & 0 \\ 0 & 0 & 1 \end{bmatrix} \begin{bmatrix} A_2 \\ B_2 \\ C_2 \end{bmatrix}$$

When this is combined with the matrix equation for first mode, it will converge to second mode.

$$\begin{bmatrix} x_1 \\ x_2 \\ x_3 \end{bmatrix} = \frac{\omega^2 m}{3k} \begin{bmatrix} 4 & 2 & 1 \\ 4 & 8 & 4 \\ 4 & 8 & 7 \end{bmatrix} \begin{bmatrix} 0 & -1.6 & -1.0 \\ 0 & 1 & 0 \\ 0 & 0 & 1 \end{bmatrix} \begin{bmatrix} x_1 \\ x_2 \\ x_3 \end{bmatrix} = \frac{\omega^2 m}{3k} \begin{bmatrix} 0 & -4.4 & -3 \\ 0 & 1.6 & 0 \\ 0 & 1.6 & 3 \end{bmatrix} \begin{bmatrix} x_1 \\ x_2 \\ x_3 \end{bmatrix}$$

Due to symmetry of the problem, the second mode is $\begin{bmatrix} 1 \\ 0 \\ -1 \end{bmatrix}$. When this is used to start the iteration process, we have

$$\begin{bmatrix} 1 \\ 0 \\ -1 \end{bmatrix} = \frac{\omega^2 m}{3k} \begin{bmatrix} 0 & -4.4 & -3 \\ 0 & 1.6 & 0 \\ 0 & 1.6 & 3 \end{bmatrix} \begin{bmatrix} 1 \\ 0 \\ -1 \end{bmatrix} = \frac{\omega^2 m}{3k} \begin{bmatrix} 3 \\ 0 \\ -3 \end{bmatrix}$$

which repeats itself. Hence

$$\begin{bmatrix} 1 \\ 0 \\ -1 \end{bmatrix} = \frac{3m\omega^2}{3k} \begin{bmatrix} 1 \\ 0 \\ -1 \end{bmatrix} \qquad \text{or} \qquad 1 = (\omega^2 m)/k \qquad \text{and} \qquad \omega_2 = \sqrt{k/m} \text{ rad/sec}$$

To obtain the third mode, write the orthogonality principle as

$$m_1 A_2 A_3 + m_2 B_2 B_3 + m_3 C_2 C_3 = 0$$
$$m_1 A_1 A_3 + m_2 B_1 B_3 + m_3 C_1 C_3 = 0$$

Substituting $A_1 = 1.0$, $B_1 = 3.2$, $C_1 = 4.0$, $A_2 = 1$, $B_2 = 0$, and $C_2 = -1$, into the orthogonality equations, we obtain

$$4m(1)A_3 + 2m(0)B_3 + m(-1)C_3 = 0$$
$$4m(1)A_3 + 2m(3.20)B_3 + m(4)C_3 = 0$$

from which $A_3 = 0.25 C_3$ and $B_3 = -0.78 C_3$. Then

$$\begin{bmatrix} A_3 \\ B_3 \\ C_3 \end{bmatrix} = \begin{bmatrix} 0 & 0 & 0.25 \\ 0 & 0 & -0.78 \\ 0 & 0 & 1.00 \end{bmatrix} \begin{bmatrix} A_3 \\ B_3 \\ C_3 \end{bmatrix}$$

and when this is combined with the matrix equation for the second mode, it will yield the third mode.

$$\begin{bmatrix} x_1 \\ x_2 \\ x_3 \end{bmatrix} = \frac{\omega^2 m}{3k} \begin{bmatrix} 0 & -4.4 & -3 \\ 0 & 1.6 & 0 \\ 0 & 1.6 & 3 \end{bmatrix} \begin{bmatrix} 0 & 0 & 0.25 \\ 0 & 0 & -.78 \\ 0 & 0 & 1.00 \end{bmatrix} \begin{bmatrix} x_1 \\ x_2 \\ x_3 \end{bmatrix} = \frac{\omega^2 m}{3k} \begin{bmatrix} 0 & 0 & 0.43 \\ 0 & 0 & -1.25 \\ 0 & 0 & 1.75 \end{bmatrix} \begin{bmatrix} x_1 \\ x_2 \\ x_3 \end{bmatrix}$$

or $$\begin{bmatrix} x_1 \\ x_2 \\ x_3 \end{bmatrix} = \frac{\omega^2 m}{3k}(1.75) \begin{bmatrix} 0 & 0 & 0.25 \\ 0 & 0 & -.72 \\ 0 & 0 & 1.00 \end{bmatrix} \begin{bmatrix} x_1 \\ x_2 \\ x_3 \end{bmatrix}$$

Assuming any arbitrary values for the third mode, it can be shown that the same third mode $\begin{bmatrix} 0.25 \\ -.72 \\ 1.00 \end{bmatrix}$ will be found. Further iteration is therefore not necessary. Thus

$$1 = (\omega^2 m/3k)(1.75) \qquad \text{or} \qquad \omega_3 = 1.32\sqrt{k/m} \text{ rad/sec}$$

23. Use matrix iteration to determine the natural frequencies of the triple pendulum as shown in Fig. 3-37 below.

From influence coefficient theory, the equations of motion are given by

$$-x_1 = \alpha_{11}m_1\ddot{x}_1 + \alpha_{12}m_2\ddot{x}_2 + \alpha_{13}m_3\ddot{x}_3$$
$$-x_2 = \alpha_{21}m_1\ddot{x}_1 + \alpha_{22}m_2\ddot{x}_2 + \alpha_{23}m_3\ddot{x}_3$$
$$-x_3 = \alpha_{31}m_1\ddot{x}_1 + \alpha_{32}m_2\ddot{x}_2 + \alpha_{33}m_3\ddot{x}_3$$

Replacing $\ddot{x}_i$ by $-\omega^2 x_i$, the equations take the form

$$x_1 = \alpha_{11}m_1x_1\omega^2 + \alpha_{12}m_2x_2\omega^2 + \alpha_{13}m_3x_3\omega^2$$
$$x_2 = \alpha_{21}m_1x_1\omega^2 + \alpha_{22}m_2x_2\omega^2 + \alpha_{23}m_3x_3\omega^2$$
$$x_3 = \alpha_{31}m_1x_1\omega^2 + \alpha_{32}m_2x_2\omega^2 + \alpha_{33}m_3x_3\omega^2$$

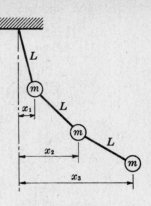

To obtain numerical values for the influence coefficients, put $L_1 = L_2 = L_3 = L$ and $m_1 = m_2 = m_3 = m$ in Problem 17, Page 89:

$$\alpha_{11} = \alpha_{12} = \alpha_{13} = L/3mg$$
$$\alpha_{21} = L/3mg, \quad \alpha_{22} = \alpha_{23} = 5L/6mg$$
$$\alpha_{31} = L/3mg, \quad \alpha_{32} = 5L/6mg, \quad \alpha_{33} = 11L/6mg$$

Fig. 3-37

In matrix notation, then, the equations become

$$\begin{bmatrix} x_1 \\ x_2 \\ x_3 \end{bmatrix} = \frac{L\omega^2}{6g}\begin{bmatrix} 2 & 2 & 2 \\ 2 & 5 & 5 \\ 2 & 5 & 11 \end{bmatrix}\begin{bmatrix} x_1 \\ x_2 \\ x_3 \end{bmatrix}$$

Begin the iteration process by an arbitrary assumption for the first mode of the system.

First iteration:

$$\begin{bmatrix} 0.2 \\ 0.6 \\ 1.0 \end{bmatrix} = \frac{L\omega^2}{6g}\begin{bmatrix} 2 & 2 & 2 \\ 2 & 5 & 5 \\ 2 & 5 & 11 \end{bmatrix}\begin{bmatrix} 0.2 \\ 0.6 \\ 1.0 \end{bmatrix} = \frac{L\omega^2}{6g}\begin{bmatrix} 3.6 \\ 8.4 \\ 14.4 \end{bmatrix} = \frac{L\omega^2}{6g}(14.4)\begin{bmatrix} 0.25 \\ 0.58 \\ 1.00 \end{bmatrix}$$

Second iteration:

$$\begin{bmatrix} 0.25 \\ 0.58 \\ 1.00 \end{bmatrix} = \frac{L\omega^2}{6g}\begin{bmatrix} 2 & 2 & 2 \\ 2 & 5 & 5 \\ 2 & 5 & 11 \end{bmatrix}\begin{bmatrix} 0.25 \\ 0.58 \\ 1.00 \end{bmatrix} = \frac{L\omega^2}{6g}\begin{bmatrix} 3.66 \\ 8.42 \\ 14.4 \end{bmatrix} = \frac{L\omega^2}{6g}(14.4)\begin{bmatrix} 0.25 \\ 0.58 \\ 1.00 \end{bmatrix}$$

Since the column repeats itself, the iteration process can be stopped. Then

$$1 = \frac{L\omega^2}{6g}(14.4) \qquad \text{or} \qquad \omega_1 = 0.65\sqrt{g/L} \text{ rad/sec}$$

To obtain the second mode, the first mode has to be suppressed during the iteration process. This is done by the use of the orthogonality principle:

$$m_1A_1A_2 + m_2B_1B_2 + m_3C_1C_2 = 0$$

Substituting the first mode into the above equation, we obtain

$$m(0.25)x_1 + m(0.58)x_2 + m(1.0)x_3 = 0 \qquad \text{or} \qquad x_1 = -2.32x_2 - 4x_3, \quad x_2 = x_2, \quad x_3 = x_3$$

and in matrix form, this becomes

$$\begin{bmatrix} x_1 \\ x_2 \\ x_3 \end{bmatrix} = \begin{bmatrix} 0 & -2.32 & -4.0 \\ 0 & 1 & 0 \\ 0 & 0 & 1 \end{bmatrix}\begin{bmatrix} x_1 \\ x_2 \\ x_3 \end{bmatrix}$$

When this is combined with the fundamental matrix equation, it will yield a matrix equation that has no first mode present:

$$\begin{bmatrix} x_1 \\ x_2 \\ x_3 \end{bmatrix} = \frac{L\omega^2}{6g}\begin{bmatrix} 2 & 2 & 2 \\ 2 & 5 & 5 \\ 2 & 5 & 11 \end{bmatrix}\begin{bmatrix} 0 & -2.3 & -4 \\ 0 & 1.0 & 0 \\ 0 & 0 & 1 \end{bmatrix}\begin{bmatrix} x_1 \\ x_2 \\ x_3 \end{bmatrix} = \frac{L\omega^2}{6g}\begin{bmatrix} 0 & -2.6 & -6 \\ 0 & 0.4 & -3 \\ 0 & 0.4 & 3 \end{bmatrix}\begin{bmatrix} x_1 \\ x_2 \\ x_3 \end{bmatrix}$$

From this matrix equation, use matrix iteration to determine the second mode.

First iteration:

$$\begin{bmatrix} -1 \\ -1 \\ 1 \end{bmatrix} = \frac{L\omega^2}{6g} \begin{bmatrix} 0 & -2.6 & -6 \\ 0 & 0.4 & -3 \\ 0 & 0.4 & 3 \end{bmatrix} \begin{bmatrix} -1 \\ -1 \\ 1 \end{bmatrix} = \frac{L\omega^2}{6g} \begin{bmatrix} -3.4 \\ -3.4 \\ 2.6 \end{bmatrix} = \frac{L\omega^2}{6g}(2.6) \begin{bmatrix} -1.3 \\ -1.3 \\ 1.0 \end{bmatrix}$$

Second iteration:

$$\begin{bmatrix} -1.3 \\ -1.3 \\ 1.0 \end{bmatrix} = \frac{L\omega^2}{6g} \begin{bmatrix} 0 & -2.6 & -6 \\ 0 & 0.4 & -3 \\ 0 & 0.4 & 3 \end{bmatrix} \begin{bmatrix} -1.3 \\ -1.3 \\ 1.0 \end{bmatrix} = \frac{L\omega^2}{6g} \begin{bmatrix} -2.6 \\ -3.5 \\ 2.5 \end{bmatrix} = \frac{L\omega^2}{6g}(2.5) \begin{bmatrix} -1.05 \\ -1.40 \\ 1.00 \end{bmatrix}$$

Third iteration:

$$\begin{bmatrix} -1.0 \\ -1.4 \\ 1.0 \end{bmatrix} = \frac{L\omega^2}{6g} \begin{bmatrix} 0 & -2.6 & -6 \\ 0 & 0.4 & -3 \\ 0 & 0.4 & 3 \end{bmatrix} \begin{bmatrix} -1.0 \\ -1.4 \\ 1.0 \end{bmatrix} = \frac{L\omega^2}{6g} \begin{bmatrix} -2.4 \\ -3.5 \\ 2.5 \end{bmatrix} = \frac{L\omega^2}{6g}(2.5) \begin{bmatrix} -1.0 \\ -1.4 \\ 1.0 \end{bmatrix}$$

Since the assumed mode in the last iteration repeats itself, the iteration process can be stopped. The mode of vibration and the natural frequency are therefore given by

$$\begin{bmatrix} -1.0 \\ -1.4 \\ 1.0 \end{bmatrix} \quad \text{and} \quad 1 = \frac{L\omega^2}{6g}(2.5) \quad \text{or} \quad \omega_2 = 1.52\sqrt{g/L} \text{ rad/sec}$$

In order to obtain the third principal mode and thus the third natural frequency of the system, both the first and second modes must not appear in the iteration process. This is again done by using the orthogonality principle expressed as

$$m_1 A_1 A_3 + m_2 B_1 B_3 + m_3 C_1 C_3 = 0, \quad m_1 A_2 A_3 + m_2 B_2 B_3 + m_3 C_2 C_3 = 0$$

For first and third modes, this becomes

$$m(0.25)x_1 + m(0.6)x_2 + m(1.0)x_3 = 0$$

and for second and third modes, we have

$$m(-1.0)x_1 + m(-1.4)x_2 + m(1.0)x_3 = 0 \quad \text{or} \quad x_1 = 8x_3, \quad x_2 = -5x_3, \quad x_3 = x_3$$

and in matrix form,

$$\begin{bmatrix} x_1 \\ x_2 \\ x_3 \end{bmatrix} = \begin{bmatrix} 0 & 0 & 8 \\ 0 & 0 & -5 \\ 0 & 0 & 1 \end{bmatrix} \begin{bmatrix} x_1 \\ x_2 \\ x_3 \end{bmatrix}$$

When this is combined with the matrix equation for second mode, we obtain the matrix equation for third mode:

$$\begin{bmatrix} x_1 \\ x_2 \\ x_3 \end{bmatrix} = \frac{L\omega^2}{6g} \begin{bmatrix} 0 & -2.6 & -6 \\ 0 & 0.4 & -3 \\ 0 & 0.4 & 3 \end{bmatrix} \begin{bmatrix} 0 & 0 & 8 \\ 0 & 0 & -5 \\ 0 & 0 & 1 \end{bmatrix} \begin{bmatrix} x_1 \\ x_2 \\ x_3 \end{bmatrix} = \frac{L\omega^2}{6g} \begin{bmatrix} 0 & 0 & 7 \\ 0 & 0 & -5 \\ 0 & 0 & 1 \end{bmatrix} \begin{bmatrix} x_1 \\ x_2 \\ x_3 \end{bmatrix}$$

Assume any convenient value for the third mode, and start the iteration process. It will be found that the same mode shape $\begin{bmatrix} 7 \\ -5 \\ 1 \end{bmatrix}$ is obtained repeatedly. This means that $\begin{bmatrix} 7 \\ -5 \\ 1 \end{bmatrix}$ is actually the third mode of the system. Thus

$$1 = L\omega^2/6g \quad \text{and} \quad \omega_3 = 2.45\sqrt{g/L} \text{ rad/sec}$$

The three natural frequencies of the triple pendulum are therefore given by

$$\omega_1 = 0.65\sqrt{g/L}, \quad \omega_2 = 1.52\sqrt{g/L}, \quad \omega_3 = 2.45\sqrt{g/L} \text{ rad/sec}$$

24. Determine the highest natural frequency of the three-degree-of-freedom spring-mass system as shown in Fig. 3-38 below. Use the inverse matrix method.

As discussed earlier, the deflection equations of the masses are

$$\begin{bmatrix} x_1 \\ x_2 \\ x_3 \end{bmatrix} = \frac{\omega^2 m}{2k} \begin{bmatrix} 1 & \frac{1}{2} & \frac{1}{2} \\ \frac{1}{2} & 1 & \frac{1}{2} \\ \frac{1}{2} & \frac{1}{2} & 1 \end{bmatrix} \begin{bmatrix} x_1 \\ x_2 \\ x_3 \end{bmatrix}$$

Using inverse matrix theory, this can be written as

$$\frac{2k}{\omega^2 m}\begin{bmatrix} x_1 \\ x_2 \\ x_3 \end{bmatrix}\begin{bmatrix} 1 & \frac{1}{2} & \frac{1}{2} \\ \frac{1}{2} & 1 & \frac{1}{2} \\ \frac{1}{2} & \frac{1}{2} & 1 \end{bmatrix}^{-1} = \begin{bmatrix} x_1 \\ x_2 \\ x_3 \end{bmatrix} \qquad (1)$$

where $[D]^{-1} = \begin{bmatrix} 1 & \frac{1}{2} & \frac{1}{2} \\ \frac{1}{2} & 1 & \frac{1}{2} \\ \frac{1}{2} & \frac{1}{2} & 1 \end{bmatrix}^{-1}$ is the inverse of matrix $[D] = \begin{bmatrix} 1 & \frac{1}{2} & \frac{1}{2} \\ \frac{1}{2} & 1 & \frac{1}{2} \\ \frac{1}{2} & \frac{1}{2} & 1 \end{bmatrix}$.

From matrix theory, Adjoint $[D]$ can be found in the following manner:

$$\text{Adjoint } [D] = \begin{bmatrix} (-1)^{1+1}\begin{vmatrix} 1 & \frac{1}{2} \\ \frac{1}{2} & 1 \end{vmatrix} & (-1)^{1+2}\begin{vmatrix} \frac{1}{2} & \frac{1}{2} \\ \frac{1}{2} & 1 \end{vmatrix} & (-1)^{1+3}\begin{vmatrix} \frac{1}{2} & \frac{1}{2} \\ 1 & \frac{1}{2} \end{vmatrix} \\ (-1)^{2+1}\begin{vmatrix} \frac{1}{2} & \frac{1}{2} \\ \frac{1}{2} & 1 \end{vmatrix} & (-1)^{2+2}\begin{vmatrix} 1 & \frac{1}{2} \\ \frac{1}{2} & 1 \end{vmatrix} & (-1)^{2+3}\begin{vmatrix} 1 & \frac{1}{2} \\ \frac{1}{2} & \frac{1}{2} \end{vmatrix} \\ (-1)^{3+1}\begin{vmatrix} \frac{1}{2} & 1 \\ \frac{1}{2} & \frac{1}{2} \end{vmatrix} & (-1)^{3+2}\begin{vmatrix} 1 & \frac{1}{2} \\ \frac{1}{2} & \frac{1}{2} \end{vmatrix} & (-1)^{3+3}\begin{vmatrix} 1 & \frac{1}{2} \\ \frac{1}{2} & 1 \end{vmatrix} \end{bmatrix} = \frac{1}{4}\begin{bmatrix} 3 & -1 & -1 \\ -1 & 3 & -1 \\ -1 & -1 & 3 \end{bmatrix}$$

and

$$|D| = \begin{vmatrix} 1 & \frac{1}{2} & \frac{1}{2} \\ \frac{1}{2} & 1 & \frac{1}{2} \\ \frac{1}{2} & \frac{1}{2} & 1 \end{vmatrix} = \frac{1}{2}$$

Hence

$$[D]^{-1} = \frac{\text{Adjoint } [D]}{|D|} = \frac{1}{2}\begin{bmatrix} 3 & -1 & -1 \\ -1 & 3 & -1 \\ -1 & -1 & 3 \end{bmatrix}$$

The inverse matrix $[D]^{-1}$ can also be found by the elementary operations as follow:

Operation	[D]			[D]$^{-1}$		
Multiply [D] by a factor of 2	1	1/2	1/2	1	0	0
	1/2	1	1/2	0	1	0
	1/2	1/2	1	0	0	1
Row (1) minus row (2)	2	1	1	2	0	0
	1	2	1	0	2	0
	1	1	2	0	0	2
Row (3) minus row (2)	1	−1	0	2	−2	0
	1	2	1	0	2	0
	1	1	2	0	0	2
Row (2) minus row (3)	1	−1	0	2	−2	0
	1	2	1	0	2	0
	0	−1	1	0	−2	2
Row (2) minus row (1)	1	−1	0	2	−2	0
	1	3	0	0	4	−2
	0	−1	1	0	−2	2
Multiply row (2) by a factor of 1/4	1	−1	0	2	−2	0
	0	4	0	−2	6	−2
	0	−1	1	0	−2	2
Add row (2) to row (1)	1	−1	0	2	−2	0
	0	1	0	−1/2	3/2	−1/2
	0	−1	1	0	−2	2
Add row (2) to row (3)	1	0	0	3/2	−1/2	−1/2
	0	1	0	−1/2	3/2	−1/2
	0	−1	1	0	−2	2
	1	0	0	3/2	−1/2	−1/2
	0	1	0	−1/2	3/2	−1/2
	0	0	1	−1/2	−1/2	3/2

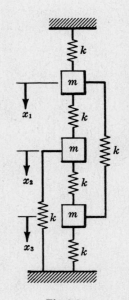

Fig. 3-38

which also gives $[D]^{-1} = \dfrac{1}{2}\begin{bmatrix} 3 & -1 & -1 \\ -1 & 3 & -1 \\ -1 & -1 & 3 \end{bmatrix}$

Substituting $[D]^{-1}$ into equation *(1)*, we have

$$\begin{bmatrix} x_1 \\ x_2 \\ x_3 \end{bmatrix} = \frac{2k}{\omega^2 m}\left(\frac{1}{2}\right)\begin{bmatrix} 3 & -1 & -1 \\ -1 & 3 & -1 \\ -1 & -1 & 3 \end{bmatrix}\begin{bmatrix} x_1 \\ x_2 \\ x_3 \end{bmatrix} \qquad (2)$$

Assume the third mode to be $\begin{bmatrix} 1 \\ -2 \\ 1 \end{bmatrix}$ and substitute this into equation *(2)* to obtain

$$\begin{bmatrix} 1 \\ -2 \\ 1 \end{bmatrix} = \frac{k}{\omega^2 m}\begin{bmatrix} 3 & -1 & -1 \\ -1 & 3 & -1 \\ -1 & -1 & 3 \end{bmatrix}\begin{bmatrix} 1 \\ -2 \\ 1 \end{bmatrix} = \frac{4k}{\omega^2 m}\begin{bmatrix} 1 \\ -2 \\ 1 \end{bmatrix}$$

The assumed mode $\begin{bmatrix} 1 \\ -2 \\ 1 \end{bmatrix}$ repeats itself. This means the assumed value is the third mode. Hence

$$1 = 4k/\omega^2 m \qquad \text{and} \qquad \omega_3 = 2\sqrt{k/m}\ \text{rad/sec}$$

25. Use the Inverse Matrix method to determine the highest natural frequency of the spring-mass system as shown in Fig. 3-39.

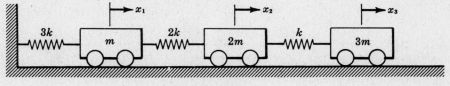

Fig. 3-39

From influence coefficient theory the equations of motion can be expressed as

$$-x_1 = \alpha_{11}m\ddot{x}_1 + \alpha_{12}2m\ddot{x}_2 + \alpha_{13}3m\ddot{x}_3$$
$$-x_2 = \alpha_{21}m\ddot{x}_1 + \alpha_{22}2m\ddot{x}_2 + \alpha_{23}3m\ddot{x}_3$$
$$-x_3 = \alpha_{31}m\ddot{x}_1 + \alpha_{32}2m\ddot{x}_2 + \alpha_{33}3m\ddot{x}_3$$

where $\alpha_{11} = \alpha_{12} = \alpha_{13} = 1/3k$; $\alpha_{21} = 1/3k$, $\alpha_{22} = \alpha_{23} = 5/6k$; $\alpha_{31} = 1/3k$, $\alpha_{32} = 5/6k$, $\alpha_{33} = 11/6k$. Replacing $\ddot{x}_i$ by $-\omega^2 x_i$, we have

$$x_1 = \alpha_{11}mx_1\omega^2 + 2\alpha_{12}mx_2\omega^2 + 3\alpha_{13}mx_3\omega^2$$
$$x_2 = \alpha_{21}mx_1\omega^2 + 2\alpha_{22}mx_2\omega^2 + 3\alpha_{23}mx_3\omega^2 \qquad (1)$$
$$x_3 = \alpha_{31}mx_1\omega^2 + 2\alpha_{32}mx_2\omega^2 + 3\alpha_{33}mx_3\omega^2$$

or in matrix notation,

$$\begin{bmatrix} x_1 \\ x_2 \\ x_3 \end{bmatrix} = \frac{\omega^2 m}{6k}\begin{bmatrix} 2 & 4 & 6 \\ 2 & 10 & 15 \\ 2 & 10 & 33 \end{bmatrix}\begin{bmatrix} x_1 \\ x_2 \\ x_3 \end{bmatrix} \qquad (2)$$

Using inverse matrix theory, equation *(2)* can be written as

$$\frac{6k}{\omega^2 m}\begin{bmatrix} x_1 \\ x_2 \\ x_3 \end{bmatrix}\begin{bmatrix} 2 & 4 & 6 \\ 2 & 10 & 15 \\ 2 & 10 & 33 \end{bmatrix}^{-1} = \begin{bmatrix} x_1 \\ x_2 \\ x_3 \end{bmatrix} \qquad (3)$$

where $[D]^{-1} = \begin{bmatrix} 2 & 4 & 6 \\ 2 & 10 & 15 \\ 2 & 10 & 33 \end{bmatrix}^{-1}$ is the inverse of $[D] = \begin{bmatrix} 2 & 4 & 6 \\ 2 & 10 & 15 \\ 2 & 10 & 33 \end{bmatrix}$

Substituting $\quad [D]^{-1} = \dfrac{1}{6}\begin{bmatrix} 5 & -2 & 0 \\ -1 & \frac{3}{2} & -\frac{1}{2} \\ 0 & -\frac{1}{3} & \frac{1}{3} \end{bmatrix}\quad$ into (3),

$$\begin{bmatrix} x_1 \\ x_2 \\ x_3 \end{bmatrix} = \frac{k}{m\omega^2}\begin{bmatrix} 5 & -2 & 0 \\ -1 & \frac{3}{2} & -\frac{1}{2} \\ 0 & -\frac{1}{3} & \frac{1}{3} \end{bmatrix}\begin{bmatrix} x_1 \\ x_2 \\ x_3 \end{bmatrix} \qquad (4)$$

Assume the third mode to be $\begin{bmatrix} 10 \\ -4 \\ 1 \end{bmatrix}$ and begin the iteration process with equation (4).

First iteration:

$$\begin{bmatrix} 10 \\ -4 \\ 1 \end{bmatrix} = \frac{k}{m\omega^2}\begin{bmatrix} 5 & -2 & 0 \\ -1 & \frac{3}{2} & -\frac{1}{2} \\ 0 & -\frac{1}{3} & \frac{1}{3} \end{bmatrix}\begin{bmatrix} 10 \\ -4 \\ 1 \end{bmatrix} = \frac{k}{m\omega^2}\begin{bmatrix} 58 \\ -16.5 \\ 1.7 \end{bmatrix} = \frac{k}{m\omega^2}(1.7)\begin{bmatrix} 34 \\ -9.7 \\ 1.0 \end{bmatrix}$$

Second iteration:

$$\begin{bmatrix} 30 \\ -10 \\ 1 \end{bmatrix} = \frac{k}{m\omega^2}\begin{bmatrix} 5 & -2 & 0 \\ -1 & \frac{3}{2} & -\frac{1}{2} \\ 0 & -\frac{1}{3} & \frac{1}{3} \end{bmatrix}\begin{bmatrix} 30 \\ -10 \\ 1 \end{bmatrix} = \frac{k}{m\omega^2}\begin{bmatrix} 170 \\ -45.5 \\ 3.7 \end{bmatrix} = \frac{k}{m\omega^2}(3.7)\begin{bmatrix} 46 \\ -12.3 \\ 1 \end{bmatrix}$$

Third iteration:

$$\begin{bmatrix} 45 \\ -11 \\ 1 \end{bmatrix} = \frac{k}{m\omega^2}\begin{bmatrix} 5 & -2 & 0 \\ -1 & \frac{3}{2} & -\frac{1}{2} \\ 0 & -\frac{1}{3} & \frac{1}{3} \end{bmatrix}\begin{bmatrix} 45 \\ -11 \\ 1 \end{bmatrix} = \frac{k}{m\omega^2}\begin{bmatrix} 247 \\ -62 \\ 4 \end{bmatrix} = \frac{k}{m\omega^2}(4)\begin{bmatrix} 61 \\ -15 \\ 1 \end{bmatrix}$$

Fourth iteration:

$$\begin{bmatrix} 60 \\ -15 \\ 1 \end{bmatrix} = \frac{k}{m\omega^2}\begin{bmatrix} 5 & -2 & 0 \\ -1 & \frac{3}{2} & -\frac{1}{2} \\ 0 & -\frac{1}{3} & \frac{1}{3} \end{bmatrix}\begin{bmatrix} 60 \\ -15 \\ 1 \end{bmatrix} = \frac{k}{m\omega^2}\begin{bmatrix} 330 \\ -83 \\ 5.4 \end{bmatrix} = \frac{k}{m\omega^2}(5.4)\begin{bmatrix} 60.7 \\ -15.3 \\ 1 \end{bmatrix}$$

The assumed column approximately repeats itself; this means the assumed value is correct. Hence

$$1 = \frac{k}{m\omega^2}(5.4) \qquad \text{and} \qquad \omega_3 = 2.36\sqrt{k/m} \text{ rad/sec}$$

THE STODOLA METHOD

26. Use the Stodola method to find the fundamental mode of vibration and its natural frequency of the spring-mass system as shown in Fig. 3-40. $k_1 = k_2 = k_3 = 1$ lb/in, $m_1 = m_2 = m_3 = 1$ lb-sec²/in.

Assume that the system is vibrating at one of its principal modes with natural frequency ω and that the motion is periodic. Then the system is acted upon by inertia forces $-m_i\ddot{x}_i$. Now

$$x_i = A_i \sin \omega t \qquad \text{and} \qquad -m_i\ddot{x}_i = \omega^2 m_i A_i$$

The Stodola method may be set up in the following tabular form as follows: Assuming an arbitrary set of values for the fundamental principal mode, the inertia force acting on each mass is equal to the product of the assumed deflection and the square of the natural frequency as shown in row 2. The spring force in row 3 is equal to the total inertia force acting on each spring. Row 4 is obtained by dividing row 3, term by term, by their respective spring constants. The calculated deflections in row 5 are found by adding the deflections due to the springs, with the mass near the

Fig. 3-40

fixed end having the least deflection and so on. The calculated deflections are then compared with the assumed deflections. This process is continued until the calculated deflections are proportional to the assumed deflections. When this is true the assumed deflections will represent the configuration of the fundamental principal mode of vibration of the system.

	k_1	m_1	k_2	m_2	k_3	m_3
1. Assumed deflection		1		1		1
2. Inertia force		ω^2		ω^2		ω^2
3. Spring force	$3\omega^2$		$2\omega^2$		ω^2	
4. Spring deflection	$3\omega^2$		$2\omega^2$		ω^2	
5. Calculated deflection		$3\omega^2$		$5\omega^2$		$6\omega^2$
		1		1.67		2
1. Assumed deflection		1		1.67		2
2. Inertia force		ω^2		$1.67\omega^2$		$2\omega^2$
3. Spring force	$4.67\omega^2$		$3.67\omega^2$		$2\omega^2$	
4. Spring deflection	$4.67\omega^2$		$3.67\omega^2$		$2\omega^2$	
5. Calculated deflection		$4.67\omega^2$		$8.34\omega^2$		$10.34\omega^2$
		1		1.79		2.21
1. Assumed deflection		1		1.79		2.21
2. Inertia force		ω^2		$1.79\omega^2$		$2.21\omega^2$
3. Spring force	$5\omega^2$		$4\omega^2$		$2.21\omega^2$	
4. Spring deflection	$5\omega^2$		$4\omega^2$		$2.21\omega^2$	
5. Calculated deflection		$5\omega^2$		$9\omega^2$		$11.21\omega^2$
		1		1.8		2.24

The assumed deflection $\begin{bmatrix} 1.00 \\ 1.79 \\ 2.21 \end{bmatrix}$ at this point is very close to the calculated deflection. Hence the

fundamental principal mode of vibration is given by

$$\begin{bmatrix} 1.00 \\ 1.80 \\ 2.24 \end{bmatrix}$$

and the fundamental natural frequency is found from
$$1.00 + 1.80 + 2.24 = (5 + 9 + 11.21)\omega^2 \qquad \text{or} \qquad \omega_1 = 0.44 \text{ rad/sec}$$

27. Use the Stodola method to determine the lowest natural frequency of the four-degree-of-freedom spring-mass system as shown in Fig. 3-41.

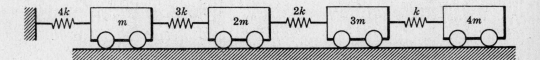

Fig. 3-41

See Problem 26 for explanation and procedure.

	$k_1 = 4k$	$m_1 = m$	$k_2 = 3k$	$m_2 = 2m$	$k_3 = 2k$	$m_3 = 3m$	$k_4 = k$	$m_4 = 4m$
Assumed deflection		4.00		3.00		2.00		1.00
Inertia force		$4\omega^2$		$6\omega^2$		$6\omega^2$		$4\omega^2$
Spring force	$20\omega^2$		$16\omega^2$		$10\omega^2$		$4\omega^2$	
Spring deflection	$5\omega^2$		$5.3\omega^2$		$5\omega^2$		$4\omega^2$	
Calculated deflection		$5\omega^2$		$10.3\omega^2$		$15.3\omega^2$		$19.3\omega^2$
Assumed deflection		1.00		2.00		3.00		4.00
Inertia force		ω^2		$4\omega^2$		$9\omega^2$		$16\omega^2$
Spring force	$30\omega^2$		$29\omega^2$		$25\omega^2$		$16\omega^2$	
Spring deflection	$7.5\omega^2$		$9.7\omega^2$		$12.5\omega^2$		$16\omega^2$	
Calculated deflection		$7.5\omega^2$		$17.2\omega^2$		$29.7\omega^2$		$45.7\omega^2$
Assumed deflection		1.00		2.00		4.00		6.00
Inertia force		ω^2		$4\omega^2$		$12\omega^2$		$24\omega^2$
Spring force	$41\omega^2$		$40\omega^2$		$36\omega^2$		$24\omega^2$	
Spring deflection	$10.25\omega^2$		$13.3\omega^2$		$18\omega^2$		$24\omega^2$	
Calculated deflection		$10.25\omega^2$		$23.55\omega^2$		$41.55\omega^2$		$65.55\omega^2$
Assumed deflection		1.00		2.2		4.00		6.4
Inertia force		ω^2		$4.4\omega^2$		$12\omega^2$		$25.6\omega^2$
Spring force	$43\omega^2$		$42\omega^2$		$37.6\omega^2$		$25.6\omega^2$	
Spring deflection	$10.75\omega^2$		$14\omega^2$		$18.8\omega^2$		$25.6\omega^2$	
Calculated deflection		$10.75\omega^2$		$24.75\omega^2$		$43.55\omega^2$		$69.15\omega^2$
		1.00		2.30		4.05		6.42

Therefore the first principal mode is given by $\begin{bmatrix} 1.00 \\ 2.30 \\ 4.05 \\ 6.42 \end{bmatrix}$ and the lowest natural frequency is obtained from

$$(1 + 2.3 + 4.05 + 6.42) = (10.75 + 24.75 + 43.55 + 69.15)\omega^2 \quad \text{or} \quad 13.77 = 148.2\omega^2$$

Hence $\qquad \omega_1^2 = 0.093 \qquad \text{and} \qquad \omega_1 = 0.306\sqrt{k/m} \text{ rad/sec}$

28. Prove that the Stodola method will converge to the fundamental mode of vibration.

The Stodola method begins with assumed deflections for the fundamental mode of a system. The corresponding inertia forces due to these assumed deflections are calculated. Compared with actual inertia forces and deflections of the system, the inertia forces just found will produce a new set of deflections which is used to start the next iteration. The process is repeated. Eventually, this process will converge to the fundamental mode; the degree of accuracy depends on the number of iterations.

The general motion of an n-degree-of-freedom system is given by

$$\begin{aligned} x_1 &= A_1 \sin(\omega_1 t + \psi_1) + A_2 \sin(\omega_2 t + \psi_2) + \cdots + A_n \sin(\omega_n t + \psi_n) \\ x_2 &= B_1 \sin(\omega_1 t + \psi_1) + B_2 \sin(\omega_2 t + \psi_2) + \cdots + B_n \sin(\omega_n t + \psi_n) \\ x_3 &= C_1 \sin(\omega_1 t + \psi_1) + C_2 \sin(\omega_2 t + \psi_2) + \cdots + C_n \sin(\omega_n t + \psi_n) \end{aligned} \qquad (1)$$

$\cdots\cdots\cdots\cdots\cdots\cdots\cdots\cdots\cdots\cdots\cdots\cdots\cdots\cdots\cdots\cdots\cdots\cdots$

Let the assumed deflections be an arbitrary superposition of all the modes of the system, with constants $a_1, a_2, \ldots, a_n$,

$$\begin{aligned} x_1 &= a_1 A_1 + a_2 A_2 + \cdots + a_n A_n \\ x_2 &= a_1 B_1 + a_2 B_2 + \cdots + a_n B_n \\ x_3 &= a_1 C_1 + a_2 C_2 + \cdots + a_n C_n \end{aligned} \qquad (2)$$

$\cdots\cdots\cdots\cdots\cdots\cdots\cdots\cdots\cdots\cdots\cdots\cdots\cdots\cdots$

The corresponding inertia forces are

$$m_1(a_1A_1 + a_2A_2 + \cdots + a_nA_n)\omega^2$$
$$m_2(a_1B_1 + a_2B_2 + \cdots + a_nB_n)\omega^2 \tag{3}$$
$$m_3(a_1C_1 + a_2C_2 + \cdots + a_nC_n)\omega^2$$
$$\cdots\cdots\cdots\cdots\cdots\cdots\cdots\cdots\cdots\cdots\cdots\cdots\cdots\cdots$$

where $m_1, m_2, \ldots, m_n$ are the masses of the system and ω is the natural frequency.

Now if the system is vibrating with all the principal modes present, the inertia forces and the corresponding deflections are

$$m_1(A_1\omega_1^2 + A_2\omega_2^2 + \cdots + A_n\omega_n^2), \quad (A_1 + A_2 + \cdots + A_n)$$
$$m_2(B_1\omega_1^2 + B_2\omega_2^2 + \cdots + B_n\omega_n^2), \quad (B_1 + B_2 + \cdots + B_n) \tag{4}$$
$$m_3(C_1\omega_1^2 + C_2\omega_2^2 + \cdots + C_n\omega_n^2), \quad (C_1 + C_2 + \cdots + C_n)$$
$$\cdots\cdots\cdots\cdots\cdots\cdots\cdots\cdots\cdots\cdots\cdots\cdots\cdots\cdots\cdots\cdots\cdots\cdots\cdots$$

Hence the inertia forces in (*3*) will produce a new set of deflections:

$$\omega^2(a_1A_1/\omega_1^2 + a_2A_2/\omega_2^2 + \cdots + a_nA_n/\omega_n^2)$$
$$\omega^2(a_1B_1/\omega_1^2 + a_2B_2/\omega_2^2 + \cdots + a_nB_n/\omega_n^2) \tag{5}$$
$$\omega^2(a_1C_1/\omega_1^2 + a_2C_2/\omega_2^2 + \cdots + a_nC_n/\omega_n^2)$$
$$\cdots\cdots\cdots\cdots\cdots\cdots\cdots\cdots\cdots\cdots\cdots\cdots\cdots\cdots$$

Now

$$x_1 = \omega^2(a_1A_1/\omega_1^2 + a_2A_2/\omega_2^2 + \cdots + a_nA_n/\omega_n^2)$$
$$x_2 = \omega^2(a_1B_1/\omega_1^2 + a_2B_2/\omega_2^2 + \cdots + a_nB_n/\omega_n^2) \tag{6}$$
$$x_3 = \omega^2(a_1C_1/\omega_1^2 + a_2C_2/\omega_2^2 + \cdots + a_nC_n/\omega_n^2)$$
$$\cdots\cdots\cdots\cdots\cdots\cdots\cdots\cdots\cdots\cdots\cdots\cdots\cdots\cdots$$

Using the deflections in (*6*) as the assumed deflections, and carrying out exactly the steps in the last iteration, we have

$$x_1 = \omega^4(a_1A_1/\omega_1^4 + a_2A_2/\omega_2^4 + \cdots + a_nA_n/\omega_n^4)$$
$$x_2 = \omega^4(a_1B_1/\omega_1^4 + a_2B_2/\omega_2^4 + \cdots + a_nB_n/\omega_n^4) \tag{7}$$
$$x_3 = \omega^4(a_1C_1/\omega_1^4 + a_2C_2/\omega_2^4 + \cdots + a_nC_n/\omega_n^4)$$
$$\cdots\cdots\cdots\cdots\cdots\cdots\cdots\cdots\cdots\cdots\cdots\cdots\cdots\cdots$$

After r iterations, the assumed deflections take the following general form:

$$x_1 = \omega^{2r}(a_1A_1/\omega_1^{2r} + a_2A_2/\omega_2^{2r} + \cdots + a_nA_n/\omega_n^{2r})$$
$$x_2 = \omega^{2r}(a_1B_1/\omega_1^{2r} + a_2B_2/\omega_2^{2r} + \cdots + a_nB_n/\omega_n^{2r}) \tag{8}$$
$$x_3 = \omega^{2r}(a_1C_1/\omega_1^{2r} + a_2C_2/\omega_2^{2r} + \cdots + a_nC_n/\omega_n^{2r})$$
$$\cdots\cdots\cdots\cdots\cdots\cdots\cdots\cdots\cdots\cdots\cdots\cdots\cdots\cdots$$

or

$$x_1 = (a_1\omega^{2r}/\omega_1^{2r})(A_1 + a_2A_2\omega_1^{2r}/a_1\omega_2^{2r} + \cdots + a_nA_n\omega_1^{2r}/a_1\omega_n^{2r})$$
$$x_2 = (a_1\omega^{2r}/\omega_1^{2r})(B_1 + a_2B_2\omega_1^{2r}/a_1\omega_2^{2r} + \cdots + a_nB_n\omega_1^{2r}/a_1\omega_n^{2r}) \tag{9}$$
$$x_3 = (a_1\omega^{2r}/\omega_1^{2r})(C_1 + a_2C_2\omega_1^{2r}/a_1\omega_2^{2r} + \cdots + a_nC_n\omega_1^{2r}/a_1\omega_n^{2r})$$
$$\cdots\cdots\cdots\cdots\cdots\cdots\cdots\cdots\cdots\cdots\cdots\cdots\cdots\cdots\cdots$$

As $\omega_1 < \omega_2 < \omega_3 < \cdots < \omega_n$, and as the number of iterations is sufficiently large or the value of r assumes a sufficient large number, the ratios of the natural frequencies become very small. In most cases, less than ten iterations are required to obtain the fundamental mode of the system. Thus for sufficiently large numbers of iterations, the calculated deflections in (*9*) become

$$\begin{aligned} x_1 &= a_1A_1\omega^{2r}/\omega_1^{2r} \\ x_2 &= a_1B_1\omega^{2r}/\omega_1^{2r} \\ x_3 &= a_1C_1\omega^{2r}/\omega_1^{2r} \\ \cdots\cdots &\cdots\cdots \end{aligned} \quad \text{or} \quad \begin{bmatrix} x_1 \\ x_2 \\ x_3 \\ \cdot \end{bmatrix} = a_1\omega^{2r}/\omega_1^{2r} \begin{bmatrix} A_1 \\ B_1 \\ C_1 \\ \cdot \end{bmatrix}$$

which approaches very closely the pure fundamental mode of vibration of the system.

Thus the Stodola method converges to the fundamental mode of vibration for an *n*-degree-of-freedom system.

THE HOLZER METHOD

29. Use the Holzer method to determine the natural frequencies of the spring-mass system as shown in Fig. 3-42. Here $m_1 = m_2 = m_3 = 1$ lb-sec^2/in.

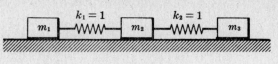

Fig. 3-42

Begin the Holzer tabulation with the column of position, indicating the masses of the system. The second column is for the values of the different masses of the system; this information is given. The third column is the product of mass and frequency squared. Displacement comes next, and is obtained from the preceding row minus the total displacement at the end of the same row. Column five is just the product of columns three and four. The total inertia force is inserted in column six. It is equal to the sum of the total inertia force in the preceding row plus the inertia force on the same row. The rest are plainly evident.

An initial displacement, usually equal to unity for convenience, is assumed. If the assumed frequency happens to be one of the natural frequencies of the systems, the final total inertia force on the system should be zero. This is because the system is having free vibration. If the final total inertia force is not equal to zero, the amount indicates the discrepancy of the assumed frequency.

Table

Position	m_i	$m_i\omega^2$	x_i	$m_i x_i \omega^2$	$\sum_1^i m_i x_i \omega^2$	k_{ij}	$\sum_1^i m_i x_i \omega^2 / k_{ij}$
Assumed frequency, $\omega = 0.5$							
1	1	0.25	1	0.25	0.25	1	0.25
2	1	0.25	0.75	0.19	0.44	1	0.44
3	1	0.25	0.31	0.07	0.51		
Assumed frequency, $\omega = 0.75$							
1	1	0.56	1	0.56	0.56	1	0.56
2	1	0.56	0.44	0.24	0.80	1	0.80
3	1	0.56	−0.36	−0.2	0.60		
Assumed frequency, $\omega = 1.0$							
1	1	1	1	1	1	1	1
2	1	1	0	0	1	1	1
3	1	1	−1	−1	0		
Assumed frequency, $\omega = 1.5$							
1	1	2.25	1.0	2.25	2.25	1	2.25
2	1	2.25	−1.25	−2.82	−0.57	1	−0.57
3	1	2.25	−0.68	−1.53	−2.10		
Assumed frequency, $\omega = 1.79$							
1	1	3.21	1	3.21	3.21	1	3.21
2	1	3.21	−2.21	−7.08	−3.87	1	−3.87
3	1	3.21	1.66	5.34	1.47		
Assumed frequency, $\omega = 2.0$							
1	1	4	1	4	4	1	4
2	1	4	−3	−12	−8	1	−8
3	1	4	5	20	12		

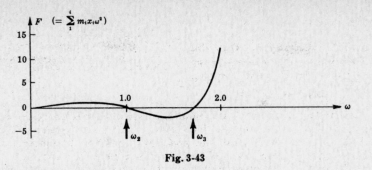

Fig. 3-43

Therefore the natural frequencies are $\omega_1 = 0$, $\omega_2 = 1.0$, $\omega_3 = 1.7$ rad/sec.

30. Use the Holzer method to determine the natural frequencies of the four-mass system as shown in Fig. 3-44, if $k = 1$ lb/in and $m = 1$ lb-sec^2/in.

See procedure given in Problem 29.

Table

Item	m	$m\omega^2$	x	$mx\omega^2$	$\Sigma mx\omega^2$	k	$\Sigma mx\omega^2/k$
Assumed frequency, $\omega = 0.2$							
1	4	.16	1	.16	.16	1	.16
2	3	.12	.84	.101	.261	2	.13
3	2	.08	.71	.056	.317	3	.105
4	1	.04	.605	.025	.342	4	.0855
5	∞	∞	.5195				
Assumed frequency, $\omega = 0.3$							
1	4	.36	1	.36	.36	1	.36
2	3	.27	.64	.173	.533	2	.267
3	2	.18	.373	.067	.600	3	.200
4	1	.09	.173	.0155	.6155	4	.1539
5	∞	∞	.0192				
Assumed frequency, $\omega = 0.4$							
1	4	.64	1	.64	.64	1	.64
2	3	.48	.36	.173	.813	2	.406
3	2	.32	−.046	−.0147	.798	3	.266
4	1	.16	−.312	−.049	.748	4	.187
5	∞	∞	−.499				
Assumed frequency, $\omega = 0.6$							
1	4	1.44	1	1.44	1.44	1	1.44
2	3	1.08	−.44	−.475	.965	2	.482
3	2	.72	−.922	−.664	.301	3	.100
4	1	.36	−1.023	−.368	−.067	4	−.017
5	∞	∞	−1.006				
Assumed frequency, $\omega = 0.8$							
1	4	2.56	1	2.56	2.56	1	2.56
2	3	1.92	−1.56	−3.00	−.44	2	−.22
3	2	1.28	−1.34	−1.72	−2.16	3	−.73
4	1	.64	−.61	−.39	−2.55	4	−.64
5	∞	∞	.03				

Fig. 3-44

Table (cont.)

Item	m	$m\omega^2$	x	$mx\omega^2$	$\Sigma mx\omega^2$	k	$\Sigma mx\omega^2/k$
Assumed frequency, $\omega = 1.0$							
1	4	4	1	4	4	1	4
2	3	3	−3	−9	−5	2	−2.5
3	2	2	−.5	−1	−6	3	−2.0
4	1	1	1.5	1.5	−4.5	4	−1.13
5	∞	∞	2.63				
Assumed frequency, $\omega = 1.5$							
1	4	9	1	9	9	1	9
2	3	6.75	−8	−54	−45	2	−22.5
3	2	4.5	14.5	65.3	20.3	3	6.77
4	1	2.25	7.73	17.4	37.7	4	9.43
5	∞	∞	−1.70				
Assumed frequency, $\omega = 1.8$							
1	4	12.96	1	12.96	12.96	1	12.96
2	3	9.72	−11.96	−116.4	−103.44	2	−51.72
3	2	6.48	39.76	257.7	154.26	3	51.42
4	1	3.24	−11.66	−37.8	116.46	4	29.12
5	∞	∞	−40.78				
Assumed frequency, $\omega = 2.0$							
1	4	16	1	16	16	1	16
2	3	12	−15	−180	−164	2	−82
3	2	8	67	536	372	3	124
4	1	4	−57	−228	144	4	36
5	∞	∞	−93				
Assumed frequency, $\omega = 2.5$							
1	4	25	1	25	25	1	25
2	3	18.75	−24	−450	−425	2	−212.5
3	2	12.5	188.5	2360	1935	3	645
4	1	6.25	−456.5	−2860	−925	4	−231
5	∞	∞	−225.5				
Assumed frequency, $\omega = 3.0$							
1	4	36	1	36	36	1	36
2	3	27	−35	−945	−909	2	−455
3	2	18	420	7560	6651	3	2220
4	1	9	−1800	−16,200	−9550	4	−2388
5	∞	∞	588				

Plot the curve with the assumed frequencies against the amplitudes of the fixed end as shown in Fig. 3-45. The natural frequencies of the system are given by the intersections of the curve with the frequency axis. The natural frequencies are

$$\omega_1 = 0.30 \text{ rad/sec}$$
$$\omega_2 = 0.81 \text{ rad/sec}$$
$$\omega_3 = 1.45 \text{ rad/sec}$$
$$\omega_4 = 2.83 \text{ rad/sec}$$

Note: Curve is not drawn to scale.

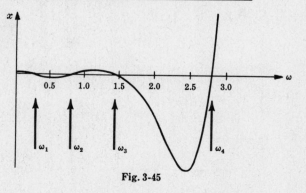

Fig. 3-45

BRANCHED SYSTEM

31. A four spring-mass branched system is shown in Fig. 3-46. If the masses are moving in the vertical direction only, derive the frequency equation of the system.

The equations of motion are given by $\Sigma F = ma$:

$$m_1 \ddot{x}_1 = -k_1 x_1 - k_2(x_1 - x_2)$$
$$m_2 \ddot{x}_2 = -k_2(x_2 - x_1) - k_3(x_2 - x_3) - k_4(x_2 - x_4)$$
$$m_3 \ddot{x}_3 = -k_3(x_3 - x_2)$$
$$m_4 \ddot{x}_4 = -k_4(x_4 - x_2)$$

Rearranging,

$$m_1 \ddot{x}_1 + (k_1 + k_2)x_1 - k_2 x_2 = 0$$
$$m_2 \ddot{x}_2 + (k_2 + k_3 + k_4)x_2 - k_3 x_3 - k_4 x_4 - k_2 x_1 = 0$$
$$m_3 \ddot{x}_3 + k_3 x_3 - k_3 x_2 = 0$$
$$m_4 \ddot{x}_4 + k_4 x_4 - k_4 x_2 = 0$$

Assume the motion is periodic and is composed of harmonic components of various amplitudes and frequencies. Let

$$x_1 = A \cos(\omega t + \psi), \qquad \ddot{x}_1 = -\omega^2 A \cos(\omega t + \psi)$$
$$x_2 = B \cos(\omega t + \psi), \qquad \ddot{x}_2 = -\omega^2 B \cos(\omega t + \psi)$$
$$x_3 = C \cos(\omega t + \psi), \qquad \ddot{x}_3 = -\omega^2 C \cos(\omega t + \psi)$$
$$x_4 = D \cos(\omega t + \psi), \qquad \ddot{x}_4 = -\omega^2 D \cos(\omega t + \psi)$$

When these relations are substituted and the term $\cos(\omega t + \psi)$ cancelled out, the differential equations of motion become a set of algebraic equations:

$$(k_1 + k_2 - m_1\omega^2)A - k_2 B = 0$$
$$-k_2 A + (k_2 + k_3 + k_4 - m_2\omega^2)B - k_3 C - k_4 D = 0$$
$$-k_3 B + (k_3 - m_3\omega^2)C = 0$$
$$-k_4 B + (k_4 - m_4\omega^2)D = 0$$

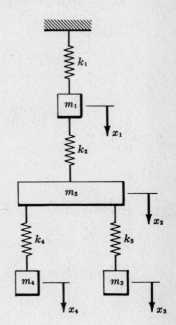

Fig. 3-46

from which the frequency equation is obtained by setting the determinant of the coefficients A, B, C, D to zero.

$$\begin{vmatrix} (k_1 + k_2 - m_1\omega^2) & -k_2 & 0 & 0 \\ -k_2 & (k_2 + k_3 + k_4 - m_2\omega^2) & -k_3 & -k_4 \\ 0 & -k_3 & (k_3 - m_3\omega^2) & 0 \\ 0 & -k_4 & 0 & (k_4 - m_4\omega^2) \end{vmatrix} = 0$$

Expand the determinant and simplify to obtain

$$\omega^8 - \left[\frac{k_1 + k_2}{m_1} + \frac{k_2 + k_3 + k_4}{m_2} + \frac{k_3}{m_3} + \frac{k_4}{m_4}\right]\omega^6$$

$$+ \left[\frac{k_1 k_2 + k_2 k_3 + k_3 k_1 + k_1 k_4 + k_2 k_4}{m_1 m_2} + \frac{k_2 k_3 + k_3 k_4}{m_2 m_3} + \frac{k_1 k_3 + k_2 k_3}{m_1 m_3}\right.$$

$$\left. + \frac{k_2 k_4 + k_3 k_4}{m_2 m_4} + \frac{k_1 k_4 + k_2 k_4}{m_1 m_4} + \frac{k_3 k_4}{m_3 m_4}\right]\omega^4$$

$$- \left[\frac{k_1 k_2 k_3 + k_2 k_3 k_4 + k_3 k_4 k_1}{m_1 m_2 m_3} + \frac{k_2 k_3 k_4}{m_2 m_3 m_4} + \frac{(k_1 + k_2)k_3 k_4}{m_3 m_4 m_1}\right.$$

$$\left. + \frac{k_4(k_1 k_2 + k_2 k_3 + k_3 k_1)}{m_4 m_1 m_2}\right]\omega^2$$

$$+ \frac{k_1 k_2 k_3 k_4}{m_1 m_2 m_3 m_4} = 0$$

32. Use the Matrix Iteration method to find the fundamental frequency of the branched system as shown in Fig. 3-47.

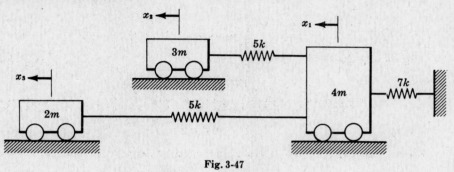

Fig. 3-47

From influence coefficient theory, the following deflection equations are obtained

$$\begin{bmatrix} x_1 \\ x_2 \\ x_3 \end{bmatrix} = \omega^2 \begin{bmatrix} \alpha_{11}m_1 & \alpha_{12}m_2 & \alpha_{13}m_3 \\ \alpha_{21}m_1 & \alpha_{22}m_2 & \alpha_{23}m_3 \\ \alpha_{31}m_1 & \alpha_{32}m_2 & \alpha_{33}m_3 \end{bmatrix} \begin{bmatrix} x_1 \\ x_2 \\ x_3 \end{bmatrix}$$

where $\alpha_{11} = \alpha_{12} = \alpha_{13} = 1/7k$, $\alpha_{21} = 1/7k$, $\alpha_{22} = 12/35k$, $\alpha_{23} = 1/7k$, $\alpha_{31} = \alpha_{32} = 1/7k$, $\alpha_{33} = 12/35k$; and $m_1 = 4m$, $m_2 = 3m$, $m_3 = 2m$.

Substituting these values into the deflection equations yields

$$\begin{bmatrix} x_1 \\ x_2 \\ x_3 \end{bmatrix} = \frac{\omega^2 m}{7k} \begin{bmatrix} 4 & 3 & 2 \\ 4 & 7.2 & 2 \\ 4 & 3 & 4.8 \end{bmatrix} \begin{bmatrix} x_1 \\ x_2 \\ x_3 \end{bmatrix}$$

Begin the iteration process with an assumed fundamental mode.

First iteration:

$$\begin{bmatrix} 0.4 \\ 0.6 \\ 1.0 \end{bmatrix} = \frac{\omega^2 m}{7k} \begin{bmatrix} 4 & 3 & 2 \\ 4 & 7.2 & 2 \\ 4 & 3 & 4.8 \end{bmatrix} \begin{bmatrix} 0.4 \\ 0.6 \\ 1.0 \end{bmatrix} = \frac{\omega^2 m}{7k} \begin{bmatrix} 5.4 \\ 7.9 \\ 8.2 \end{bmatrix} = \frac{\omega^2 m}{7k}(8.2) \begin{bmatrix} 0.66 \\ 0.96 \\ 1.00 \end{bmatrix}$$

Second iteration:

$$\begin{bmatrix} 0.6 \\ 0.9 \\ 1.0 \end{bmatrix} = \frac{\omega^2 m}{7k} \begin{bmatrix} 4 & 3 & 2 \\ 4 & 7.2 & 2 \\ 4 & 3 & 4.8 \end{bmatrix} \begin{bmatrix} 0.6 \\ 0.9 \\ 1.0 \end{bmatrix} = \frac{\omega^2 m}{7k} \begin{bmatrix} 7.1 \\ 10.9 \\ 9.9 \end{bmatrix} = \frac{\omega^2 m}{7k}(9.9) \begin{bmatrix} 0.8 \\ 1.1 \\ 1.0 \end{bmatrix}$$

Third iteration:

$$\begin{bmatrix} 0.8 \\ 1.1 \\ 1.0 \end{bmatrix} = \frac{\omega^2 m}{7k} \begin{bmatrix} 4 & 3 & 2 \\ 4 & 7.2 & 2 \\ 4 & 3 & 4.8 \end{bmatrix} \begin{bmatrix} 0.8 \\ 1.1 \\ 1.0 \end{bmatrix} = \frac{\omega^2 m}{7k} \begin{bmatrix} 8.5 \\ 13.1 \\ 11.3 \end{bmatrix} = \frac{\omega^2 m}{7k}(11.3) \begin{bmatrix} 0.75 \\ 1.16 \\ 1.00 \end{bmatrix}$$

Fourth iteration:

$$\begin{bmatrix} 0.8 \\ 1.2 \\ 1.0 \end{bmatrix} = \frac{\omega^2 m}{7k} \begin{bmatrix} 4 & 3 & 2 \\ 4 & 7.2 & 2 \\ 4 & 3 & 4.8 \end{bmatrix} \begin{bmatrix} 0.8 \\ 1.2 \\ 1.0 \end{bmatrix} = \frac{\omega^2 m}{7k} \begin{bmatrix} 8.8 \\ 13.8 \\ 11.6 \end{bmatrix} = \frac{\omega^2 m}{7k}(11.6) \begin{bmatrix} 0.76 \\ 1.19 \\ 1.00 \end{bmatrix}$$

The assumed mode $\begin{bmatrix} 0.8 \\ 1.2 \\ 1.0 \end{bmatrix}$ almost repeats itself in the last iteration, i.e. $\begin{bmatrix} 0.76 \\ 1.19 \\ 1.00 \end{bmatrix}$. Thus approximately,

$$\begin{bmatrix} 0.8 \\ 1.2 \\ 1.0 \end{bmatrix} = \frac{\omega^2 m}{7k}(11.6) \begin{bmatrix} 0.8 \\ 1.2 \\ 1.0 \end{bmatrix}$$

or $\qquad\qquad 1 = (\omega^2 m/7k)(11.6) \qquad$ and so $\qquad \omega_1 = 0.78\sqrt{k/m}$ rad/sec

33. Use the Stodola method to determine the lowest natural frequency of the branched system as shown in Fig. 3-48.

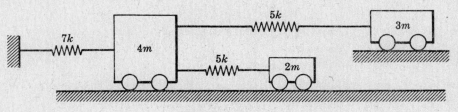

Fig. 3-48

See Problem 26, Page 98, for an explanation of the Stodola method.

	$k_1 = 7k$	$m_1 = 4m$	$k_2 = 5k$	$m_2 = 3m$	$k_3 = 5k$	$m_3 = 2m$
Assumed deflection		1		1		1
Inertia force		$4\omega^2$		$3\omega^2$		$2\omega^2$
Spring force	$9\omega^2$		$3\omega^2$		$2\omega^2$	
Spring deflection	$1.3\omega^2$		$0.6\omega^2$		$0.4\omega^2$	
Calculated deflection		$1.3\omega^2$		$1.9\omega^2$		$1.7\omega^2$
		1		1.46		1.31
Assumed deflection		1		1.4		1.3
Inertia force		$4\omega^2$		$4.2\omega^2$		$2.6\omega^2$
Spring force	$10.8\omega^2$		$4.2\omega^2$		$2.6\omega^2$	
Spring deflection	$1.54\omega^2$		$0.84\omega^2$		$0.52\omega^2$	
Calculated deflection		$1.54\omega^2$		$2.38\omega^2$		$2.06\omega^2$
		1		1.54		1.34
Assumed deflection		1		1.52		1.34
Inertia force		$4\omega^2$		$4.56\omega^2$		$2.68\omega^2$
Spring force	$11.24\omega^2$		$4.56\omega^2$		$2.68\omega^2$	
Spring deflection	$1.61\omega^2$		$0.92\omega^2$		$0.53\omega^2$	
Calculated deflection		$1.61\omega^2$		$2.53\omega^2$		$2.14\omega^2$
		1		1.56		1.32
Assumed deflection		1		1.56		1.32
Inertia force		$4\omega^2$		$4.68\omega^2$		$2.64\omega^2$
Spring force	$11.32\omega^2$		$4.68\omega^2$		$2.64\omega^2$	
Spring deflection	$1.62\omega^2$		$0.93\omega^2$		$0.53\omega^2$	
Calculated deflection		$1.62\omega^2$		$2.55\omega^2$		$2.15\omega^2$
		1		1.57		1.33

The assumed deflection $\begin{bmatrix} 1.00 \\ 1.56 \\ 1.32 \end{bmatrix}$ at this point is quite close to the calculated deflection. Hence the

fundamental principal mode of vibration is given by $\begin{bmatrix} 1.00 \\ 1.57 \\ 1.33 \end{bmatrix}$ and the lowest natural frequency is found from

$$(1 + 1.57 + 1.33) = (1.62 + 2.55 + 2.15)\omega^2 \qquad \text{or} \qquad \omega_1 = 0.79\sqrt{k/m} \text{ rad/sec}$$

THE MECHANICAL IMPEDANCE METHOD

34. Use the Mechanical Impedance method to find the frequency equation of the spring-mass system as shown in Fig. 3-49.

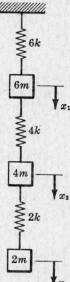

Fig. 3-49

The mechanical impedances are k and $-m\omega^2$ for spring and mass. For junction x_1, this becomes

$$(6k + 4k - 6m\omega^2)x_1$$

where the slippage term is $4kx_2$. Since there is no force acting on junction x_1, one equation will be obtained as

$$(6k + 4k - 6m\omega^2)x_1 - 4kx_2 = 0$$

Similarly, for junction x_2 the equation is

$$(4k + 2k - 4m\omega^2)x_2 - 4kx_1 - 2kx_3 = 0$$

where $4kx_1$ and $2kx_3$ are the slippage terms.

For junction x_3, this is given by

$$(2k - 2m\omega^2)x_3 - 2kx_2 = 0$$

Rearrange the equations to get

$$\begin{aligned}(10k - 6m\omega^2)x_1 - 4kx_2 &= 0\\ -4kx_1 + (6k - 4m\omega^2)x_2 - 2kx_3 &= 0\\ -2kx_2 + (2k - 2m\omega^2)x_3 &= 0\end{aligned}$$

Hence the frequency equation is given by

$$\begin{vmatrix} (10k - 6m\omega^2) & -4k & 0\\ -4k & (6k - 4m\omega^2) & -2k\\ 0 & -2k & (2k - 2m\omega^2)\end{vmatrix} = 0$$

35. Use the Mechanical Impedance method to determine the steady state vibrations of the masses of the system as shown in Fig. 3-50. Let $k_1 = k_2 = k_3 = k_4 = 1$ lb/in, $c_1 = c_2 = c_3 = c_4 = 1$ sec-lb/in, $m_1 = m_2 = m_3 = 1$ lb-sec²/in, and $\omega = 1$ rad/sec.

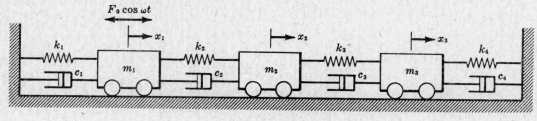

Fig. 3-50

Writing the impedance for junction x_1 and its amplitude, we obtain

$$(k_1 + ic_1\omega - m_1\omega^2 + k_2 + ic_2\omega)x_1$$

and the slippage terms for the junction x_1 are $k_2x_2 + ic_2\omega x_2$; hence the first equation is given by

$$(k_1 + k_2 + ic_1\omega + ic_2\omega - m_1\omega^2)x_1 - k_2x_2 - ic_2\omega x_2 = F_0$$

Similarly, the equations for junctions x_2 and x_3 are

$$(k_2 + k_3 + ic_2\omega + ic_3\omega - m_2\omega^2)x_2 - k_2x_1 - ic_2\omega x_1 - k_3x_3 - ic_3\omega x_3 = 0$$
$$(k_3 + k_4 + ic_3\omega + ic_4\omega - m_3\omega^2)x_3 - k_3x_2 - ic_3x_2\omega = 0$$

Substituting the given values into the equations of motion, we get

$$\begin{aligned}(1 + 2i)x_1 - (1 + i)x_2 &= F_0\\ -(1 + i)x_1 + (1 + 2i)x_2 - (1 + i)x_3 &= 0\\ -(1 + i)x_2 + (1 + 2i)x_3 &= 0\end{aligned}$$

Solving by Cramer's rule,

$$x_1 = \frac{\begin{vmatrix} F_0 & -(1+i) & 0 \\ 0 & (1+2i) & -(1+i) \\ 0 & -(1+i) & (1+2i) \end{vmatrix}}{\begin{vmatrix} (1+2i) & -(1+i) & 0 \\ -(1+i) & (1+2i) & -(1+i) \\ 0 & -(1+i) & (1+2i) \end{vmatrix}}$$

$$x_2 = \frac{\begin{vmatrix} (1+2i) & F_0 & 0 \\ -(1+i) & 0 & -(1+i) \\ 0 & 0 & (1+2i) \end{vmatrix}}{\begin{vmatrix} (1+2i) & -(1+i) & 0 \\ -(1+i) & (1+2i) & -(1+i) \\ 0 & -(1+i) & (1+2i) \end{vmatrix}}$$

$$x_3 = \frac{\begin{vmatrix} (1+2i) & -(1+i) & F_0 \\ -(1+i) & (1+2i) & 0 \\ 0 & -(1+i) & 0 \end{vmatrix}}{\begin{vmatrix} (1+2i) & -(1+i) & 0 \\ -(1+i) & (1+2i) & -(1+i) \\ 0 & -(1+i) & (1+2i) \end{vmatrix}}$$

Expand the determinants to obtain

$$x_1 = \frac{F_0(1+2i)(1+2i) - F_0(1+i)^2}{(1+2i)^3 - (1+2i)(1+i)^2 - (1+2i)(1+i)^2}$$

$$x_2 = \frac{F_0(1+i)(1+2i)}{(1+2i)^3 - (1+2i)(1+i)^2 - (1+2i)(1+i)^2}$$

$$x_3 = \frac{F_0(1+i)^2}{(1+2i)^3 - (1+2i)(1+i)^2 - (1+2i)(1+i)^2}$$

The equation for x_1 simplifies to

$$x_1 = \frac{F_0(3-2i)}{(3+6i)} = \frac{13F_0}{(-3+24i)}$$

Therefore the numerical value of the amplitude of x_1 is

$$\frac{13F_0}{\sqrt{9+24^2}} = 0.54F_0 \quad \text{and the phase angle} \quad \phi = \tan^{-1}(24/-3) = -82.9°$$

Thus the steady state response of the mass m_1 is

$$x_1(t) = 0.54F_0 \cos(\omega t - 82.9°)$$

Similar expressions can be found for masses m_2 and m_3 as

$$x_2(t) = 0.47F_0 \cos(\omega t - 45°)$$
$$x_3(t) = 0.29F_0 \cos(\omega t - 26.7°)$$

THE ORTHOGONALITY PRINCIPLE

36. Show that the orthogonality principle holds for Problem 8, Page 80.

For three-degree-of-freedom systems, the orthogonality principle can be written as

$$m_1 A_1 A_2 + m_2 B_1 B_2 + m_3 C_1 C_2 = 0$$
$$m_1 A_2 A_3 + m_2 B_2 B_3 + m_3 C_2 C_3 = 0$$
$$m_1 A_3 A_1 + m_2 B_3 B_1 + m_3 C_3 C_1 = 0$$

where m's are the masses; A's, B's and C's are the amplitudes of vibration of the system.

The general motion of the problem is

$$x_1(t) = \tfrac{1}{3} - \tfrac{1}{2}\sin(t + \pi/2) + \tfrac{1}{6}\sin(\sqrt{3}\,t + \pi/2)$$
$$x_2(t) = \tfrac{1}{3} - \tfrac{1}{3}\sin(\sqrt{3}\,t + \pi/2)$$
$$x_3(t) = \tfrac{1}{3} + \tfrac{1}{2}\sin(t + \pi/2) + \tfrac{1}{6}\sin(\sqrt{3}\,t + \pi/2)$$

Substituting the corresponding amplitudes of vibration into the equations for the principle of orthogonality, we obtain

$$m(\tfrac{1}{3})(-\tfrac{1}{2}) + m(\tfrac{1}{3})(0) + m(\tfrac{1}{3})(\tfrac{1}{2}) = 0$$
$$m(-\tfrac{1}{2})(\tfrac{1}{6}) + m(0)(\tfrac{1}{3}) + m(\tfrac{1}{2})(\tfrac{1}{6}) = 0$$
$$m(\tfrac{1}{6})(\tfrac{1}{3}) + m(-\tfrac{1}{3})(\tfrac{1}{3}) + m(\tfrac{1}{6})(\tfrac{1}{3}) = 0$$

Hence the orthogonality principle is completely satisfied.

Supplementary Problems

37. Derive the equations of motion of the system as shown in Fig. 3-51 below. The connecting rods are weightless and restrict motion to the plane of the paper.

Ans. $4m\ddot{\theta}_1 + 2k\theta_1 - k\theta_2 = 0$

$\quad\quad 4m\ddot{\theta}_2 + 2k\theta_2 - k\theta_3 - k\theta_1 = 0$

$\quad\quad 4m\ddot{\theta}_3 + 2k\theta_3 - k\theta_2 = 0$

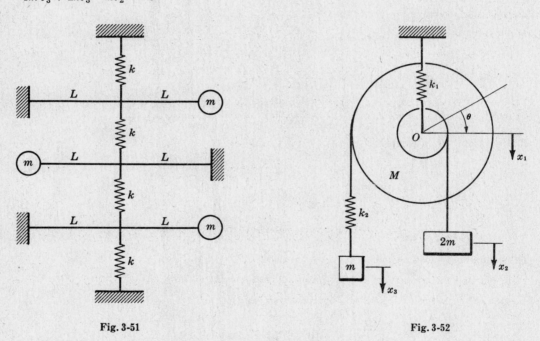

Fig. 3-51   Fig. 3-52

38. The circular homogeneous cylinder of total mass M and radius $2a$ is suspended by a spring of stiffness k_1, and is free to rotate with respect to its center of mass O as shown in Fig. 3-52 above. Derive the equations of motion.

Ans. $3M\ddot{x}_1 + (k_1 + 9k_2)x_1 - 2M\ddot{x}_2 - 6k_2x_2 - 3k_2x_3 = 0$

$\quad\quad (2M + 2m)\ddot{x}_2 + 4k_2x_2 + 2k_2x_3 - 2M\ddot{x}_1 - 6k_2x_1 = 0$

$\quad\quad m\ddot{x}_3 + k_2x_3 - 3k_2x_1 + 2k_2x_2 = 0$

39. Calculate the natural frequencies of the system as shown in Fig. 3-53.

Ans. $\omega_1 = 0.39\sqrt{k/m}, \quad \omega_2 = 1.47\sqrt{k/m},$
$\quad\quad \omega_3 = 2.36\sqrt{k/m}$ rad/sec

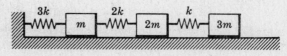

Fig. 3-53

40. The equivalent spring for the cantilever is $K = 10$ lb/in, and $k = 1$ lb/in, $m = 1$ lb-sec^2/in. Calculate the natural frequencies of the system as shown in Fig. 3-54 below.

   Ans. $\omega_1 = 3.16$, $\omega_2 = 3.34$, $\omega_3 = 3.62$ rad/sec

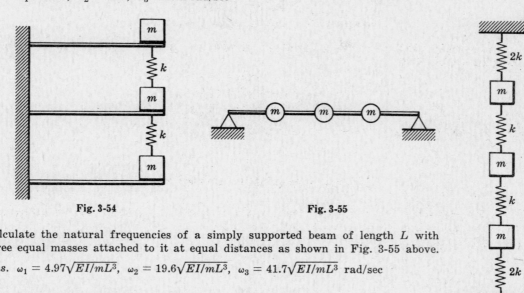

Fig. 3-54   Fig. 3-55

41. Calculate the natural frequencies of a simply supported beam of length L with three equal masses attached to it at equal distances as shown in Fig. 3-55 above.

   Ans. $\omega_1 = 4.97\sqrt{EI/mL^3}$, $\omega_2 = 19.6\sqrt{EI/mL^3}$, $\omega_3 = 41.7\sqrt{EI/mL^3}$ rad/sec

42. Determine the general motion of the system as shown in Fig. 3-56.

   Ans. $x_1(t) = A_1 \sin(\sqrt{k/m}\, t + \phi_1) + A_2 \sin(\sqrt{3k/m}\, t + \phi_2) + A_3 \sin(\sqrt{4k/m}\, t + \phi_3)$

   $x_2(t) = 2A_1 \sin(\sqrt{k/m}\, t + \phi_1) - A_3 \sin(\sqrt{4k/m}\, t + \phi_3)$

   $x_3(t) = A_1 \sin(\sqrt{k/m}\, t + \phi_1) - A_2 \sin(\sqrt{3k/m}\, t + \phi_2) + A_3 \sin(\sqrt{4k/m}\, t + \phi_3)$

Fig. 3-56

43. Fig. 3-57 below shows an excitation force $F_0 \sin \omega t$ applied to the center mass of the system. Determine the amplitude of the steady state response of the first mass of the system.

   Ans. $x_1(t) = \dfrac{F_0 k(2k - m\omega^2)}{\omega^6 - (6k/m)\omega^4 + (10k^2/m^2)\omega^2 - 4k^3/m^3}$

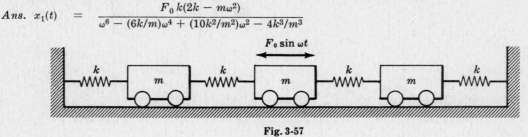

Fig. 3-57

44. Determine the natural frequencies of the spring-mass system as shown in Fig. 3-58 below; $m = k = 1$.

   Ans. $\omega_1 = 0.62$, $\omega_2 = 1.18$, $\omega_3 = 1.62$, $\omega_4 = 1.9$ rad/sec

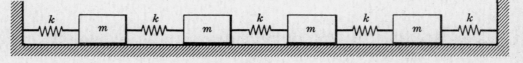

Fig. 3-58

45. Calculate the natural frequencies of a simply-supported beam of length L with four equal masses attached to it at equal distances as shown in Fig. 3-59.

   Ans. $\omega_1 = 4.97\sqrt{EI/mL^3}$, $\omega_2 = 19.6\sqrt{EI/mL^3}$, $\omega_3 = 42.1\sqrt{EI/mL^3}$, $\omega_4 = 55.4\sqrt{EI/mL^3}$ rad/sec

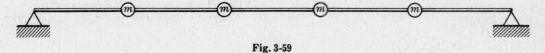

Fig. 3-59

INFLUENCE COEFFICIENTS

46. Find the influence coefficients of the spring-mass system as shown in Fig. 3-60.
    Ans. $\alpha_{11} = 1/2k$, $\alpha_{12} = 1/2k$, $\alpha_{21} = 1/2k$, $\alpha_{22} = 3/2k$

47. Determine the influence coefficients of the spring-mass system as shown in Fig. 3-61
    below.   *Ans.* $\alpha_{11} = 1/3k$, $\alpha_{12} = 1/3k$, $\alpha_{13} = 1/3k$
                      $\alpha_{21} = 1/3k$, $\alpha_{22} = 5/6k$, $\alpha_{23} = 5/6k$
                      $\alpha_{31} = 1/3k$, $\alpha_{32} = 5/6k$, $\alpha_{33} = 11/6k$

48. A simply-supported beam of length L has three equal masses attached to it at equal
    distances as shown in Fig. 3-62 below. Determine the influence coefficients.
    Ans. $\alpha_{11} = 3L^3/256EI$ $\alpha_{12} = 3.67L^3/256EI$ $\alpha_{13} = 2.33L^3/256EI$
           $\alpha_{21} = 3.67L^3/256EI$ $\alpha_{22} = 5.33L^3/256EI$ $\alpha_{23} = 3.67L^3/256EI$
           $\alpha_{31} = 2.33L^3/256EI$ $\alpha_{32} = 3.67L^3/256EI$ $\alpha_{33} = 3L^3/256EI$

Fig. 3-60

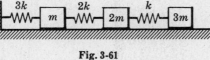

Fig. 3-61   Fig. 3-62

MATRICES

49. Solve Problem 5, Page 78, by the matrix method.

50. Determine the dynamic matrix of the branched-
    system as shown in Fig. 3-63. What are the
    principal coordinates?

    Ans. $[C] = \begin{bmatrix} k/m & -2k/3m & -k/3m \\ -2k/m & 2k/m & 0 \\ -k/m & 0 & 2k/m \end{bmatrix}$

    $p_1 = 0.36x_1 - 0.13x_2 + 0.06x_3$
    $p_2 = 0.2x_2 + 0.33x_3$
    $p_3 = -0.36x_1 - 0.13x_2 + 0.14x_3$

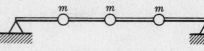

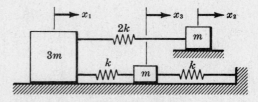

Fig. 3-63

51. Using the inverse matrix method, calculate the highest natural frequency of the system as shown in
    Fig. 3-55.

52. Use matrix iteration to determine the natural frequencies of the system as shown in Fig. 3-4.

THE STODOLA METHOD

53. Use the Stodola method to find the fundamental frequency of the system as shown in Fig. 3-64.
    m_1 and m_2 weigh 4 lb and 6 lb respectively; $k_1 = 5$, $k_2 = 2$, $k_3 = 3$ lb/in; the pulley weighs 50 lb and has
    a radius of gyration of 6 in.   *Ans.* $\omega = 9.7$ rad/sec

54. Use the Stodola method to find the fundamental frequency of the spring-mass system as shown in
    Fig. 3-65. $m_1 = 6$, $m_2 = 8$, $m_3 = 10$ lb-sec²/in; $k_1 = 3$, $k_2 = 6$, $k_3 = 8$, $k_4 = 4$, $k_5 = 2$, $k_6 = 5$ lb/in.
    Ans. $\omega = 0.61$ rad/sec

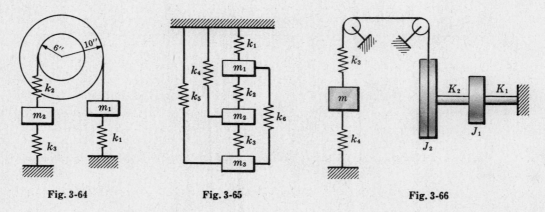

Fig. 3-64   Fig. 3-65   Fig. 3-66

55. Find the fundamental frequency of the system as shown in Fig. 3-66 above by the Stodola method. Assume the shafts and pulleys are weightless; $K_1 = 20$, $K_2 = 50$ in-lb/rad; $k_3 = 5$, $k_4 = 4$, lb/in; $m = 3$ lb-sec^2/in; $J_1 = 100$, $J_2 = 400$. *Ans.* $\omega = 5.25$ rad/sec

56. A locomotive weighing 64,400 lb is coupled to three freight cars as shown in Fig. 3-67. The first and third cars weigh 32,200 lb each, and the second weighs 16,100 lb. The coupling springs are each equal to K, where $K = 10,000$ lb/in. What is the lowest natural frequency? *Ans.* $\omega = 7.4$ rad/sec

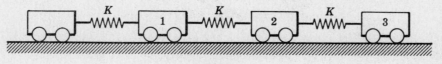

Fig. 3-67

THE HOLZER METHOD

57. For the spring-mass system as shown in Fig. 3-68, find all the frequencies if $k_1 = k_7 = 0$, $k_2 = k_3 = k_4 = k_5 = k_6 = k$, and $m_1 = m_2 = m_3 = m_4 = m_5 = m_6 = m$.
Ans. $\omega_1 = 0$, $\omega_2 = 0.52\sqrt{k/m}$, $\omega_3 = \sqrt{k/m}$, $\omega_4 = 1.4\sqrt{k/m}$, $\omega_5 = 1.73\sqrt{k/m}$, $\omega_6 = 1.93\sqrt{k/m}$ rad/sec

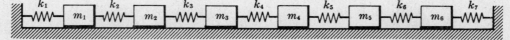

Fig. 3-68

58. Use the Holzer method to determine the natural frequencies of the spring-mass system as shown in Fig. 3-68. $k_1 = 0$, and all the other springs are equal to k; all masses have equal mass m.
Ans. $\omega_1 = 0.24\sqrt{k/m}$, $\omega_2 = 0.71\sqrt{k/m}$, $\omega_3 = 1.14\sqrt{k/m}$, $\omega_4 = 1.49\sqrt{k/m}$,
$\omega_5 = 1.77\sqrt{k/m}$, $\omega_6 = 1.94\sqrt{k/m}$ rad/sec

59. If all the springs are equal to k, and all the masses are equal to m, find the natural frequencies of the system as shown in Fig. 3-68.
Ans. $\omega_1 = 0.45\sqrt{k/m}$, $\omega_2 = 0.87\sqrt{k/m}$, $\omega_3 = 1.25\sqrt{k/m}$, $\omega_4 = 1.56\sqrt{k/m}$,
$\omega_5 = 1.80\sqrt{k/m}$, $\omega_6 = 1.95\sqrt{k/m}$ rad/sec

60. Verify the result of Problem 56 by the Holzer method.

BRANCHED SYSTEMS

61. Assuming all the contacting surfaces are frictionless, find the influence coefficients of the system as shown in Fig. 3-69.
Ans. $\alpha_{11} = 2/k$, $\alpha_{12} = 2/k$, $\alpha_{13} = 1/k$, $\alpha_{21} = 2/k$, $\alpha_{22} = 1/2k$, $\alpha_{23} = 1/k$, $\alpha_{31} = 1/k$, $\alpha_{32} = 1/k$, $\alpha_{33} = 1/k$

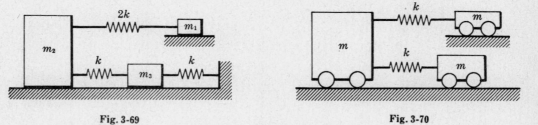

Fig. 3-69   Fig. 3-70

62. Determine the frequencies of oscillation for the system as shown in Fig. 3-70.
Ans. $\omega_1 = 0$, $\omega_2 = \sqrt{k/m}$, $\omega_3 = \sqrt{3k/m}$ rad/sec

63. Derive the frequency equation for the system as shown in Fig. 3-7 by the Mechanical Impedance method.

64. Prove that the orthogonality principle for n-degree-of-freedom systems can be expressed as

$$\sum_{i=1}^{n} m_i A_i^r A_i^s = 0$$

where A_i are the amplitudes of vibration at principal modes, and r and s are the principal modes of the system.

65. Check the general motion of the system given in Problem 42, Page 111, by the orthogonality principle.

Chapter 4

Torsional Vibration

INTRODUCTION

Torsional vibration is periodic angular motion of elastic shafts with circular rotors rigidly attached to them.

Since there are close resemblances and similarities between rectilinear and torsional vibrations, the theory and analysis discussed for the former can be applied equally well to the latter. The following table gives the analogy between the two types of vibration.

Analogy between Rectilinear and Torsional Vibrations

	Rectilinear vibration		Torsional vibration	
	Symbol	Unit	Symbol	Unit
Time	t	sec	t	sec
Displacement	x	in	θ	rad
Velocity	$\dot{x}$	in/sec	$\dot{\theta}$	rad/sec
Acceleration	$\ddot{x}$	in/sec^2	$\ddot{\theta}$	rad/sec^2
Spring constant	k	lb/in	K	in-lb/rad
Damping coefficient	c	lb-sec/in	η	in-lb-sec/rad
Damping factor	ζ	dimensionless	ζ	dimensionless
Mass	m	lb-sec^2/in	J	lb-in-sec^2/rad
Force or torque	$F = m\ddot{x}$	lb	$T = J\ddot{\theta}$	in-lb
Momentum	$m\dot{x}$	lb-sec	$J\dot{\theta}$	in-lb-sec
Impulse	Ft	lb-sec	Tt	in-lb-sec
Kinetic energy	$\frac{1}{2}m\dot{x}^2$	lb-in	$\frac{1}{2}J\dot{\theta}^2$	lb-in
Potential energy	$\frac{1}{2}kx^2$	lb-in	$\frac{1}{2}K\theta^2$	lb-in
Work	$\int F\,dx$	lb-in	$\int T\,d\theta$	lb-in
Natural frequency	$\omega_n = \sqrt{k/m}$	rad/sec	$\omega_n = \sqrt{K/J}$	rad/sec
Equation of motion	$m\ddot{x} + c\dot{x} + kx = F_0 \sin \omega t$		$J\ddot{\theta} + \eta\dot{\theta} + K\theta = T_0 \sin \omega t$	
Initial conditions	$x(0) = x_0, \quad \dot{x}(0) = \dot{x}_0$		$\theta(0) = \theta_0, \quad \dot{\theta}(0) = \dot{\theta}_0$	
Transient responses	$x_c = Ae^{-\zeta\omega_n t} \sin(\omega_d t + \phi)$ $\omega_d = \sqrt{1 - \zeta^2}\,\omega_n$		$\theta_c = Ae^{-\zeta\omega_n t} \sin(\omega_d t + \phi)$ $\omega_d = \sqrt{1 - \zeta^2}\,\omega_n$	
Steady state responses	$x_p = X \sin(\omega t - \psi)$ $X = \dfrac{F_0}{\sqrt{(k - m\omega^2)^2 + (c\omega)^2}}$		$\theta_p = \Phi \sin(\omega t - \psi)$ $\Phi = \dfrac{T_0}{\sqrt{(K - J\omega^2)^2 + (\eta\omega)^2}}$	

Solved Problems

1. A circular disc of moment of inertia J is attached to the lower end of an elastic vertical shaft. If the mass of the shaft is small, and the shaft has torsional stiffness K, derive the differential equation of motion for the free torsional vibration of the disc. What is its natural frequency?

The Energy method:

The configuration of the system can be described by the amount of twist of the shaft, or the angle θ as shown in Fig. 4-1. Since the disc is rigidly attached to the end of the shaft, the angular position of the disc at any instant is given by the angle θ.

The kinetic energy of the system is given by
$$\text{K.E.} = \tfrac{1}{2}J\dot{\theta}^2$$
and the potential energy by
$$\text{P.E.} = \tfrac{1}{2}K\theta^2$$
Now
$$\frac{d}{dt}(\text{K.E.} + \text{P.E.}) = 0 \quad \text{or} \quad (J\ddot{\theta}\dot{\theta} + K\theta\dot{\theta}) = 0$$
Since $\dot{\theta}$ is not always zero,
$$J\ddot{\theta} + K\theta = 0 \quad \text{and} \quad \omega_n = \sqrt{K/J} \text{ rad/sec}$$

K

J

θ

Fig. 4-1

Newton's law of motion:

Because the shaft is elastic, the torque exerted by the shaft on the disc must be proportional but opposite in direction to the angle of twist θ, i.e.,
$$J\ddot{\theta} = -K\theta \quad \text{or} \quad J\ddot{\theta} + K\theta = 0$$
where K is the torsional stiffness of the shaft, defined as torque in in-lb necessary to produce 1 radian of twist. Hence
$$\omega_n = \sqrt{K/J} \text{ rad/sec}$$

2. In the determination of the moment of inertia of flywheels, equal weights are placed on the diameter of the flywheel at equal distance from the center O. The natural frequencies of oscillation of the flywheel at the end of an elastic shaft are measured with and without the added weights. Derive an expression for the moment of inertia I of the flywheel.

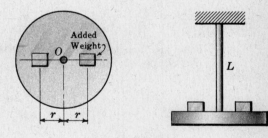

Added Weight

O

r r

L

Fig. 4-2

For free torsional vibration, the natural frequency is given by
$$\omega_1^2 = JG/ILl \text{ (rad/sec)}^2$$

where I is the mass moment of inertia of the flywheel in in-lb-sec^2, G the shear modulus of elasticity, J the rectangular moment of inertia in in^4, and L the length of the shaft.

When the equal weights are added as shown, the natural frequency becomes
$$\omega_2^2 = JG/(I + 2mr^2)L$$

Comparing the two natural frequencies, we obtain
$$\omega_1^2/\omega_2^2 = \frac{JG/IL}{JG/(I+2mr^2)L} = \frac{I+2mr^2}{I}$$
from which
$$I = \frac{2mr^2\omega_2^2}{(\omega_1^2 - \omega_2^2)} \text{ in-lb-sec}^2$$

3. Determine the natural frequencies of the system as shown in Fig. 4-3. Assume that the mass of the shaft is small compared to the masses of the rotors.

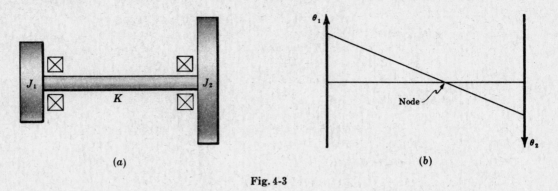

(a)   (b)

Fig. 4-3

When equal and opposite torques are applied to the two rotors and suddenly removed, the two rotors will execute free torsional vibration. The two rotors move in opposite direction.

Let θ_1 and θ_2 be the angles of rotation of rotors A and B respectively. Applying the torque equation, the equations of motion are given by

$$J_1 \ddot{\theta}_1 + K(\theta_1 + \theta_2) = 0$$
$$J_2 \ddot{\theta}_2 + K(\theta_2 + \theta_1) = 0$$

where K is the torsional stiffness of the shaft, and J_1 and J_2 are the mass moments of inertia of the rotors.

Assume the motion is periodic and is composed of harmonic components of various amplitudes and frequencies. Let

$$\theta_1 = A \sin(\omega t + \phi) \qquad \text{and} \qquad \ddot{\theta}_1 = -\omega^2 A \sin(\omega t + \phi)$$
$$\theta_2 = B \sin(\omega t + \phi) \qquad \text{and} \qquad \ddot{\theta}_2 = -\omega^2 B \sin(\omega t + \phi)$$

Substituting these values into the equations of motion and simplifying, we obtain

$$(K - J_1\omega^2)A + KB = 0$$
$$KA + (K - J_2\omega^2)B = 0$$

These are homogeneous linear algebraic equations in A and B. The solution that $A = B = 0$ simply defines the equilibrium position of the system. The other solution is obtained by equating to zero the determinant of the coefficients of A and B, i.e.,

$$\begin{vmatrix} (K - J_1\omega^2) & K \\ K & (K - J_2\omega^2) \end{vmatrix} = 0$$

Expand the determinant to obtain the frequency equation $\omega^2[\omega^2 J_1 J_2 - K(J_1 + J_2)] = 0$; hence

$$\omega_1 = 0 \qquad \text{and} \qquad \omega_2 = \sqrt{K(J_1 + J_2)/J_1 J_2} \text{ rad/sec}$$

The amplitude ratios are obtained from the algebraic equations and are given by

$$\frac{A}{B} = \frac{-K}{(K - J_1\omega^2)} = \frac{-(K - J_2\omega^2)}{K} \qquad \text{and} \qquad \frac{A_2}{B_2} = \frac{-KJ_2}{KJ_2 - K(J_1 + J_2)} = \frac{J_2}{J_1}$$

The second mode shape is plotted in Fig. 4-3(b). At the nodal point, the shaft is immovable.

The natural frequency can also be found directly by solving simultaneously the two equations of motion. Multiplying the first equation of motion by J_2 and the second by J_1, we obtain

$$J_1 J_2 \ddot{\theta}_1 + KJ_2(\theta_1 + \theta_2) = 0$$
$$J_1 J_2 \ddot{\theta}_2 + KJ_1(\theta_1 + \theta_2) = 0$$

Adding,

$$J_1 J_2(\ddot{\theta}_1 + \ddot{\theta}_2) + K(J_1 + J_2)(\theta_1 + \theta_2) = 0 \qquad \text{and hence} \qquad \omega_n = \sqrt{K(J_1 + J_2)/J_1 J_2} \text{ rad/sec}$$

4. A steel shaft is made up of two shafts of lengths L_1 and L_2, and diameters d_1 and d_2, as shown in Fig. 4-4. Find the equivalent shaft for this given steel shaft.

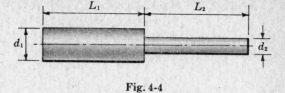

Fig. 4-4

The equivalent shaft is defined as a shaft of length L and constant diameter d but having the same torsional stiffness K as the given shaft.

Consider the action of an applied torque to one end of the given shaft. The torque will be transmitted from the end of shaft d_1, through the connecting point of the two shafts, to the other end of shaft d_2. The total angle of twist of the entire shaft is given by

$$\theta = \theta_1(\text{twist of shaft of diameter } d_1) + \theta_2(\text{twist of shaft of diameter } d_2)$$

$$= \frac{32TL_1}{\pi d_1^4 G} + \frac{32TL_2}{\pi d_2^4 G}$$

where T is the applied torque, and G is the shear modulus of elasticity.

The total angle of twist of the entire shaft can then be expressed as

$$\theta = \frac{32T}{\pi G d_1^4}\left[L_1 + \frac{d_1^4}{d_2^4}L_2\right] \quad \text{or} \quad \frac{32T}{\pi G d_2^4}\left[L_2 + \frac{d_2^4}{d_1^4}L_1\right]$$

The equivalent shaft is therefore given by a shaft of constant diameter d_1 and length $(L_1 + d_1^4 L_2/d_2^4)$ or a shaft of constant diameter d_2 and length $(L_2 + d_2^4 L_1/d_1^4)$.

5. Determine the frequency equation and the general motion of the two-degree-of-freedom torsional system as shown in Fig. 4-5.

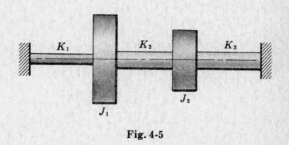

Fig. 4-5

Let θ_1 and θ_2 be the angular displacements of the rotors J_1 and J_2 respectively. Applying the torque equation $\Sigma M = J\ddot{\theta}$, we obtain

$$J_1\ddot{\theta}_1 = -K_1\theta_1 - K_2(\theta_1 - \theta_2)$$
$$J_2\ddot{\theta}_2 = -K_2(\theta_2 - \theta_1) - K_3\theta_2$$

Rearranging, the equations of motion become

$$J_1\ddot{\theta}_1 + (K_1 + K_2)\theta_1 - K_2\theta_2 = 0$$
$$J_2\ddot{\theta}_2 + (K_2 + K_3)\theta_2 - K_2\theta_1 = 0$$

Assume that the motion is periodic and is composed of harmonic motions of various amplitudes and frequencies. Let one of these components be

$$\theta_1 = A\cos(\omega t + \phi), \qquad \ddot{\theta}_1 = -\omega^2 A\cos(\omega t + \phi)$$
$$\theta_2 = B\cos(\omega t + \phi), \qquad \ddot{\theta}_2 = -\omega^2 B\cos(\omega t + \phi)$$

where A, B, and ϕ are arbitrary constants and ω one of the natural frequencies of the system. Substituting these values into the equations of motion, we obtain

$$(K_1 + K_2 - J_1\omega^2)A - K_2 B = 0$$
$$-K_2 A + (K_2 + K_3 - J_2\omega^2)B = 0$$

These are homogeneous linear algebraic equations in A and B. The solution that $A = B = 0$ simply defines the equilibrium condition of the system. The other solution is obtained by equating to zero the determinant of the coefficients of A and B, i.e.,

$$\begin{vmatrix} (K_1 + K_2 - J_1\omega^2) & -K_2 \\ -K_2 & (K_2 + K_3 - J_2\omega^2) \end{vmatrix} = 0$$

Expand the determinant to obtain

$$\omega^4 - \left[\frac{K_1 + K_2}{J_1} + \frac{K_2 + K_3}{J_2}\right]\omega^2 + \frac{K_1 K_2 + K_2 K_3 + K_3 K_1}{J_1 J_2} = 0$$

which is the frequency equation of the system. The two natural frequencies of the system are found by solving this frequency equation.

Therefore the general motion of the system is composed of two harmonic motions of frequencies ω_1 and ω_2, i.e.,

$$\theta_1(t) = A_1 \cos(\omega_1 t + \psi_1) + A_2 \cos(\omega_2 t + \psi_2)$$
$$\theta_2(t) = B_1 \cos(\omega_1 t + \psi_1) + B_2 \cos(\omega_2 t + \psi_2)$$

where the A's, B's, and ψ's are arbitrary constants. But the amplitude ratios are determined from the algebraic equations as

$$\frac{A_1}{B_1} = \frac{K_2}{K_1 + K_2 - J_1\omega_1^2} = \frac{K_2 + K_3 - J_2\omega_1^2}{K_2} = \frac{1}{\lambda_1}$$
$$\frac{A_2}{B_2} = \frac{K_2}{K_1 + K_2 - J_1\omega_2^2} = \frac{K_2 + K_3 - J_2\omega_2^2}{K_2} = \frac{1}{\lambda_2}$$

Hence the general motion finally becomes

$$\theta_1(t) = A_1 \cos(\omega_1 t + \psi_1) + A_2 \cos(\omega_2 t + \psi_2)$$
$$\theta_2(t) = \lambda_1 A_1 \cos(\omega_1 t + \psi_1) + \lambda_2 A_2 \cos(\omega_2 t + \psi_2)$$

where the four constants A_1, A_2, ψ_1, and ψ_2 are to be evaluated by the four initial conditions $\theta_1(0)$, $\dot{\theta}_1(0)$, $\theta_2(0)$, and $\dot{\theta}_2(0)$.

6. If the moments of inertia of the gears are negligible and $J_1 = 2J_2$, $K_1 = K_2 = K$, and the gear reduction ratio $n = 3$, determine the frequency of the torsional vibration.

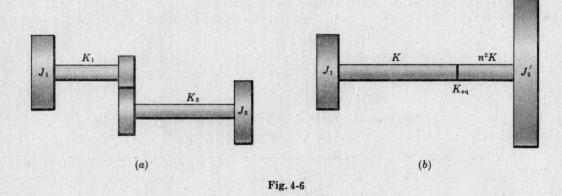

(a)   (b)

Fig. 4-6

Let θ_1 and θ_2 be the angular displacements of the rotors J_1 and J_2 respectively. Because of the gears, θ_2 is equal to $n\theta_1$.

The total energy of the system consists of both kinetic and potential energy, and remains constant.

$$\text{K.E.} = \tfrac{1}{2}J_1\dot{\theta}_1^2 + \tfrac{1}{2}J_2\dot{\theta}_2^2$$
$$\text{P.E.} = \tfrac{1}{2}K_1\theta_1^2 + \tfrac{1}{2}K_2\theta_2^2$$

or

$$\text{K.E.} = \tfrac{1}{2}J_1\dot{\theta}_1^2 + \tfrac{1}{2}J_2(n\dot{\theta}_1)^2$$
$$\text{P.E.} = \tfrac{1}{2}K_1\theta_1^2 + \tfrac{1}{2}K_2(n\theta_1)^2$$

The total energy will be the same when the geared system is replaced by its equivalent system. This is done by replacing J_2 with $J_2' = n^2 J_2$, and K_2 with $K_2 = n^2 K$. Therefore the equivalent system is as shown in Fig. 4-6(b).

The natural frequency of the equivalent system is given by

$$\omega_n = \sqrt{K_{eq}(J_1 + J_2')/J_1 J_2'}$$

where $J_2' = n^2 J_2 = n^2 J_1/2 = 4.5 J_1$ and $K_{eq} = n^2 K^2/(K + n^2 K) = 0.9K$. Hence $\omega_n = \sqrt{1.05K/J_1}$ rad/sec.

7. A uniform shaft of length $4L$ is supported at three different points as shown in Fig. 4-7. Two rotors of equal weight are attached to the shaft at distance L from the ends. What are the critical speeds of the shaft?

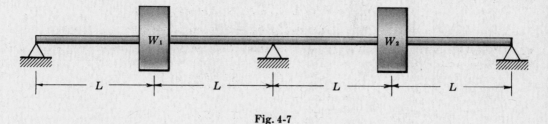

Fig. 4-7

Let y_1 and y_2 be the deflections of the rotors W_1 and W_2 respectively, and $\alpha_{11}, \alpha_{12}, \alpha_{21}, \alpha_{22}$ be the influence coefficients. Applying d'Alembert's principle, we obtain

$$y_1 = -\alpha_{11}(W_1/g)\ddot{y}_1 - \alpha_{12}(W_2/g)\ddot{y}_2$$
$$y_2 = -\alpha_{21}(W_1/g)\ddot{y}_1 - \alpha_{22}(W_2/g)\ddot{y}_2$$

Assume the motion is periodic, and let

$$y_1 = A\cos(\omega t + \phi), \qquad \ddot{y}_1 = -\omega^2 A\cos(\omega t + \phi)$$
$$y_2 = B\cos(\omega t + \phi), \qquad \ddot{y}_2 = -\omega^2 B\cos(\omega t + \phi)$$

Substituting these values into the equations yields

$$(1 - \alpha_{11}W_1\omega^2/g)A - (W_2\alpha_{12}\omega^2/g)B = 0$$
$$-(W_1\alpha_{21}\omega^2/g)A + (1 - \alpha_{22}W_2\omega^2/g)B = 0$$

These are homogeneous linear algebraic equations in A and B. The solution that $A = B = 0$ simply defines the equilibrium position of the system. The other solution is obtained by equating to zero the determinant of the coefficients of A and B, i.e.,

$$\begin{vmatrix} (1 - \alpha_{11}W_1\omega^2/g) & -(W_2\alpha_{12}\omega^2/g) \\ -(W_1\alpha_{21}\omega^2/g) & (1 - \alpha_{22}W_2\omega^2/g) \end{vmatrix} = 0$$

Expanding the determinant gives

$$(\alpha_{11}\alpha_{22} - \alpha_{12}\alpha_{21})\frac{W_1 W_2}{g}\omega^4 - (\alpha_{11}W_1/g + \alpha_{22}W_2/g)\omega^2 + 1 = 0$$

which is the frequency equation of the system.

Due to the symmetry of the problem, $\alpha_{11} = \alpha_{22}$ and $\alpha_{12} = \alpha_{21}$. And from Strength of Materials, it can be shown that they have the following values:

$$\alpha_{11} = \frac{23L^3}{192EI} \qquad \text{and} \qquad \alpha_{12} = \frac{-9L^3}{192EI}$$

Substitute these values for the influence coefficients into the frequency equation and solve to obtain

$$\omega_1 = \sqrt{\frac{6EIg}{WL^3}} \qquad \text{and} \qquad \omega_2 = \sqrt{\frac{13.9EIg}{WL^3}} \text{ rad/sec}$$

But the natural frequencies of the shaft are also the critical speeds of the shaft. Hence the critical speeds of the shaft are

$$f_1 = 0.4\sqrt{\frac{EIg}{WL^3}} \qquad \text{and} \qquad f_2 = 0.6\sqrt{\frac{EIg}{WL^3}} \text{ cycles/sec}$$

8. Use the Holzer method to determine the natural frequencies for torsional vibration of the system as shown in Fig. 4-8. $J_1 = J_2 = J_3 = 1$ and $K_1 = K_2 = 1$.

As discussed in Chapter 3, the Holzer method is conveniently used in tabular form as follows:

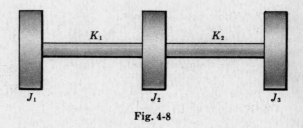

Fig. 4-8

Table

Item	J_i	$J_i\omega^2$	θ_i	$J_i\omega^2\theta_i$	$\Sigma J_i\omega^2\theta_i$	K_{ij}	$\Sigma J_i\omega^2\theta_i/K_{ij}$
Assumed frequency, $\omega = 0.5$							
1	1	0.25	1	0.25	0.25	1	0.25
2	1	0.25	0.75	0.19	0.44	1	0.44
3	1	0.25	0.31	0.07	0.51		
Assumed frequency, $\omega = 0.75$							
1	1	0.56	1	0.56	0.56	1	0.56
2	1	0.56	0.44	0.24	0.80	1	0.80
3	1	0.56	−.36	−.20	0.60		
Assumed frequency, $\omega = 1.0$							
1	1	1	1	1	1	1	1
2	1	1	0	0	1	1	1
3	1	1	−1	−1	0		
Assumed frequency, $\omega = 1.5$							
1	1	2.25	1.0	2.25	2.25	1	2.25
2	1	2.25	−1.25	−2.82	−.57	1	−.57
3	1	2.25	−.68	−1.53	−2.1		
Assumed frequency, $\omega = 1.79$							
1	1	3.21	1	3.21	3.21	1	3.21
2	1	3.21	−2.21	−7.08	−3.87	1	−3.87
3	1	3.21	1.66	5.43	1.56		

Plotting the values for the assumed frequency against the remaining torque, we obtain

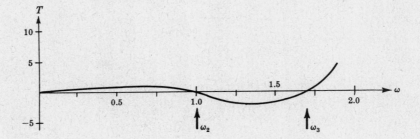

Thus the natural frequencies are $\omega_1 = 0$, $\omega_2 = 1.0$, $\omega_3 = 1.7$ rad/sec.

9. Use the Holzer method to determine the natural frequencies for torsional vibration of the four-degree-of-freedom system as shown in Fig. 4-9.

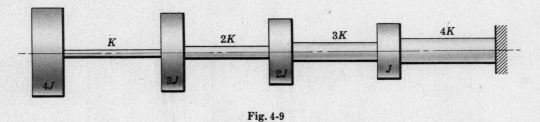

Fig. 4-9

As discussed in Problem 29 on Page 102, the Holzer method is conveniently used in the following tabular form:

Table

Item	J_i	$J_i\omega^2$	θ_i	$J_i\omega^2\theta_i$	$\Sigma J_i\omega^2\theta_i$	K_{ij}	$\Sigma J_i\omega^2\theta_i/K_{ij}$
Assumed frequency, $\omega = 0.2$							
1	4	0.16	1.0	0.16	0.16	1	0.16
2	3	0.12	0.84	0.101	0.26	2	0.13
3	2	0.08	0.71	0.056	0.317	3	0.105
4	1	0.04	0.59	0.024	0.342	4	0.086
5	∞	∞	0.50				
Assumed frequency, $\omega = 0.3$							
1	4	0.36	1.0	0.36	0.36	1	0.36
2	3	0.27	0.64	0.173	0.533	2	0.267
3	2	0.18	0.373	0.067	0.600	3	0.200
4	1	0.09	0.173	0.016	0.616	4	0.154
5	∞	∞	0.019				
Assumed frequency, $\omega = 0.4$							
1	4	0.64	1.0	0.64	0.64	1	0.64
2	3	0.48	0.36	0.173	0.813	2	0.406
3	2	0.32	−.046	−.015	0.798	3	0.266
4	1	0.16	−.312	−.050	0.748	4	0.187
5	∞	∞	−.499				
Assumed frequency, $\omega = 0.6$							
1	4	1.44	1.0	1.44	1.44	1	1.44
2	3	1.08	−.44	−.475	0.965	2	0.482
3	2	0.72	−.922	−.664	0.301	3	0.100
4	1	0.36	−1.023	−.368	−.067	4	−.017
5	∞	∞	−1.006				
Assumed frequency, $\omega = 0.8$							
1	4	2.56	1.0	2.56	2.56	1	2.56
2	3	1.92	−1.56	−3.0	−.44	2	−.22
3	2	1.28	−1.34	−1.72	−2.16	3	−.73
4	1	0.64	−.61	−.39	−2.55	4	−.64
5	∞	∞	0.03				
Assumed frequency, $\omega = 1.0$							
1	4	4	1.0	4	4	1	4
2	3	3	−3.0	−9.0	−5.0	2	−2.5
3	2	2	−0.5	−1.0	−6.0	3	−2.0
4	1	1	1.5	1.5	−4.5	4	−1.13
5	∞	∞	2.63				
Assumed frequency, $\omega = 1.5$							
1	4	9	1.0	9	9	1	9
2	3	6.75	−8.0	−54	−45	2	−22.5
3	2	4.50	14.5	65.3	20.3	3	6.77
4	1	2.25	7.73	17.4	37.7	4	9.43
5	∞	∞	−1.70				
Assumed frequency, $\omega = 1.8$							
1	4	12.96	1.0	12.96	12.96	1	12.96
2	3	9.72	−11.96	−116.4	−103.44	2	−51.72
3	2	6.48	39.76	257.7	154.26	3	51.42
4	1	3.24	−11.66	−37.8	116.46	4	29.12
5	∞	∞	−40.78				

Table (cont.)

Item	J_i	$J_i\omega^2$	θ_i	$J_i\omega^2\theta_i$	$\Sigma J_i\omega^2\theta_i$	K_{ij}	$\Sigma J_i\omega^2\theta_i/K_{ij}$
Assumed frequency, $\omega = 2.0$							
1	4	16	1.0	16	16	1	16
2	3	12	−15	−180	−164	2	−82
3	2	8	67	536	372	3	124
4	1	4	−57	−228	144	4	36
5	∞	∞	−93				
Assumed frequency, $\omega = 2.5$							
1	4	25	1.0	25	25	1	25
2	3	18.75	−24	−450	−425	2	−212.5
3	4	12.5	188.5	2360	1935	3	645
4	1	6.25	−456.5	−2860	−925	4	−231
5	∞	∞	−225.5				
Assumed frequency, $\omega = 3.0$							
1	4	36	1.0	36	36	1	36
2	3	27	−35	−945	−909	2	−455
3	2	18	420	7560	6651	3	2220
4	1	9	−1800	−16,200	−9549	4	−2389
5	∞	∞	589				

Plotting the values of the assumed frequency against the remaining amplitudes θ_i in column 4 as in previous problems, the natural frequencies are found to be

$$\omega_1 = 0.3\sqrt{K/J}, \quad \omega_2 = 0.81\sqrt{K/J}, \quad \omega_3 = 1.45\sqrt{K/J}, \quad \omega_4 = 2.83\sqrt{K/J} \text{ rad/sec}$$

10. Three circular rotors are rigidly connected to the solid steel shaft as shown in Fig. 4-10. Use the Stodola method to find the fundamental frequency of the system. $J_1 = 3, J_2 = 0.5, J_3 = 1.0$ in-lb-sec^2/rad; $K_1 = 2(10)^6, K_2 = 3(10)^6, K_3 = 4(10)^6$ in-lb/rad; and the mass of the shaft is negligible.

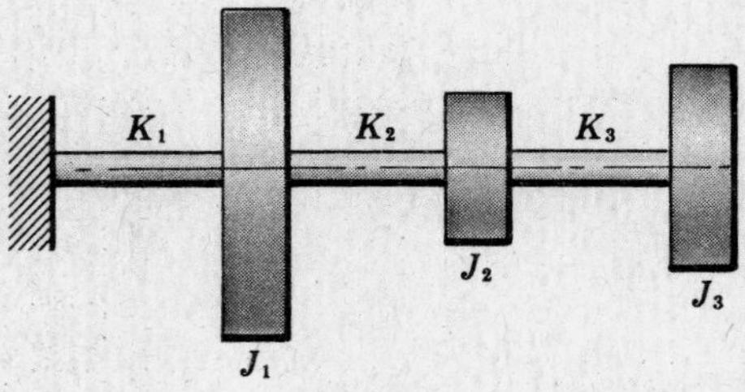

As discussed in Problem 26, Page 98, the Stodola method is conveniently used in the following tabular form:

Fig. 4-10

Table

	K_1	J_1	K_2	J_2	K_3	J_3	Factor
Assumed deflection		1		1.5		2	a
Inertia torque		3		0.75		2	$a\omega^2$
Shaft torque	5.75		2.75		2		$a\omega^2$
Shaft twist	2.88		0.92		0.5		$a\omega^2(10)^{-6}$
Calculated deflection		2.88		3.8		4.3	$a\omega^2(10)^{-6}$
		1		1.32		1.49	a
Assumed deflection		1		1.3		1.5	a
Inertia torque		3		0.65		1.5	$a\omega^2$
Shaft torque	5.15		2.15		1.5		$a\omega^2$
Shaft twist	2.58		0.72		0.38		$a\omega^2(10)^{-6}$
Calculated deflection		2.58		3.3		3.68	$a\omega^2(10)^{-6}$
		1		1.28		1.43	a

Table (cont.)

	K_1	J_1	K_2	J_2	K_3	J_3	Factor
Assumed deflection		1		1.28		1.43	a
Inertia torque		3		0.64		1.43	$a\omega^2$
Shaft torque	5.07		2.07		1.43		$a\omega^2$
Shaft twist	2.54		0.69		0.36		$a\omega^2(10)^{-6}$
Calculated deflection		2.54		3.23		3.59	$a\omega^2(10)^{-6}$
		1		1.274		1.429	a

The resulting configuration of the last iteration is fairly close to that of the assumed values. Hence the computation is stopped and the fundamental natural frequency is calculated.

$$(2.54 + 3.23 + 3.59)a\omega^2(10)^{-6} = (1 + 1.274 + 1.429)a \quad \text{and} \quad \omega_1 = 626 \text{ rad/sec}$$

11. A harmonic torque $T_0 \sin \omega t$ is applied to the first rotor of a uniform shaft as shown in Fig. 4-11. Find the steady state vibration of the system. $J_1 = 5$, $J_2 = 10$, $J_3 = 15$ in-lb-sec^2/rad; $K_1 = 10(10)^6$, $K_2 = 20(10)^6$ in-lb/rad, $\omega = 1000$ rad/sec.

The equations of motion are

$$J_1 \ddot{\theta}_1 + K_1(\theta_1 - \theta_2) = T_0 \sin \omega t$$
$$J_2 \ddot{\theta}_2 + K_2(\theta_2 - \theta_3) + K_1(\theta_2 - \theta_1) = 0$$
$$J_3 \ddot{\theta}_3 + K_2(\theta_3 - \theta_2) = 0$$

Assume that the motion is periodic and is composed of harmonic components of various amplitudes and frequencies. Let

$$\theta_1 = A \sin \omega t, \quad \ddot{\theta}_1 = -\omega^2 A \sin \omega t$$
$$\theta_2 = B \sin \omega t, \quad \ddot{\theta}_2 = -\omega^2 B \sin \omega t$$
$$\theta_3 = C \sin \omega t, \quad \ddot{\theta}_3 = -\omega^2 C \sin \omega t$$

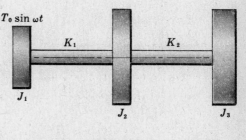

Fig. 4-11

Substituting these expressions into the equations of motion and simplifying, we obtain

$$(K_1 - J_1\omega^2)A - K_1 B = T_0$$
$$-K_1 A + (K_1 + K_2 - J_2\omega^2)B - K_2 C = 0$$
$$-K_2 B + (K_2 - J_3\omega^2)C = 0$$

Solving by Cramer's rule, the coefficients A, B, C are found to be

$$A = 1.5(10)^{-7} T_0, \quad B = -0.25(10)^{-7} T_0, \quad C = -(10)^{-7} T_0$$

and hence the steady state vibration is

$$\theta_1(t) = 1.5(10)^{-7} T_0 \sin \omega t$$
$$\theta_2(t) = -0.25(10)^{-7} T_0 \sin \omega t$$
$$\theta_3(t) = -(10)^{-7} T_0 \sin \omega t$$

12. Determine the lowest natural frequency of the branched torsional system as shown in Fig. 4-12 below. $J_1 = 10$, $J_2 = 15$, $J_3 = 20$, $J_4 = 10$, $J_5 = 10$, $J_6 = 20$ in-lb-sec^2/rad; $K_1 = 100$, $K_2 = 200$, $K_3 = 200$, $K_4 = 100$, $K_5 = 150$ in-lb/rad.

The natural frequencies can be found by the Holzer method.

Unit angular displacement is assumed for the discs at the ends of the branches. Proceeding to the junction, the resulting angular displacements of the branches should be the same. If this is not true, proper adjustment must be made of the assumed values until the resulting angular displacements at the junction are the same. The sum of the inertia torques of the branches is then equal to the torque acting on the main shaft.

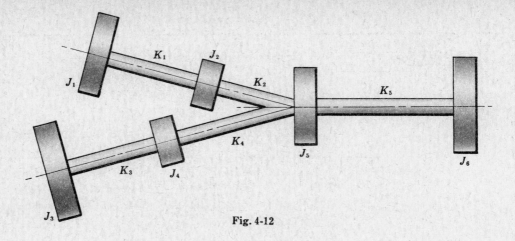

Fig. 4-12

Table

Item	J_i	$J_i\omega^2$	θ_i	$J_i\omega^2\theta_i$	$\Sigma\,\omega^2 J_i\theta_i$	K_{ij}	$\Sigma J_i\omega^2\theta_i/K_{ij}$
Assumed frequency, $\omega^2 = 1.0$							
1	10	10	1	10	10	100	0.1
2	15	15	0.9	13.5	23.5	200	0.12
5	10	10	0.78	7.8	31.3		
3	20	20	1	20	20	200	0.1
4	10	10	0.9	9	29	100	0.29
5	10	10	0.61	6.1	35.1		

At the junction, disc J_5 cannot have amplitudes of 0.78 and 0.61 at the same time, so let (0.78/0.61) be the new assumed value for disc J_3.

Item	J_i	$J_i\omega^2$	θ_i	$J_i\omega^2\theta_i$	$\Sigma\,\omega^2 J_i\theta_i$	K_{ij}	$\Sigma J_i\omega^2\theta_i/K_{ij}$
3	20	20	1.28	25.6	25.6	200	0.13
4	10	10	1.15	11.5	37.1	100	0.37
5	10	10	0.78	7.8	68.4	150	0.456

The torque acting on shaft K_5 equals the sum of the inertia torques developed by discs J_1, J_2, J_3, J_4 and J_5, i.e. $23.5 + 37.1 + 7.8 = 68.4$.

Item	J_i	$J_i\omega^2$	θ_i	$J_i\omega^2\theta_i$	$\Sigma\,\omega^2 J_i\theta_i$	K_{ij}	$\Sigma J_i\omega^2\theta_i/K_{ij}$
6	20	20	0.32	6.4	74.8		
Assumed frequency, $\omega^2 = 2.0$							
1	10	20	1.0	20.0	20.0	100	0.2
2	15	30	0.8	24.0	44.0	200	0.22
5	10	20	0.58				
3	20	40	1.0	40.0	40.0	200	0.2
4	10	20	0.8	16.0	56.0	100	0.56
5	10	20	0.24				

At the junction, disc J_5 cannot have amplitudes of 0.58 and 0.24 at the same time. Therefore let (0.58/0.24) be the new assumed value for disc J_3.

Item	J_i	$J_i\omega^2$	θ_i	$J_i\omega^2\theta_i$	$\Sigma\,\omega^2 J_i\theta_i$	K_{ij}	$\Sigma J_i\omega^2\theta_i/K_{ij}$
3	20	40	2.42	96.8	96.8	200	0.48
4	10	20	1.94	38.8	135.6	100	1.356
5	10	20	0.58	11.6	191.2	150	1.28

The torque acting on shaft K_5 equals to the sum of the inertia torques developed by discs J_1, J_2, J_3, J_4, and J_5, i.e. $44 + 135.6 + 11.6 = 191.2$.

Item	J_i	$J_i\omega^2$	θ_i	$J_i\omega^2\theta_i$	$\Sigma\,\omega^2 J_i\theta_i$	K_{ij}	$\Sigma J_i\omega^2\theta_i/K_{ij}$
6	20	40	-0.7	-28	163.2		

Table (cont.)

Item	J_i	$J_i\omega^2$	θ_i	$J_i\omega^2\theta_i$	$\Sigma\,\omega^2 J_i\theta_i$	K_{ij}	$\Sigma J_i\omega^2\theta_i/K_{ij}$
Assumed frequency, $\omega^2 = 3.0$							
1	10	30	1.0	30	30	100	0.3
2	15	45	0.7	31.5	61.5	200	0.31
5	10	30	0.39				
3	20	60	1.0	60	60	200	0.3
4	10	30	0.7	21	81	100	0.81
5	10	30	−.11	−3.3	77.7		

Since disc J_5 cannot have amplitudes of 0.39 and −0.11 at the same time, let $(0.39/-.11)$ be the new assumed value for disc J_3.

Item	J_i	$J_i\omega^2$	θ_i	$J_i\omega^2\theta_i$	$\Sigma\,\omega^2 J_i\theta_i$	K_{ij}	$\Sigma J_i\omega^2\theta_i/K_{ij}$
3	20	60	−3.55	−213	−213	200	−1.06
4	10	30	−2.49	−74.9	−287.9	100	−2.88
5	10	30	0.39	11.7	−214.7	150	−1.43

The torque acting on shaft K_5 equals to the sum of the inertia torques developed by discs $J_1, J_2, J_3, J_4,$ and J_5, i.e. $61.5 + (-287.9) + 11.7 = -214.7$.

Item	J_i	$J_i\omega^2$	θ_i	$J_i\omega^2\theta_i$	$\Sigma\,\omega^2 J_i\theta_i$	K_{ij}	$\Sigma J_i\omega^2\theta_i/K_{ij}$
6	20	60	1.82	109	−105.7		

Plotting the values of the assumed frequency against the remaining torques, we obtain

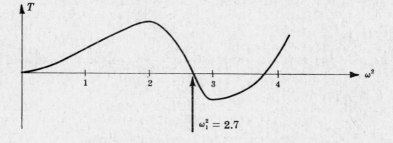

and thus the fundamental natural frequency of the system is $\omega_1^2 = 2.7$ or $\omega_1 = 1.64$ rad/sec.

13. Three rotors of weight 1000 lb, 1500 lb, and 2800 lb respectively, are attached to a steel shaft of diameter 4 in. as shown in Fig. 4-13. The distances between the rotors are 26 in. If G for steel is $12(10)^6$ lb/in², find the natural frequencies of vibration.

Employ the torque equation $\Sigma M = J\ddot\theta$. The differential equations of motion for the system are

$$J_1\ddot\theta_1 + K(\theta_1 - \theta_2) = 0$$
$$J_2\ddot\theta_2 + K(\theta_2 - \theta_3) + K(\theta_2 - \theta_1) = 0$$
$$J_3\ddot\theta_3 + K(\theta_3 - \theta_2) = 0$$

Assume that the motion is periodic and is composed of harmonic components of various amplitudes and frequencies. Let

$$\theta_1 = A\cos(\omega t + \phi), \qquad \ddot\theta_1 = -\omega^2 A\cos(\omega t + \phi)$$
$$\theta_2 = B\cos(\omega t + \phi), \qquad \ddot\theta_2 = -\omega^2 B\cos(\omega t + \phi)$$
$$\theta_3 = C\cos(\omega t + \phi), \qquad \ddot\theta_3 = -\omega^2 C\cos(\omega t + \phi)$$

Substituting these values into the equations of motion yields

$$(K - J_1\omega^2)A - KB = 0$$
$$-KA + (2K - J_2\omega^2)B - KC = 0$$
$$-KB + (K - J_3\omega^2)C = 0$$

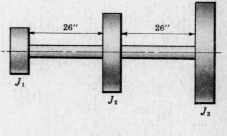

Fig. 4-13

The frequency equation is obtained by equating to zero the determinant of the coefficients of A, B, and C, i.e.,

$$\begin{vmatrix} (K - J_1\omega^2) & -K & 0 \\ -K & (2K - J_2\omega^2) & -K \\ 0 & -K & (K - J_3\omega^2) \end{vmatrix} = 0$$

Expand the determinant and simplify to obtain

$$\omega^2\{\omega^4 - [K/J_3 + 2K/J_2 + K/J_1]\omega^2 + K^2/J_1J_2 + K^2/J_2J_3 + K^2/J_3J_1\} = 0$$

where $\omega_1^2 = 0$ corresponds to the rigid body movement of the shaft without any angular displacement.

Substituting $K = \pi d^4 G/32L = 11.2(10)^6$, $J_1 = \frac{1}{2}Wr^2/g = 210$, $J_2 = 600$, and $J_3 = 1450$ into the frequency equation, we obtain

$$\omega_3 = 67.2 \quad \text{and} \quad \omega_2 = 23.7 \text{ cycles/sec}$$

Supplementary Problems

14. The ends of the shaft with a heavy disc of moment of inertia J are built in as shown in Fig. 4-14. Find the natural frequency of torsional vibration of the disc.

Ans. $\omega_n = \sqrt{\dfrac{\pi d^4 G(L_1 + L_2)}{32JL_1L_2}}$ rad/sec

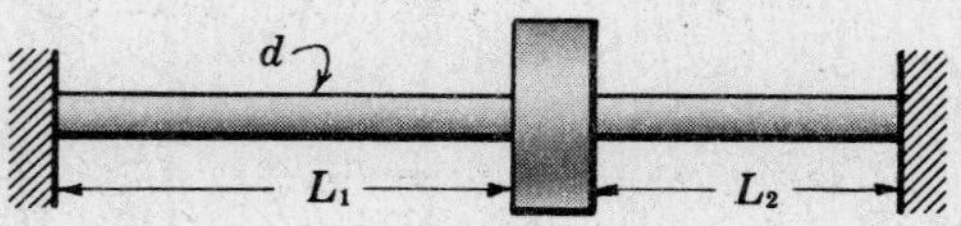

Fig. 4-14

15. Find the equivalent shaft for the system as shown in Fig. 4-15 below.
Ans. $d = d_1$, $L = L_1 + (d_1/d_2)^4(a_2/a_1)^2 L_2$

16. An external torque $T_0 \sin \omega t$ is acting on rotor J_2. Determine the steady state responses of the system as shown in Fig. 4-16 below.

Ans. $\theta_1 = \dfrac{K_2 T_0 \sin \omega t}{(K_1 + K_2 - J_1\omega^2)(K_2 - J_2\omega^2) - K_2^2}$, $\qquad \theta_2 = \dfrac{(K_1 + K_2 - J_1\omega^2)T_0 \sin \omega t}{(K_1 + K_2 - J_1\omega^2)(K_2 - J_2\omega^2) - K_2^2}$

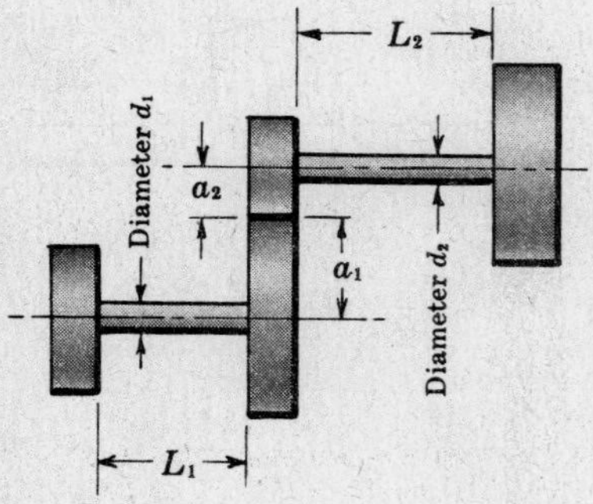

Fig. 4-15

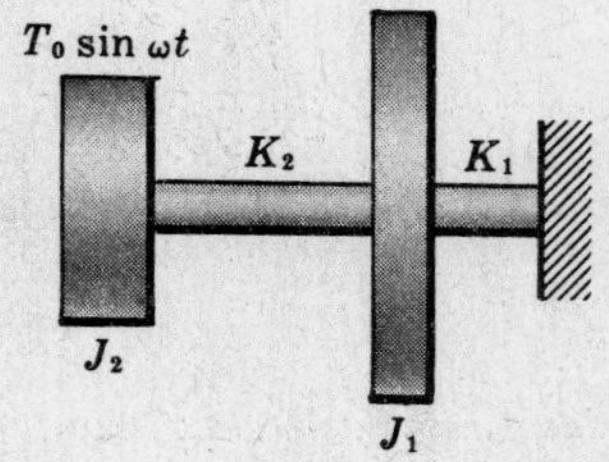

Fig. 4-16

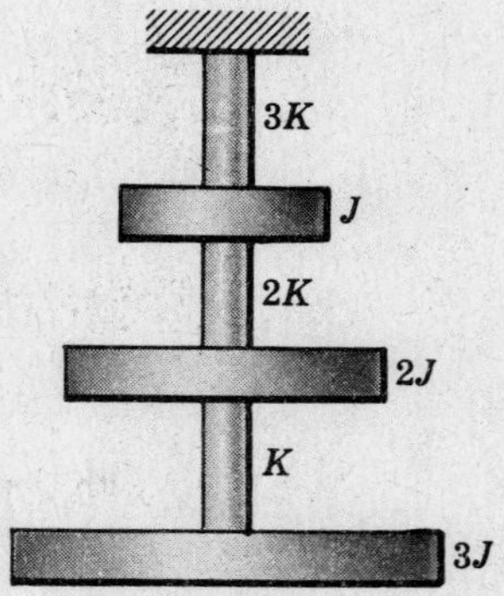

Fig. 4-17

17. Calculate the natural frequencies of the torsional system as shown in Fig. 4-17 above.
Ans. $\omega_1 = 0.39\sqrt{K/J}$, $\omega_2 = 1.47\sqrt{K/J}$, $\omega_3 = 2.36\sqrt{K/J}$ rad/sec

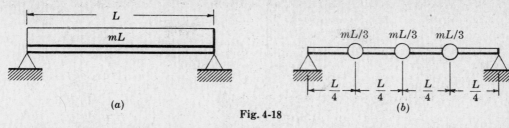

(a)

Fig. 4-18

(b)

18. A beam is loaded uniformly and simply supported as shown in Fig. 4-18(a). Determine the fundamental natural frequency by the "lumped" configuration as shown in Fig. 4-18(b).

Ans. $f_1 = 1.36\sqrt{EI/mL^4}$ cycles/sec

19. Calculate the natural frequencies of the torsional system as shown in Fig. 4-19. The shaft carrying three rotors is fixed at both ends.

Ans. $\omega_1 = 0.54\sqrt{K/J}$, $\omega_2 = 1.17\sqrt{K/J}$,
$\omega_3 = 1.82\sqrt{K/J}$ rad/sec

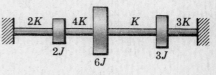

Fig. 4-19

20. If $J = K = 1$, determine the general motion of the system as shown in Fig. 4-20 if an initial angular displacement of 1 radian is applied to the first disc.

Ans. $\theta_1(t) = \frac{1}{3} - \frac{1}{2}\cos t + \frac{1}{6}(\cos\sqrt{3}\,t)$
$\theta_2(t) = \frac{1}{3} - \frac{1}{3}(\cos\sqrt{3}\,t)$
$\theta_3(t) = \frac{1}{3} + \frac{1}{2}\cos t + \frac{1}{6}(\cos\sqrt{3}\,t)$

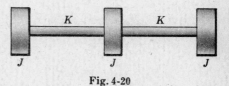

Fig. 4-20

21. Use the Holzer method to determine the natural frequencies of the system as shown in Fig. 4-21. The system is fixed at both ends and is having torsional vibration. $K = 1000$ in-lb/rad and $J = 10$ in-lb-sec²/rad.

Ans. $\omega_1 = 7.66$, $\omega_2 = 14.12$, $\omega_3 = 18.57$ rad/sec

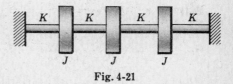

Fig. 4-21

22. Use the Holzer method to determine the natural frequencies of the system as shown in Fig. 4-22. $K = 10(10)^6$ in-lb/rad and $J = 10^3$ in-lb-sec²/rad.   *Ans.* $\omega_1 = 46$, $\omega_2 = 100$, $\omega_3 = 134$ rad/sec

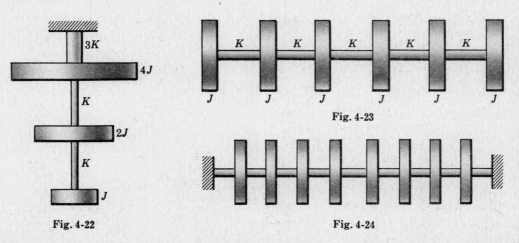

Fig. 4-22

Fig. 4-23

Fig. 4-24

23. Use the Holzer method to determine the natural frequencies of the system as shown in Fig. 4-23.

Ans. $\omega_1 = 0$, $\omega_2 = 0.518\sqrt{K/J}$, $\omega_3 = \sqrt{K/J}$, $\omega_4 = 1.41\sqrt{K/J}$, $\omega_5 = 1.73\sqrt{K/J}$, $\omega_6 = 1.93\sqrt{K/J}$ rad/sec

24. A uniform shaft with eight equal rotors attached at equal distances is shown in Fig. 4-24. Use the Stodola method to determine the lowest natural frequency of the system. Verify your answer by the Holzer method.

Ans. $\omega_1 = 0.347\sqrt{K/J}$ rad/sec

Chapter 5

Vibrations of Continuous Media

INTRODUCTION

Mechanical systems that have their masses and elastic forces *distributed*, such as cables, rods, beams, plates, etc., rather than *"lumped" together* in concentrated masses by springs belong to this class of vibrations of continuous media.

These systems consist of an infinitely large number of particles, and hence require an infinitely large number of coordinates to specify their configurations. Consequently, mechanical systems of this class have an *infinite number of natural frequencies and natural modes of vibration.*

In general, vibrations of continuous media are governed by partial differential equations; and in their analysis, all materials are assumed to be homogeneous and isotropic and to obey Hooke's law.

LONGITUDINAL VIBRATION OF BARS

The differential equation of motion is

$$\frac{\partial^2 u}{\partial t^2} = a^2 \frac{\partial^2 u}{\partial x^2}$$

where u = displacement of any cross-section,
$a^2 = Eg/\gamma$, E = modulus of elasticity of bar, γ = specific weight of bar,
x = coordinate along the longitudinal axis.

The general solution is given by

$$u(x,t) = \sum_{i=1,2,\ldots}^{\infty} (A_i \cos p_i t + B_i \sin p_i t)\left(C_i \cos\frac{p_i x}{a} + D_i \sin\frac{p_i x}{a}\right)$$

where A_i, B_i, C_i, D_i are constants to be determined by the initial and boundary conditions, and p_i the natural frequencies of the system.

TRANSVERSE VIBRATION OF BEAMS

The differential equation of motion is

$$\frac{\partial^2 y}{\partial t^2} + a^2 \frac{\partial^4 y}{\partial x^4} = 0$$

where y = deflection of the beam,
x = coordinate along the longitudinal axis of the beam,
$a^2 = EIg/A\gamma$, A = area of cross section, γ = specific weight of beam,

and EI is known as the flexural rigidity of the beam. The general solution is given by

$$y(x,t) = (A \cos pt + B \sin pt)(C_1 \cos kx + C_2 \sin kx + C_3 \cosh kx + C_4 \sinh kx)$$

where p = natural frequency, $k^2 = p/a$,

and the constants A, B are to be evaluated by the two initial conditions, and C_1, C_2, C_3, C_4 by the four boundary conditions of the problem.

128

THE ORTHOGONALITY PRINCIPLE

The orthogonality principle of the normal functions for the longitudinal vibration of bars is given by

$$\int_0^L X_i X_j \, dx \;=\; \frac{a^2}{p_j^2 - p_i^2}\left[X_j \frac{dX_i}{dx} - X_i \frac{dX_j}{dx} \right]_0^L$$

where X_i and X_j are normal functions of vibration.

The orthogonality principle of the normal functions for the transverse vibration of beams is given by

$$\int_0^L X_i X_j \, dx \;=\; \frac{a^2}{p_j^2 - p_i^2}\left[X_i \frac{d^3X_j}{dx^3} - X_j \frac{d^3X_i}{dx^3} + \frac{dX_j}{dx}\frac{d^2X_i}{dx^2} - \frac{dX_i}{dx}\frac{d^2X_j}{dx^2} \right]_0^L$$

where X_i and X_j are normal functions.

TORSIONAL VIBRATIONS OF CIRCULAR SHAFTS

The differential equation of motion is

$$\frac{\partial^2 \theta}{\partial t^2} \;=\; a^2 \frac{\partial^2 \theta}{\partial x^2}$$

where θ = angular twist of the shaft,

x = coordinate along the longitudinal axis of the shaft,

$a^2 = Gg/\gamma$, G = shear modulus of elasticity,

γ = weight per unit volume of the shaft.

As the equation of motion for the torsional vibrations of circular shafts is the same as the equation of motion for the longitudinal vibrations of bars, any of the problems already solved in longitudinal vibration may be applied equally well to the shafts in torsional vibration.

Solved Problems

LONGITUDINAL VIBRATION OF BARS

1. Derive the differential equation of motion for the longitudinal vibration of uniform bars and investigate its general solution.

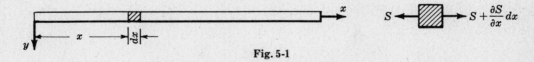

Fig. 5-1

Let u be the displacement of any cross section dx of the bar. Then the strain ϵ_x at any point x is

$$\frac{\partial u}{\partial x} \;=\; \epsilon_x$$

For an elastic bar, the stress is $\sigma_x = E\epsilon_x$, where E is the modulus of elasticity. Thus the tensile force at x is

$$S \;=\; \int_A \sigma_x \, dA \;=\; EA\frac{\partial u}{\partial x}$$

and the inertia force is $\rho A \, dx \frac{\partial^2 u}{\partial t^2}$, where $\rho = \gamma/g$ is the density of the bar and A the area of cross section. Balancing the two forces, we have

$$S + \frac{\partial S}{\partial x} dx \;=\; S + \frac{\gamma A}{g}\frac{\partial^2 u}{\partial t^2} dx \qquad \text{or} \qquad \frac{\partial^2 u}{\partial t^2} \;=\; a^2 \frac{\partial^2 u}{\partial x^2}$$

where $a^2 = Eg/\gamma$. This is the differential equation of motion for the longitudinal vibration of uniform bars.

For the solution of this partial differential equation, let us look for a solution in the form of $u(x, t) = X(x)\,T(t)$. Substituting this expression into the equation of motion yields

$$a^2 \frac{d^2X/dx^2}{X} = \frac{d^2T/dt^2}{T}$$

Since the left-hand side is a function of x alone, and the right-hand side a function of t alone, each side must be equal to a constant. Let this constant be $-p^2$. This leads to two ordinary differential equations:

$$\ddot{T} + p^2 T = 0 \qquad \text{and} \qquad d^2X/dx^2 + (p/a)^2 X = 0$$

the solutions of which are well known:

$$T(t) = A \cos pt + B \sin pt$$
$$X(x) = C \cos (p/a)x + D \sin (p/a)x$$

As $X(x)$ is a function of x alone and determines the shape of the normal mode of vibration under consideration, it is called a *normal function*. Thus the general solution to the partial differential equation is

$$u(x, t) = \sum_{i=1,2,\ldots}^{\infty} (A_i \cos p_i t + B_i \sin p_i t)\left(C_i \cos \frac{p_i x}{a} + D_i \sin \frac{p_i x}{a}\right)$$

where A_i, B_i, C_i, D_i are arbitrary constants determined by the initial and boundary conditions of the problem, and p_i the natural frequencies of the system.

2. A rectangular bar of length L and uniform cross section is having longitudinal vibrations. Derive the frequency equation if both ends of the bar are free as shown in Fig. 5-2.

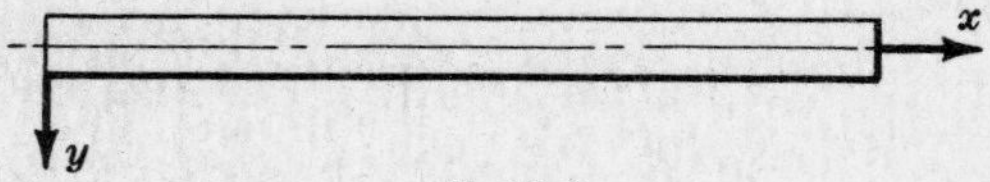

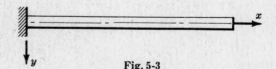

Fig. 5-2

The general solution for the longitudinal vibration of uniform bars is given by

$$u(x, t) = \sum_{i=1,2,\ldots}^{\infty} (A_i \cos p_i t + B_i \sin p_i t)\left(C_i \cos \frac{p_i x}{a} + D_i \sin \frac{p_i x}{a}\right)$$

where $a = \sqrt{Eg/\gamma}$ and $p_i =$ the natural frequencies of the system.

The forces at the ends of the given bar during vibration should equal zero, i.e. the boundary conditions of the problem are $(\partial u/\partial x)_{x=0} = (\partial u/\partial x)_{x=L} = 0$. Substituting these boundary conditions into the general solution, we get

$$(A_i \cos p_i t + B_i \sin p_i t)\left(-\frac{p_i}{a} C_i \sin \frac{p_i x}{a} + \frac{p_i}{a} D_i \cos \frac{p_i x}{a}\right) = 0 \qquad \text{or} \qquad D_i = 0$$

and

$$(A_i \cos p_i t + B_i \sin p_i t)\left(-\frac{p_i}{a} C_i \sin \frac{p_i L}{a}\right) = 0 \qquad \text{or} \qquad \sin\left(\frac{p_i L}{a}\right) = 0$$

This is the frequency equation from which the natural frequencies of the beam having free longitudinal vibration can be determined.

3. Derive an expression for the free longitudinal vibration of a uniform bar of length L, one end of which is fixed and the other end free as shown in Fig. 5-3.

Fig. 5-3

The general solution for the free longitudinal vibration of uniform bars is given by

$$u(x, t) = \sum_{i=1,2,\ldots}^{\infty} (A_i \cos p_i t + B_i \sin p_i t)\left(C_i \cos \frac{p_i x}{a} + D_i \sin \frac{p_i x}{a}\right)$$

The tensile force at the free end of this bar is equal to zero while the displacement at the fixed end of the bar is also equal to zero; i.e. the boundary conditions of the problem are $(u)_{x=0} = 0$, $(\partial u/\partial x)_{x=L} = 0$. Substituting these boundary conditions into the general solution, we obtain

$$(A_i \cos p_i t + B_i \sin p_i t)(C_i) = (u)_{x=0} = 0 \qquad \text{or} \qquad C_i = 0$$

and

$$(A_i \cos p_i t + B_i \sin p_i t)\left(\frac{p_i D_i}{a} \cos \frac{p_i L}{a}\right) = (u)_{x=L} = 0$$

or $\cos \dfrac{p_i L}{a} = 0$ as D_i cannot be equal to zero.   Hence $p_i = \dfrac{i\pi a}{2L}$ where $i = 1, 3, 5, \ldots$.   Then the general expression for longitudinal vibration is

$$u(x, t) = \sum_{i=1,3,5,\ldots}^{\infty} \sin \frac{i\pi x}{2L}\left(A_i \cos \frac{i\pi a t}{2L} + B_i \sin \frac{i\pi a t}{2L}\right)$$

where $a = \sqrt{Eg/\gamma}$.

4. Determine the normal functions for free longitudinal vibration of a bar of length L and uniform cross-section. Both ends of the bar are fixed.

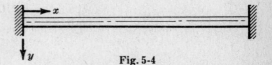

Fig. 5-4

For longitudinal vibration of uniform bars, the general solution is

$$u(x, t) = \sum_{i=1,2,\ldots}^{\infty} (A_i \cos p_i t + B_i \sin p_i t)\left[C_i \cos \frac{p_i}{a}x + D_i \sin \frac{p_i}{a}x\right]$$

The displacements of this bar at its ends are equal to zero, i.e. the boundary conditions are $(u)_{x=0} = (u)_{x=L} = 0$. Substituting these boundary conditions into the general solution, we have

$$(u)_{x=0} = \sum_{i=1,2,\ldots}^{\infty} T_i [C_i \cos (p_i/a)x + D_i \sin (p_i/a)x] = 0 \qquad \text{or} \qquad C_i = 0$$

and

$$(u)_{x=L} = \sum_{i=1,2,\ldots}^{\infty} T_i [D_i \sin (p_i/a)x] = 0$$

or $\sin (p_i L/a) = 0$, and $p_i = i\pi a/L$, where $i = 1, 2, 3, \ldots$. Hence the normal function is

$$X_i(x) = D_i \sin i\pi x/L, \qquad i = 1, 2, 3, \ldots$$

5. A uniform bar of length L is initially compressed by equal forces at both ends as shown in Fig. 5-5. If these compressive forces are suddenly removed, find the vibration produced.

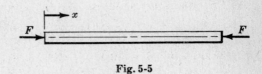

Fig. 5-5

Let ϵ denote the unit compression at time $t = 0$. Then the initial conditions are
$$(u)_{t=0} = \epsilon L/2 - \epsilon x, \qquad (\dot{u})_{t=0} = 0$$

For longitudinal vibration of uniform bars, the general solution is

$$u(x, t) = \sum_{i=1,2,3,\ldots}^{\infty} (A_i \cos p_i t + B_i \sin p_i t)[C_i \cos (p_i/a)x + D_i \sin (p_i/a)x]$$

and for both ends free, this becomes (see Problem 2 on Page 130)

$$u(x, t) = \sum_{i=1,2,\ldots}^{\infty} \cos \frac{i\pi x}{L}\left(A_i \cos \frac{i\pi a t}{L} + B_i \sin \frac{i\pi a t}{L}\right)$$

Substituting the initial conditions into the general solution, we have

$$(u)_{t=0} = \sum_{i=1,2,\ldots}^{\infty} A_i \cos \frac{i\pi x}{L} = \frac{\epsilon L}{2} - \epsilon x, \qquad (\dot{u})_{t=0} = \sum_{i=1,2,\ldots}^{\infty} B_i \frac{i\pi a}{L} \cos \frac{i\pi x}{L} = 0$$

Hence $B_i = 0$ and

$$\int_0^L A_i \cos \frac{i\pi x}{L} \cos \frac{i\pi x}{L}\, dx = \int_0^L \left(\frac{\epsilon L}{2} - \epsilon x\right) \cos \frac{i\pi x}{L}\, dx$$

or

$$A_i(L/2) = \int_0^L \left(\frac{\epsilon L}{2} - \epsilon x\right) \cos \frac{i\pi x}{L}\, dx$$

from which $A_i = \displaystyle\int_0^L \epsilon \cos \frac{i\pi x}{L}\, dx - \frac{2\epsilon}{L} \int_0^L x \cos \frac{i\pi x}{L}\, dx = \frac{2L\epsilon}{i^2 \pi^2}(1 - \cos i\pi)$

Then $A_i = 4\epsilon L/i^2\pi^2$ for $i = $ odd, and $A_i = 0$ for $i = $ even. Thus the vibration produced is given by

$$u(x, t) = \frac{4\epsilon L}{\pi^2} \sum_{i=1,3,\ldots}^{\infty} \frac{(\cos i\pi x/L)(\cos i\pi a t/L)}{i^2}$$

6. A uniform bar of length L is initially stretched by an axial force P_0 applied at the free end as shown in Fig. 5-6. If this axial force is suddenly removed at time $t = 0$, find the resulting longitudinal vibration of the bar.

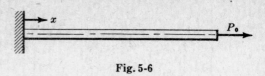

Fig. 5-6

Let ϵ denote the unit elongation at time $t = 0$. Then the initial conditions are

$$(u)_{t=0} = \epsilon x, \qquad (\dot{u})_{t=0} = 0$$

As discussed in Problem 3 on Page 130, the general solution for uniform bars with one end fixed and one end free is

$$u(x, t) = \sum_{i=1,3,\ldots}^{\infty} \sin\frac{i\pi x}{2L}\left(A_i \cos\frac{i\pi at}{2L} + B_i \sin\frac{i\pi at}{2L}\right)$$

Substituting the initial conditions into the general solution, we have

$$(u)_{t=0} = \sum_{i=1,3,\ldots}^{\infty} A_i \sin\frac{i\pi x}{2L} = \epsilon x, \qquad (\dot{u})_{t=0} = \sum_{i=1,3,\ldots}^{\infty} \frac{i\pi a}{2L}B_i \cos\frac{i\pi at}{2L} = 0$$

Hence $B_i = 0$ and

$$\int_0^L A_i \sin\frac{i\pi x}{2L} \sin\frac{i\pi x}{2L}dx = \int_0^L \epsilon x \sin\frac{i\pi x}{2L}dx$$

from which

$$A_i = \frac{2\epsilon}{L}\int_0^L x \sin\frac{i\pi x}{2L}dx = \frac{8\epsilon L}{i^2\pi^2}\sin\frac{i\pi x}{2L}\Big]_0^L = \frac{8\epsilon L}{i^2\pi^2}(-1)^{(i-1)/2}$$

where $i = 1, 3, 5, \ldots$. Thus the longitudinal vibration of the bar is

$$u(x, t) = \frac{8\epsilon L}{\pi^2}\sum_{i=1,3,\ldots}^{\infty} \frac{(-1)^{(i-1)/2}}{i^2}\sin\frac{i\pi x}{2L}\cos\frac{i\pi at}{2L}$$

7. A uniform bar of length L is built-in at both ends as shown in Fig. 5-7. The bar is set into longitudinal motion by giving it a constant velocity V_0 at all points along the bar in the x-direction. Find the resulting longitudinal vibration of the bar.

Fig. 5-7

The general solution for the longitudinal vibration of uniform bars with both ends fixed is

$$u(x, t) = \sum_{i=1,2,\ldots}^{\infty} \sin\frac{i\pi x}{L}\left(A_i \cos\frac{i\pi a}{L}t + B_i \sin\frac{i\pi a}{L}t\right)$$

The initial conditions for this problem are: $(u)_{t=0} = 0$, $(\dot{u})_{t=0} = V_0$. Then

$$u(x, 0) = \sum_{i=1,2,\ldots}^{\infty} A_i \sin\frac{i\pi x}{L} = 0 \qquad \text{or} \qquad A_i = 0$$

$$\dot{u}(x, t) = \sum_{i=1,2,\ldots}^{\infty} B_i \frac{i\pi a}{L} \sin\frac{i\pi x}{L} \cos\frac{i\pi a}{L}t$$

and

$$\dot{u}(x, 0) = \sum_{i=1,2,\ldots}^{\infty} B_i \frac{i\pi a}{L} \sin\frac{i\pi x}{L} = V_0$$

from which

$$\frac{i\pi a}{L}\int_0^L B_i \sin\frac{i\pi x}{L} \sin\frac{i\pi x}{L}dx = \int_0^L V_0 \sin\frac{i\pi x}{L}dx$$

or

$$B_i = \frac{2}{i\pi a}\int_0^L V_0 \sin\frac{i\pi x}{L}dx = \frac{2V_0 L}{i^2\pi^2 a}(1 - \cos i\pi)$$

Then $B_i = \dfrac{4V_0 L}{i^2\pi^2 a}$ when $i = $ odd, and $B_i = 0$ when $i = $ even. Thus the resulting motion is

$$u(x, t) = \frac{4V_0 L}{\pi^2 a}\sum_{i=1,3,\ldots}^{\infty} \frac{1}{i^2}\sin\frac{i\pi x}{L}\sin\frac{i\pi at}{L}$$

8. A bar of length L is fixed at one end and connected at the other end by a spring of constant k as shown in Fig. 5-8. Derive the frequency equation of the system.

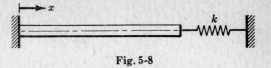

Fig. 5-8

The general expression for longitudinal vibration of uniform bars with one end fixed and one end free is

$$u(x, t) = \sum_{i=1,3,\ldots}^{\infty} \sin\left[\frac{p_i x}{a}\right](A_i \cos p_i t + B_i \sin p_i t)$$

where u is the displacement of any cross section of the bar in the x-direction.

The boundary conditions are: $(u)_{x=0} = 0$, $-k(u)_{x=L} = AE\left(\dfrac{\partial u}{\partial x}\right)_{x=L}$. From the second boundary condition, we have

$$-k \sin\left(\frac{p_i L}{a}\right) = \frac{AE p_i}{a}\cos\left(\frac{p_i L}{a}\right) \qquad \text{or} \qquad \tan\left(\frac{p_i L}{a}\right) = -\frac{AE p_i}{ka}$$

This is the frequency equation of the system. And if k of the spring is very small compared to that of the bar, $\tan (p_i L/a) \to \infty$, i.e.,

$$p_i L/a = i\pi/2, \quad i = 1, 3, 5, \ldots$$

and $p = i\pi a/2L$ is the normal function for longitudinal vibration of uniform bars with one end fixed and the other end free.

9. Derive the differential equation of motion for longitudinal vibration of uniform bars with internal damping given by the expression $\sigma_x = E\epsilon_x + c\dot{\epsilon}_x$, where c is the internal damping coefficient. Solve the resulting equation.

Note that for uniform bars without damping, stress $\sigma_x = E\epsilon_x$, and $E\dfrac{\partial^2 u}{\partial x^2} = \rho \dfrac{\partial^2 u}{\partial t^2}$. Therefore for the case with internal damping the equation of motion is given by

$$\left(E + c\frac{\partial}{\partial t}\right)\frac{\partial^2 u}{\partial x^2} = \rho\frac{\partial^2 u}{\partial t^2} \qquad \text{or} \qquad \frac{E}{\rho}\frac{\partial^2 u}{\partial x^2} + \frac{c}{\rho}\frac{\partial^3 u}{\partial t\,\partial x^2} = \frac{\partial^2 u}{\partial t^2}$$

As for the case without damping, let $u_i = X_i T_i$ and $E/\rho = a^2$. Then

$$a^2 X_i'' T_i + \frac{c}{\rho} X_i'' \dot{T}_i = X_i \ddot{T}_i \qquad \text{where} \qquad X'' = d^2X/dx^2, \quad \ddot{T} = d^2T/dt^2$$

Rearranging, $\quad X_i''\left(a^2 T_i + \dfrac{c}{\rho}\dot{T}_i\right) = X_i \ddot{T}_i \qquad$ or $\qquad \dfrac{X_i''}{X_i} = \dfrac{\ddot{T}_i}{a^2 T_i + (c/\rho)\dot{T}_i}$

Since the left-hand side of the equation is a function of x alone, and the right-hand side a function of t alone, each side must equal to a constant. This leads to two ordinary differential equations with $-(p_i^2/a_i^2)$ as the constant for the above equation:

$$(1) \quad X_i'' + \frac{p_i^2}{a^2}X_i = 0 \qquad\qquad (2) \quad \ddot{T}_i + \frac{c}{\rho}\frac{p_i^2}{a^2}\dot{T}_i + p_i^2 T_i = 0$$

From equation (1),

$$X_i = C_i \cos (p_i/a)x + D_i \sin (p_i/a)x$$

and from equation (2), letting $\alpha_i = cp_i^2/2E$, we obtain

$$\ddot{T}_i + 2\alpha_i \dot{T}_i + p_i^2 T_i = 0$$

the solution of which can be expressed in the form

$$T_i = A_i e^{\lambda_1 t} + B_i e^{\lambda_2 t}$$

where $\lambda = -\alpha_i \pm \sqrt{\alpha_i^2 - p_i^2}$. The solution can be further simplified to

$$T_i = e^{-\alpha_i t}(A_i \cos \beta_i t + B_i \sin \beta_i t)$$

where $\beta_i = \sqrt{p_i^2 - \alpha_i^2}$. Thus the general expression for free longitudinal vibration of uniform bars with internal damping is

$$u(x, t) = \sum_{i=1,2,\ldots}^{\infty} e^{-\alpha_i t}(A_i \cos \beta_i t + B_i \sin \beta_i t)\left(C_i \cos\frac{p_i x}{a} + D_i \sin\frac{p_i x}{a}\right)$$

10. A bar of length L is fixed at one end and has a concentrated mass attached at the other end as shown in Fig. 5-9. Derive the frequency equation.

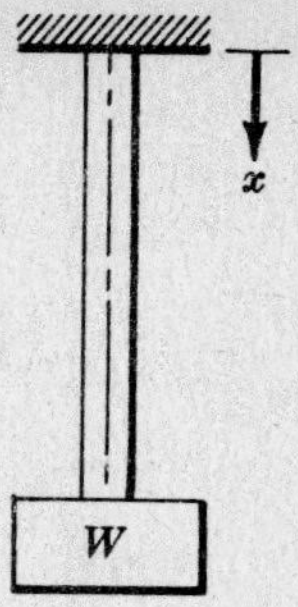

The general expression for longitudinal vibration of uniform bars is given by

$$u(x, t) = \sum_{i=1,2,\ldots}^{\infty} (A_i \cos p_i t + B_i \sin p_i t)\left(C_i \cos \frac{p_i x}{a} + D_i \sin \frac{p_i x}{a}\right)$$

There is no displacement at the fixed end, and a dynamic force in the bar at the free end is equal to the inertia force of the concentrated mass, i.e. the boundary conditions are

$$(u)_{x=0} = 0, \qquad AE\left(\frac{\partial u}{\partial x}\right)_{x=L} = -\frac{W}{g}\left(\frac{\partial^2 u}{\partial t^2}\right)_{x=L}$$

where A is the cross sectional area of the bar.

Fig. 5-9

From the first of these boundary conditions,

$$u(0, t) = \sum_{i=1,2,\ldots}^{\infty} T_i C_i = 0 \qquad \text{or} \qquad C_i = 0$$

where $T_i = (A_i \cos p_i t + B_i \sin p_i t)$. From the second boundary condition,

$$\frac{AEp_i}{a} \cos \frac{p_i L}{a} = \frac{W}{g} p_i^2 \sin \frac{p_i L}{a} \qquad \text{or} \qquad \frac{AL\gamma}{W} = \frac{p_i L}{a} \tan \frac{p_i L}{a}$$

where γ is the weight per unit volume of the bar, and $a^2 = Eg/\gamma$. This is then the frequency equation of the system.

When $AL\gamma/W \to \infty$, i.e. when the mass of the weight is very small compared to the mass of the bar, the frequency equation becomes $\cos p_i L/a = 0$. The system becomes that of a bar fixed at one end and free at the other end.

When W is large compared to the weight of the bar, it can be shown that $p_1 = \sqrt{AEg/WL}$. This corresponds to the natural frequency of a simple single-degree-of-freedom spring-mass system where $m = W/g$ and $k = AE/L$.

11. Derive the orthogonality principle of the normal functions for longitudinal vibration of uniform bars.

Let X_i and X_j be two normal functions corresponding to normal modes of vibration of uniform bars. The general differential equation of motion is

$$a^2 \frac{\partial^2 u}{\partial x^2} = \frac{\partial^2 u}{\partial t^2}$$

where one solution is $u_i = X_i T_i$ and the other is $u_j = X_j T_j$. As discussed earlier, these lead to the following ordinary differential equations:

$$a^2 X_i'' + p_i^2 X_i = 0, \qquad a^2 X_j'' + p_j^2 X_j = 0$$

where $X'' = d^2 X/dx^2$. Multiplying the first of these equations by X_j and the second by X_i, subtracting one from the other and integrating, we have

$$a^2 X_i'' X_j + p_i^2 X_i X_j = 0, \qquad a^2 X_j'' X_i + p_j^2 X_j X_i = 0$$

and

$$\int_0^L X_i X_j \, dx = -\frac{a^2}{p_i^2 - p_j^2} \int_0^L (X_i'' X_j - X_i X_j'') \, dx$$

or

$$\int_0^L X_i X_j \, dx = -\frac{a^2}{p_i^2 - p_j^2}\left[X_i' X_j - X_i X_j'\right]_0^L$$

(1) For bars with both ends free, the boundary conditions are

$$(\partial u/\partial x)_{x=0} = (\partial u/\partial x)_{x=L} = 0 \qquad \text{or} \qquad (X')_{x=0} = (X')_{x=L} = 0$$

Therefore

$$\int_0^L X_i X_j \, dx = 0, \qquad i \neq j$$

(2) For bars with one end fixed and the other end free, the boundary conditions are

$$(u)_{x=0} = (\partial u/\partial x)_{x=L} = 0 \qquad \text{or} \qquad (X)_{x=0} = (X')_{x=L} = 0$$

Therefore

$$\int_0^L X_i X_j \, dx = 0, \qquad i \neq j$$

(3) For bars with both ends fixed, the boundary conditions are

$$(u)_{x=0} = (u)_{x=L} = 0 \qquad \text{or} \qquad (X)_{x=0} = (X)_{x=L} = 0$$

Therefore

$$\int_0^L X_i X_j \, dx = 0, \qquad i \neq j$$

As $\displaystyle\int_0^L X_i X_j \, dx = 0$ is true for each of the above three cases, it is true for longitudinal vibration of uniform bars in general. This is then the orthogonality principle of the normal functions.

12. A uniform bar of length L is fixed at one end and the free end is stretched uniformly to L_0 and released at $t=0$. Find the resulting longitudinal vibration.

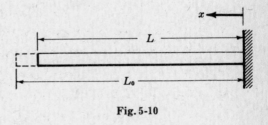

Fig. 5-10

The general solution for bars with one end fixed and the other end free is given by

$$u(x, t) = \sum_{i=1,3,\ldots}^{\infty} \sin\frac{i\pi x}{2L}\left(A_i \cos\frac{i\pi at}{2L} + B_i \sin\frac{i\pi at}{2L}\right)$$

where $a^2 = Eg/\gamma$, and A_i, B_i are constants determined by the initial conditions of the problem:

$$(1) \quad u(x,0) = \frac{L_0 - L}{L}x \qquad (2) \quad \left(\frac{\partial u}{\partial t}\right)_{t=0} = 0$$

Using initial condition (2), we have

$$(\dot{u})_{t=0} = \sum_{i=1,3,\ldots}^{\infty} \frac{i\pi a}{2L} B_i \sin\frac{i\pi x}{2L} = 0 \qquad \text{or} \qquad B_i = 0$$

From initial condition (1),

$$u(x,0) = \sum_{i=1,3,\ldots}^{\infty} A_i \sin\frac{i\pi x}{2L} = \frac{L_0 - L}{L}x$$

Multiply both sides of the above equation by $\sin i\pi x/2L$ and obtain

$$\int_0^L A_i \sin\frac{i\pi x}{2L}\sin\frac{i\pi x}{2L}\,dx = \frac{L_0-L}{L}\int_0^L x\sin\frac{i\pi x}{2L}\,dx$$

or

$$A_i = \frac{2(L_0-L)}{L^2}\int_0^L x\sin\frac{i\pi x}{2L}\,dx = \frac{8(L_0-L)}{i^2\pi^2}(-1)^{(i-1)/2}$$

where $i = 1, 3, 5, \ldots$. Hence the longitudinal vibration of the bar is

$$u(x,t) = \frac{8(L_0-L)}{\pi^2}\sum_{i=1,3,\ldots}^{\infty}(-1)^{(i-1)/2}\frac{1}{i^2}\sin\frac{i\pi x}{2L}\cos\frac{i\pi at}{2L}$$

13. Determine the longitudinal forced vibration of a uniform bar of length L subjected to a sinusoidal force $F_0 \sin \omega t$ at the free end as shown in Fig. 5-11.

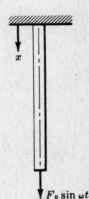

The equation of motion for longitudinal vibration of uniform bars is given by

$$\frac{\partial^2 u}{\partial t^2} = a^2\frac{\partial^2 u}{\partial x^2}$$

where u is the displacement of any cross section of the bar, and $a^2 = E/\rho$.

Let $u(x,t) = X(x)\sin \omega t$ be the general form of solution for the steady state forced vibration of the bar. Substitute this expression into the equation of motion and obtain

$$\frac{d^2X}{dx^2} + \frac{\omega^2 X}{a^2} = 0$$

The solution for X may be written as

$$X(x) = A_1 \cos\frac{\omega}{a}x + A_2 \sin\frac{\omega}{a}x$$

Fig. 5-11

and hence
$$u(x, t) = \left(A_1 \cos \frac{\omega}{a} x + A_2 \sin \frac{\omega}{a} x\right) \sin \omega t$$

The boundary conditions for this problem are

at $x = 0$, $(u) = 0$ and so $A_1 = 0$

at $x = L$, $AE \dfrac{\partial u}{\partial x} = F_0 \sin \omega t$

or $\dfrac{AE\omega A_2}{a} \cos \dfrac{\omega L}{a} \sin \omega t = F_0 \sin \omega t$ which gives $A_2 = \dfrac{F_0 a}{AE\omega} \sec \dfrac{\omega L}{a}$

Therefore the forced vibration of the bar is

$$u(x, t) = \frac{F_0 a}{AE\omega} \sec \frac{\omega L}{a} \sin \frac{\omega}{a} x \sin \omega t$$

14. A uniform bar of length L is free at one end and is forced to follow a sinusoidal movement $U_0 \sin \omega t$ at the other as shown in Fig. 5-12. Find the steady state vibration.

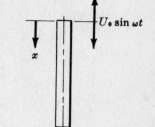

The general differential equation of motion for longitudinal vibration of uniform bars is given by

$$\frac{\partial^2 u}{\partial t^2} = a^2 \frac{\partial^2 u}{\partial x^2}$$

where u is the displacement of any cross section of the bar.

The boundary conditions for this problem are:
$$(u)_{x=0} = U_0 \sin \omega t, \qquad (\partial u/\partial x)_{x=L} = 0$$

Fig. 5-12

Since we are interested in the steady state vibration, let $u(x, t) = X(x) \sin \omega t$ be the general form of solution. Substitute this expression for u into the partial differential equation of motion and obtain

$$-X(x)\omega^2 \sin \omega t = a^2 \sin \omega t \frac{d^2 X}{dx^2} \qquad \text{or} \qquad \frac{d^2 X}{dx^2} + \frac{\omega^2 X}{a^2} = 0$$

The solution may be written as

$$X(x) = A_1 \cos \frac{\omega x}{a} + A_2 \sin \frac{\omega x}{a}$$

and hence
$$u(x, t) = \left(A_1 \cos \frac{\omega x}{a} + A_2 \sin \frac{\omega x}{a}\right) \sin \omega t$$

From the first boundary condition,
$$u(0, t) = A_1 \sin \omega t = U_0 \sin \omega t \qquad \text{and so} \qquad A_1 = U_0$$
and from the second boundary condition,

$$(\partial u/\partial x)_{x=L} = \frac{\omega}{a}\left[-U_0 \sin \frac{\omega L}{a} + A_2 \cos \frac{\omega L}{a}\right] \sin \omega t = 0 \qquad \text{or} \qquad A_2 = U_0 \tan \frac{\omega L}{a}$$

Therefore the steady state vibration is given by

$$u(x, t) = U_0 \left[\cos \frac{\omega x}{a} + \tan \frac{\omega L}{a} \sin \frac{\omega x}{a}\right] \sin \omega t$$

It is clear that resonance will occur if the forcing frequency $\omega = \pi a/2L, 3\pi a/2L, 5\pi a/2L, \ldots$. The amplitude of the steady state vibration $u(x, t)$ will then theoretically approach infinity.

TRANSVERSE VIBRATIONS OF BARS

15. Derive the differential equation of motion for the transverse vibration of beams.

The differential equation of motion for the transverse vibration of beams can be derived from the deflection curve of a beam discussed in Strength of Materials:

$$EI \frac{d^2 y}{dx^2} = -M \tag{1}$$

where EI is known as the flexural rigidity of the beam, M is the bending moment at any cross section, and y the deflection of the beam. Assuming EI is constant, and differentiating equation (1) twice, we obtain

$$EI\frac{d^3y}{dx^3} = -Q \qquad (2)$$

$$EI\frac{d^4y}{dx^4} = w \qquad (3)$$

where Q is the shearing force, w the load intensity.

For free transverse vibration of beams without external loading, it is necessary to consider the inertia forces, $-(\gamma A/g)\,\partial^2 y/\partial t^2$, as the load intensity along the entire length of the beam. Thus equation (3) becomes

$$EI\frac{\partial^4 y}{\partial x^4} = -\frac{\gamma A}{g}\frac{\partial^2 y}{\partial t^2} \qquad (4)$$

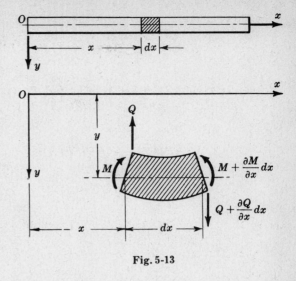

Fig. 5-13

where partial derivatives are used because y is a function of x and t. Since $a^2 = EIg/A\gamma$, (4) becomes

$$\frac{\partial^2 y}{\partial t^2} + a^2\frac{\partial^4 y}{\partial x^4} = 0 \qquad (5)$$

This is then the differential equation of motion for the transverse vibration of beams with constant cross section where stiffness and transverse inertia of the beam have been included. Other effects, such as external damping, elastic foundation, axial forces, and rotatory inertia can also be incorporated if desired.

16. Determine the normal modes of transverse vibration of a simply supported beam of length L and uniform cross section as shown in Fig. 5-14.

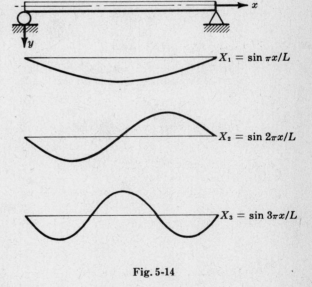

The differential equation of motion for the transverse vibration of beams of uniform cross section is given by

$$\frac{\partial^2 y}{\partial t^2} + a^2\frac{\partial^4 y}{\partial x^4} = 0$$

Assume $y = X(x) \cdot T(t)$. Substituting this expression into the equation of motion yields

$$X\ddot{T} + a^2 X^{iv} T = 0$$

where $X^{iv} = d^4X/dx^4$. This can be written as

$$X^{iv}/X = -\ddot{T}/a^2 T$$

Since the left-hand side is a function of x alone, and the right-hand side a function of t alone, each side must be equal to a constant. Let this constant be p^2/a^2. This leads to two ordinary differential equations:

Fig. 5-14

$$(1)\quad X^{iv} - (p^2/a^2)X = 0 \qquad\qquad (2)\quad \ddot{T} + p^2 T = 0$$

The solution for equation (2) is well known,

$$T(t) = A\cos pt + B\sin pt$$

and the solution for equation (1) is of this form $X = e^{\lambda x}$, where $\lambda = k,\ -k,\ ik,\ -ik$ and $k^4 = p^2/a^2$. The general solution may be written in the form

$$X(x) = C_1\cos kx + C_2\sin kx + C_3\cosh kx + C_4\sinh kx$$

or $\quad C_1(\cos kx + \cosh kx) + C_2(\sin kx + \sinh kx) + C_3(\cos kx - \cosh kx) + C_4(\sin kx - \sinh kx)$

In this case of a simply supported beam, the displacement and the bending moment are both equal to zero at each end of the beam, i.e. the four boundary conditions are

$$(X)_{x=0} = 0, \qquad (X)_{x=L} = 0, \qquad (d^2X/dx^2)_{x=0} = 0, \qquad (d^2X/dx^2)_{x=L} = 0$$

From these four boundary conditions, the four unknown constants C_1, C_2, C_3, and C_4 can be evaluated. Using the first form of the general solution, we have

$$(X)_{x=0} = C_1 + C_3 = 0, \qquad (d^2X/dx^2)_{x=0} = -C_1 + C_3 = 0$$

and from these two equations, $C_1 = C_3 = 0$. Now

$$(X)_{x=L} = C_2 \sin kL + C_4 \sinh kL = 0$$
$$(d^2X/dx^2)_{x=L} = -k^2C_2 \sin kL + k^2C_4 \sinh kL = 0$$

and so $C_4 = 0$ and $\sin kL = 0$. Then

$$\sqrt{p/a}\, L = i\pi \qquad \text{where} \quad i = 1, 2, 3, \dots$$

The natural frequencies are thus given by $p_i = i^2\pi^2 a/L^2$, and the normal function is $X_i = \sin i\pi x/L$. The first three normal modes of vibration are shown in Fig. 5-14 above.

17. List the boundary conditions for the two cantilevers as shown in Fig. 5-15(a) and (b).

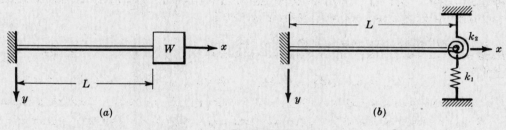

Fig. 5-15

(a) The displacement and slope are equal to zero at the built-in end whereas at the free end the inertia force is equal to the shearing force, i.e.,

$$(y)_{x=0} = 0 \qquad (\partial y/\partial x)_{x=0} = 0$$

$$EI\frac{\partial^3 y}{\partial x^3}\bigg]_{x=L} = -\frac{W}{g}\frac{\partial^2 y}{\partial t^2}\bigg]_{x=L} \qquad EI\frac{\partial^2 y}{\partial x^2}\bigg]_{x=L} = 0$$

if rotatory inertia is neglected.

(b) Similarly, the displacement and slope are zero at the built-in end whereas the shearing force and bending moment are balanced by spring forces, i.e.,

$$(y)_{x=0} = 0 \qquad (\partial y/\partial x)_{x=0} = 0$$

$$EI\frac{\partial^2 y}{\partial x^2}\bigg]_{x=L} = -k_2\frac{\partial y}{\partial x}\bigg]_{x=L} \qquad EI\frac{\partial^3 y}{\partial x^3}\bigg]_{x=L} = k_1(y)_{x=L}$$

18. Find the frequency equation of a beam with both ends free and having transverse vibration.

In this problem, the bending moment and the shearing force are both zero at each end of the beam, i.e.,

$$(d^2X/dx^2)_{x=0} = 0, \qquad (d^3X/dx^3)_{x=0} = 0 \qquad (d^2X/dx^2)_{x=L} = 0, \qquad (d^3X/dx^3)_{x=L} = 0$$

For this case, it is convenient to take the general solution of the normal function for transverse vibration of bars in the following form:

$$X = C_1(\cos kx + \cosh kx) + C_2(\cos kx - \cosh kx) + C_3(\sin kx + \sinh kx) + C_4(\sin kx - \sinh kx)$$

From the first two boundary conditions, $C_2 = C_4 = 0$ and hence
$$X = C_1(\cos kx + \cosh kx) + C_3(\sin kx + \sinh kx)$$

From the other two boundary conditions,
$$(-\cos kL + \cosh kL)C_1 + (-\sin kL + \sinh kL)C_3 = 0$$
$$(\sin kL + \sinh kL)C_1 + (-\cos kL + \cosh kL)C_3 = 0$$

Solving for the constants C_1 and C_3, we obtain the following determinant:
$$\begin{vmatrix} (-\cos kL + \cosh kL) & (-\sin kL + \sinh kL) \\ (\sin kL + \sinh kL) & (-\cos kL + \cosh kL) \end{vmatrix} = 0$$

Expand the determinant to obtain
$$(-\cos kL + \cosh kL)^2 - (\sinh^2 kL - \sin^2 kL) = 0$$

or
$$-2\cos kL \cosh kL + \cos^2 kL + \cosh^2 kL - \sinh^2 kL + \sin^2 kL = 0$$

But
$$\cosh^2 kL - \sinh^2 kL = 1, \qquad \cos^2 kL + \sin^2 kL = 1$$

and so the frequency equation is
$$\cos kL \cosh kL = 1$$

19. Derive the frequency equation of a beam of length L with one end built-in and the other end simply supported as shown in Fig. 5-16.

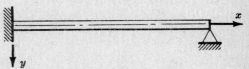

Fig. 5-16

The deflections are zero at both ends, while the slope at $x = 0$ and the bending moment at $x = L$ are also zero, i.e.,
$$(X)_{x=0} = 0, \qquad (X)_{x=L} = 0 \qquad (dX/dx)_{x=0} = 0, \qquad (d^2X/dx^2)_{x=L} = 0$$

Take the general solution of the normal function for the transverse vibration of uniform bars as
$$X = C_1(\cos kx + \cosh kx) + C_2(\cos kx - \cosh kx) + C_3(\sin kx + \sinh kx) + C_4(\sin kx - \sinh kx)$$

From $(X)_{x=0} = 0$, $C_1 = 0$; and from $(dX/dx)_{x=0} = 0$, $C_3 = 0$. Then
$$X = C_2(\cos kx - \cosh kx) + C_4(\sin kx - \sinh kx)$$

Substituting boundary conditions $(X)_{x=L} = 0$ and $(d^2X/dx^2)_{x=L} = 0$ into the above expression gives
$$(\cos kL - \cosh kL)C_2 + (\sin kL - \sinh kL)C_4 = 0$$
$$-(\cos kL + \cosh kL)C_2 - (\sin kL + \sinh kL)C_4 = 0$$

The nontrivial solution for C_2 and C_4 is found by equating to zero the determinant of the above equations, i.e.,
$$\begin{vmatrix} (\cos kL - \cosh kL) & (\sin kL - \sinh kL) \\ -(\cos kL + \cosh kL) & -(\sin kL + \sinh kL) \end{vmatrix} = 0$$

Expand to obtain
$$\sin kL \cosh kL - \cos kL \sinh kL = 0$$

and the frequency equation is therefore
$$\tan kL = \tanh kL$$

20. Prove that the normal functions of transverse vibration of bars are orthogonal.

The general equation for the transverse vibration of bars is given by
$$\frac{\partial^2 y}{\partial t^2} + a^2 \frac{\partial^4 y}{\partial x^4} = 0 \tag{1}$$

and the corresponding solution is
$$y = X(A\cos pt + B\sin pt) \tag{2}$$

where X is the normal function. And one of the resulting equations is (see Problem 16)
$$\frac{d^4X}{dx^4} = \frac{p^2}{a^2}X \tag{3}$$

If X_i and X_j are two normal functions corresponding to two normal modes of vibration, we have the following two equations:

$$\frac{d^4X_i}{dx^4} = \frac{p_i^2}{a^2}X_i \tag{4}$$

$$\frac{d^4X_j}{dx^4} = \frac{p_j^2}{a^2}X_j \tag{5}$$

Multiplying equation (4) by X_j and (5) by X_i, and subtracting (4) from (5), we have

$$\frac{p_j^2 - p_i^2}{a^2} \int_0^L X_iX_j\,dx = \int_0^L \left(X_i\frac{d^4X_j}{dx^4} - X_j\frac{d^4X_i}{dx^4} \right) dx \tag{6}$$

from which

$$\frac{p_j^2 - p_i^2}{a^2} \int_0^L X_iX_j\,dx = \left[X_i\frac{d^3X_j}{dx^3} - X_j\frac{d^3X_i}{dx^3} + \frac{dX_j}{dx}\frac{d^2X_i}{dx^2} - \frac{dX_i}{dx}\frac{d^2X_j}{dx^2} \right]_0^L \tag{7}$$

The boundary conditions are:

(a) Free end: bending moment and shear force are equal to zero, i.e.,

$$d^2X/dx^2 = 0, \quad d^3X/dx^3 = 0$$

(b) Simply supported end: bending moment and deflection are equal to zero, i.e.,

$$d^2X/dx^2 = 0, \quad X = 0$$

(c) Built-in end: deflection and slope are equal to zero, i.e.,

$$X = 0, \quad dX/dx = 0$$

Substituting the above boundary conditions into equation (7), we obtain

$$\int_0^L X_iX_j\,dx = 0, \quad i \neq j$$

This is known as the condition of orthogonality of the normal functions. In other words, the normal functions of transverse vibration for the above-mentioned cases are orthogonal.

21. Show that the normal functions are orthogonal for the beam as shown in Fig. 5-17.

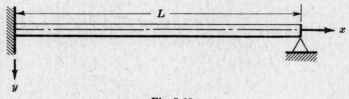

Fig. 5-17

The condition of orthogonality of the normal modes can be expressed as

$$\int_0^L X_iX_j\,dx = \frac{-a^2}{p_i^2 - p_j^2}\left[X_i\frac{d^3X_j}{dx^3} - X_j\frac{d^3X_i}{dx^3} + \frac{dX_j}{dx}\frac{d^2X_i}{dx^2} - \frac{dX_i}{dx}\frac{d^2X_j}{dx^2} \right]_0^L$$

In this case, at $x = L$,

$$X_i = X_j = 0, \quad d^2X_i/dx^2 = d^2X_j/dx^2 = 0$$

and at $x = 0$,

$$X_i = X_j = 0, \quad dX_i/dx = dX_j/dx = 0$$

Substituting these boundary conditions into the above condition of orthogonality, we find that the normal functions are orthogonal, i.e.,

$$\int_0^L X_iX_j\,dx = 0$$

22. Investigate the orthogonality condition of the normal functions of the beam as shown in Fig. 5-18 below.

The orthogonality condition for the transverse vibration of beam is given by

$$\int_0^L X_i X_j \, dx = \frac{a^2}{p_j^2 - p_i^2}\left[X_i \frac{d^3 X_j}{dx^3} - X_j \frac{d^3 X_i}{dx^3} + \frac{dX_j}{dx}\frac{d^2 X_i}{dx^2} - \frac{dX_i}{dx}\frac{d^2 X_j}{dx^2} \right]_0^L$$

where X_i and X_j are normal functions.

At $x = L$, the bending moment is equal to zero, i.e. $d^2X_i/dx^2 = d^2X_j/dx^2 = 0$; and the deflection and slope are equal to zero at the fixed end, i.e. $dX_i/dx = dX_j/dx = X_i = X_j = 0$. Hence the orthogonality condition becomes

$$\int_0^L X_i X_j \, dx = \frac{a^2}{p_j^2 - p_i^2}\left[X_i \frac{d^3 X_j}{dx^3} - X_j \frac{d^3 X_i}{dx^3} \right]_{x=L}$$

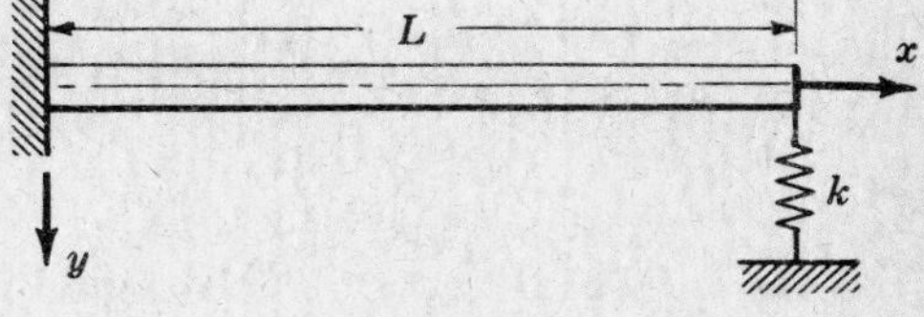

But at the free end of the beam, the shearing force is equal to the spring force, i.e.,

$$kX_i = EI(d^3X_i/dx^3) \qquad \text{and} \qquad kX_j = EI(d^3X_j/dx^3)$$

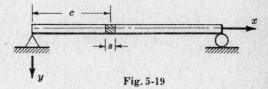

Fig. 5-18

or $X_i = (EI/k)(d^3X_i/dx^3)$ and $X_j = (EI/k)(d^3X_j/dx^3)$, which yields $\int_0^L X_i X_j \, dx = 0$. Thus the normal functions are orthogonal.

23. A simply supported beam of length L is as shown in Fig. 5-19. At time $t = 0$, a very short portion s of the beam is hit by a hammer, giving an initial velocity V_0 to that portion of the beam. Find the resulting transverse vibrations of the beam.

Fig. 5-19

The general expression for free transverse vibration of simply-supported beams is given by

$$y(x,t) = \sum_{i=1,2,\ldots}^\infty \sin\frac{i\pi x}{L}(A_i \cos p_i t + B_i \sin p_i t)$$

where the constants A_i and B_i are to be evaluated by the initial conditions, and $p_i = ai^2\pi^2/L^2$. Substituting $t = 0$ in the above expression and in the derivative of the above expression with respect to t, we obtain

$$(y)_{t=0} = \sum_{i=1,2,\ldots}^\infty A_i \sin\frac{i\pi x}{L}, \qquad (\dot{y})_{t=0} = \sum_{i=1,2,\ldots}^\infty p_i B_i \sin\frac{i\pi x}{L}$$

Now the initial conditions are

$$y = 0 \text{ at } t = 0 \qquad \dot{y} = V_0, \ (c - s/2) \leqq x \leqq (c + s/2)$$

and hence $A_i = 0$ and

$$V_0 \Big]_{(c-s/2)}^{(c+s/2)} = \sum_{i=1,2,\ldots}^\infty p_i B_i \sin\frac{i\pi x}{L}$$

or

$$\int_0^L p_i B_i \sin\frac{i\pi x}{L}\sin\frac{i\pi x}{L}\,dx = \int_{(c-s/2)}^{(c+s/2)} V_0 \sin\frac{i\pi x}{L}\,dx$$

and $B_i = \dfrac{2V_0 L}{i\pi L p_i}\sin\dfrac{i\pi c}{L}\left(2\sin\dfrac{i\pi s}{2L} \right)$. Therefore the vibration of the beam is

$$y(x,t) = \frac{4V_0}{i\pi}\sum_{i=1,2,\ldots}^\infty \frac{1}{p_i}\sin\frac{i\pi c}{L}\sin\frac{i\pi x}{L}\sin p_i t \sin\frac{i\pi s}{2L}$$

24. The response of a simply supported beam is given for a pulsating force $F_0 \sin \omega t$ applied at mid-span as

$$y(x,t) = \frac{2gF_0}{A\gamma L}\sum_{i=1,2,\ldots}^\infty \frac{\sin i\pi/2}{(p_i^2 - \omega^2)}\sin\frac{i\pi x}{L}\left(\sin\omega t - \frac{\omega}{p_i}\sin p_i t \right)$$

in which $p_i = i^2\pi^2 a/L^2$. Determine the beam response if the load is applied for a half cycle only.

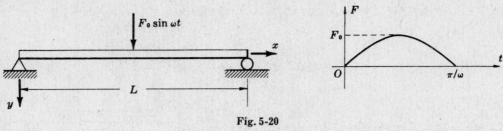

Fig. 5-20

After the pulsating load is removed, i.e. at the end of the π/ω period, the beam is having free vibration with initial conditions governed by the applied force at the end of the half cycle.

The general expression for transverse vibration of a beam is given by

$$y(x, T) = \sum_{i = 1, 2, \ldots}^{\infty} \sin\frac{i\pi x}{L}(A_i \cos p_i T + B_i \sin p_i T)$$

where T is the time after the force is removed, or $T = t - \pi/\omega$. And from the given initial conditions,

$$y(x, \pi/\omega) = -\frac{2gF_0\omega}{A\gamma L} \sum_{i = 1, 2, \ldots}^{\infty} \frac{\sin i\pi/2}{(p_i^2 - \omega^2)p_i} \sin\frac{p_i\pi}{\omega} \sin\frac{i\pi x}{L}$$

$$\dot{y}(x, \pi/\omega) = -\frac{2gF_0\omega}{A\gamma L} \sum_{i = 1, 2, \ldots}^{\infty} \frac{\sin i\pi/2}{(p_i^2 - \omega^2)} \left(1 + \cos\frac{p_i\pi}{\omega}\right) \sin\frac{i\pi x}{L}$$

which means
$$y(x, T)_{T=0} = y(x, \pi/\omega), \qquad \dot{y}(x, T)_{T=0} = \dot{y}(x, \pi/\omega)$$

and
$$y(x, T)_{T=0} = \sum_{i = 1, 2, \ldots}^{\infty} A_i \sin\frac{i\pi x}{L}, \qquad \dot{y}(x, T)_{T=0} = \sum_{i = 1, 2, \ldots}^{\infty} p_i B_i \sin\frac{i\pi x}{L}$$

Then
$$\frac{-2gF_0\omega}{A\gamma L} \sum_{i = 1, 2, \ldots}^{\infty} \frac{\sin i\pi/2}{(p_i^2 - \omega^2)p_i} \sin\frac{p_i\pi}{\omega} \sin\frac{i\pi x}{L} = \sum_{i = 1, 2, \ldots}^{\infty} A_i \sin\frac{i\pi x}{L}$$

$$\frac{-2gF_0\omega}{A\gamma L} \sum_{i = 1, 2, \ldots}^{\infty} \frac{\sin i\pi/2}{(p_i^2 - \omega^2)} \left(1 + \cos\frac{p_i\pi}{\omega}\right) \sin\frac{i\pi x}{L} = \sum_{i = 1, 2, \ldots}^{\infty} p_i B_i \sin\frac{i\pi x}{L}$$

which give
$$A_i = -\frac{2gF_0\omega}{A\gamma L} \sum_{i = 1, 2, \ldots}^{\infty} \frac{\sin i\pi/2}{(p_i^2 - \omega^2)p_i} \sin\left(\frac{p_i\pi}{\omega}\right)$$

$$B_i = -\frac{2gF_0\omega}{A\gamma L} \sum_{i = 1, 2, \ldots}^{\infty} \frac{\sin i\pi/2}{(p_i^2 - \omega^2)p_i} \left(1 + \cos\frac{p_i\pi}{\omega}\right)$$

and hence the free vibration of the beam is given by

$$y(x, T) = \sum_{i = 1, 2, \ldots}^{\infty} \sin\frac{i\pi x}{L}(A_i \cos p_i T + B_i \sin p_i T)$$

where A_i and B_i have just been determined.

25. A simple beam of length L is hinged at one end and the other end is dropped from rest through a height h as shown in Fig. 5-21. If the beam turns as a rigid body until impact, and if there is no loss of energy and no rebound at supports, find the resulting free vibration.

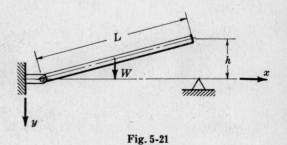

Fig. 5-21

After the impact, the beam is a simply supported one with solution

$$y(x, t) = \sum_{i = 1, 2, 3, \ldots}^{\infty} \sin\frac{i\pi x}{L}(A_i \sin p_i t + B_i \cos p_i t)$$

and initial conditions

$$(1) \quad (y)_{t=0} = 0 \qquad (2) \quad (dy/dt)_{t=0} = \omega x$$

where ω is the angular velocity of the beam upon impact. From (1), $B_i = 0$; and from (2),

$$\omega x = \sum_{i=1,2,\ldots}^{\infty} A_i p_i \sin \frac{i\pi x}{L}$$

or

$$\int_0^L \omega x \sin \frac{i\pi x}{L} dx = A_i p_i \int_0^L \sin \frac{i\pi x}{L} \sin \frac{i\pi x}{L} dx$$

from which

$$A_i = -\frac{2\omega L}{i\pi p_i} \cos i\pi$$

Assume the center of gravity is at the center of the beam; then from energy consideration, i.e. Δ K.E. $= \Delta$ P.E.,

$$\tfrac{1}{2}I\omega^2 = \tfrac{1}{2}Mgh, \qquad \tfrac{1}{2}(\tfrac{1}{3}ML^2)\omega^2 = \tfrac{1}{2}Mgh, \qquad \omega = \sqrt{3gh/L^2}$$

Therefore the free vibration of the beam upon impact is given by

$$y(x, t) = \frac{2\sqrt{3gh}}{\pi} \sum_{i=1,2,\ldots}^{\infty} \frac{1}{ip_i}(-\cos i\pi) \sin \frac{i\pi x}{L} \sin p_i t$$

26. A simply supported beam of length L carries a uniform load of intensity W as shown in Fig. 5-22. Find the resulting vibration when the load is suddenly removed.

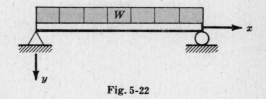

For simply supported beams, the general expression for free transverse vibration is

Fig. 5-22

$$y(x, t) = \sum_{i=1,2,\ldots}^{\infty} \sin \frac{i\pi x}{L}(A_i \cos p_i t + B_i \sin p_i t)$$

Substituting $t = 0$ into the above expression and into the derivative of the above expression with respect to t, we obtain

$$(y)_{t=0} = \sum_{i=1,2,\ldots}^{\infty} A_i \sin \frac{i\pi x}{L} \qquad (\dot{y})_{t=0} = \sum_{i=1,2,\ldots}^{\infty} p_i B_i \sin \frac{i\pi x}{L} = 0$$

and thus $B_i = 0$. Since at $t = 0$, $EI(d^4y/dx^4) = W$,

$$EI \sum_{i=1,2,\ldots}^{\infty} A_i(i\pi/L)^4 \sin \frac{i\pi x}{L} = W$$

or $\quad A_i = \dfrac{2L^3}{EIi^4\pi^4} \displaystyle\int W \sin \frac{i\pi x}{L} dx = \dfrac{2L^4W}{EIi^5\pi^5}(1 - \cos i\pi) = \dfrac{4L^4W}{EIi^5\pi^5}$ where $i = 1, 3, 5, \ldots$

The vibration of the beam is therefore

$$y(x, t) = \frac{4WL^4}{\pi^5 EI} \sum_{i=1,3,\ldots}^{\infty} \frac{1}{i^5} \cos p_i t \sin \frac{i\pi x}{L}$$

27. A concentrated load F_0 is applied at a point on a simply supported beam as shown in Fig. 5-23. What vibrations of the beam will ensue if the load F_0 is suddenly removed?

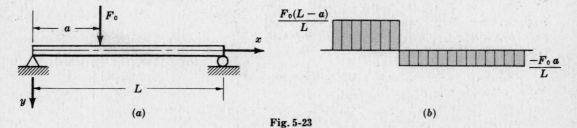

Fig. 5-23

The general expression for free transverse vibration of simply supported beams is

$$y(x, t) = \sum_{i=1,2,\ldots}^{\infty} \sin\frac{i\pi x}{L}(A_i \cos p_i t + B_i \sin p_i t)$$

Substituting $t = 0$ into the above expression and into the derivative of the above expression with respect to time t, we obtain

$$(y)_{t=0} = \sum_{i=1,2,\ldots}^{\infty} A_i \sin\frac{i\pi x}{L} \qquad (\dot{y})_{t=0} = \sum_{i=1,2,\ldots}^{\infty} p_i B_i \sin\frac{i\pi x}{L}$$

Now from the initial condition $\dot{y} = 0$, $B_i = 0$; then

$$y(x, t) = \sum_{i=1,2,\ldots}^{\infty} A_i \sin\frac{i\pi x}{L} \cos p_i t$$

$$\frac{d^3 y}{dx^3} = -\sum_{i=1,2,\ldots}^{\infty} A_i (i\pi/L)^3 \cos\frac{i\pi x}{L} \cos p_i t$$

But $d^3 y/dx^3 = -Q/EI$, where Q is the shearing force along the beam. Equating the two expressions gives

$$\sum_{i=1,2,\ldots}^{\infty} A_i (i\pi/L)^3 \cos\frac{i\pi x}{L} = Q/EI$$

where

$$Q = \begin{cases} F_0 (L-a)/L & 0 \le x \le a \\ -F_0 a/L & a \le x \le L \end{cases}$$

Multiply both sides of the above expression by $\cos i\pi x/L$ and integrate to obtain

$$A_i = \frac{2L^2}{i^3\pi^3} \int_0^L (Q/EI) \cos\frac{i\pi x}{L} dx$$

Substitute the expression for Q into the equation for A_i to get

$$A_i = \frac{2L^2}{i^3\pi^3}\left[\int_0^a \frac{F_0(L-a)/L}{EI} \cos\frac{i\pi x}{L} dx - \int_a^L \frac{F_0 a/L}{EI} \cos\frac{i\pi x}{L} dx\right] = \frac{2L^3 F_0}{i^4\pi^4 EI} \sin\frac{i\pi a}{L}$$

and hence the vibration of the beam is

$$y(x, t) = \frac{2F_0 L^3}{\pi^4 EI} \sum_{i=1,2,\ldots}^{\infty} \frac{1}{i^4} \sin\frac{i\pi a}{L} \sin\frac{i\pi x}{L} \cos p_i t$$

28. A uniformly distributed forcing function of intensity $F_0 \sin \omega t$ is applied to the simply supported beam of length L as shown in Fig. 5-24. Determine the steady state vibration of the beam.

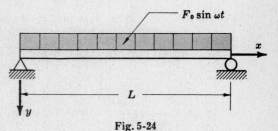

Fig. 5-24

The general equation for transverse vibration of bars with external forcing function is given by

$$\frac{\partial^2 y}{\partial t^2} + a^2\frac{\partial^4 y}{\partial x^4} = \frac{F_0 g}{A\gamma} \sin \omega t$$

where $a^2 = EIg/A\gamma$, A = cross-sectional area, γ = specific weight.

Assume a particular solution in the form

$$y(x, t) = X(x) \sin \omega t$$

Substitute this into the general equation to obtain

$$\frac{d^4 X}{dx^4} - \frac{\omega^2}{a^2} X = \frac{F_0 g}{A\gamma a^2}$$

the complete solution of which is

$$X(x) = C_1 \cos\sqrt{\omega/a}\, x + C_2 \sin\sqrt{\omega/a}\, x + C_3 \cosh\sqrt{\omega/a}\, x + C_4 \sinh\sqrt{\omega/a}\, x - F_0 g/A\gamma\omega^2$$

In this case the boundary conditions are

$$(X)_{x=0} = 0, \quad (d^2X/dx^2)_{x=0} = 0 \qquad (X)_{x=L} = 0, \quad (d^2X/dx^2)_{x=L} = 0$$

From the first two boundary conditions, $C_1 = C_3 = F_0 g/2A\gamma\omega^2$; and the other two boundary conditions give

$$C_2 = \frac{F_0 g}{2A\gamma\omega^2} \tan (\sqrt{\omega/a}\, L/2), \qquad C_4 = \frac{-F_0 g}{2A\gamma\omega^2} \tanh (\sqrt{\omega/a}\, L/2)$$

Since $y(x, t) = X \sin \omega t$, the expression for the steady state vibration is

$$y(x, t) = \frac{F_0 g}{A\gamma\omega^2} \left\{ \frac{\cos [\sqrt{\omega/a}\, (L/2 - x)]}{2 \cos (\sqrt{\omega/a}\, L/2)} + \frac{\cosh [\sqrt{\omega/a}\, (L/2 - x)]}{2 \cosh (\sqrt{\omega/a}\, L/2)} - 1 \right\} \sin \omega t$$

29. Determine the frequency equation of the beam as shown in Fig. 5-25.

Let X_1 and X_2 be the normal functions of the beam. The general solution for the normal functions can be expressed as

$$X_1 = A_1 \cos kx_1 + B_1 \cosh kx_1 + C_1 \sin kx_1 + D_1 \sinh kx_1$$

$$X_2 = A_2 \cos kx_2 + B_2 \cosh kx_2 + C_2 \sin kx_2 + D_2 \sinh kx_2$$

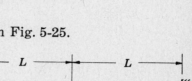

Fig. 5-25

The boundary conditions are:

at $x_1 = 0$, $X_1 = 0$ (1)   at $x_1 = L$, $X_1 = 0$ (5)

         $dX_1/dx_1 = 0$ (2) $dX_1/dx_1 = -dX_2/dx_2$ (6)

at $x_2 = 0$, $X_2 = 0$ (3)   at $x_2 = L$, $X_2 = 0$ (7)

         $dX_2/dx_2 = 0$ (4) $d^2X_2/dx_2^2 = -d^2X_1/dx_1^2$ (8)

From conditions (1) and (3), $A_1 = -B_1$ and $A_2 = -B_2$. Put these expressions into (2) and (4) to obtain $C_1 = -D_1$ and $C_2 = -D_2$.

Using (5) and (7), we have

$$A_1(\cos kL - \cosh kL) + C_1(\sin kL - \sinh kL) = 0$$
$$A_2(\cos kL - \cosh kL) + C_2(\sin kL - \sinh kL) = 0$$

from which A_1 and A_2 can be expressed in terms of C_1 and C_2 respectively, i.e.,

$$A_1 = \frac{-C_1(\sin kL - \sinh kL)}{(\cos kL - \cosh kL)} \qquad A_2 = \frac{-C_2(\sin kL - \sinh kL)}{(\cos kL - \cosh kL)}$$

From conditions (6) and (8), we have

$$-A_1(\sin kL + \sinh kL) + C_1(\cos kL - \cosh kL) = A_2(\sin kL + \sinh kL) - C_2(\cos kL - \cosh kL)$$
$$-A_1(\cos kL + \cosh kL) - C_1(\sin kL + \sinh kL) = A_2(\cos kL + \cosh kL) + C_2(\sin kL + \sinh kL)$$

Substituting the expressions for A_1 and A_2 into the above equations and solving for C_1 and C_2, we find

$$(\sin kL \cosh kL - \cos kL \sinh kL)(1 - \cos kL \cosh kL) = 0$$

and hence the frequency equations are given by

$$\cos kL \cosh kL = 1, \qquad \tan kL = \tanh kL$$

30. List the boundary conditions for the continuous beam as shown in Fig. 5-26.

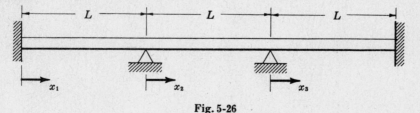

Fig. 5-26

Assume that the flexural rigidity of the beam is the same for all spans. The deflection is equal to zero at the supports and both ends, i.e.,

$$(X_1)_{x_1=0} = (X_1)_{x_1=L} = (X_2)_{x_2=0} = (X_2)_{x_2=L} = (X_3)_{x_3=0} = (X_3)_{x_3=L} = 0$$

The slope is equal to zero at the ends,

$$(dX_1/dx_1)_{x_1=0} = (dX_3/dx_3)_{x_3=L} = 0$$

and
$$(dX_1/dx_1)_{x_1=L} = (dX_2/dx_2)_{x_2=0}, \qquad (dX_2/dx_2)_{x_2=L} = (dX_3/dx_3)_{x_3=0}$$

The bending moments at the supports can be expressed as

$$(d^2X_2/dx_2^2)_{x_2=0} = (d^2X_1/dx_1^2)_{x_1=L}, \qquad (d^2X_2/dx_2^2)_{x_2=L} = (d^2X_3/dx_3^2)_{x_3=0}$$

Thus there are twelve boundary conditions for a three-span beam.

TORSIONAL VIBRATIONS

31. Derive the differential equation of motion for the free torsional vibration of a circular shaft if the mass of the shaft is not small and hence cannot be neglected.

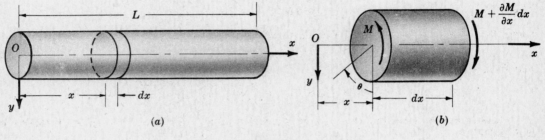

Fig. 5-27

Assume that the distortion of the shaft during torsional vibration is small, so that its cross sections remain plane and its radii remain straight.

As shown in Fig. 5-27(b), and from Strength of Materials, the twisting moment acting on any cross section is equal to

$$GI_p \frac{\partial \theta}{\partial x} \qquad \text{and} \qquad GI_p \left(\frac{\partial \theta}{\partial x} + \frac{\partial^2 \theta}{\partial x^2} dx \right)$$

where $\partial\theta/\partial x$ = angle of twist per unit length of shaft,

　　G = modulus of elasticity in shear,

　　I_p = polar moment of inertia of the cross section.

　　(GI_p is sometimes called the torsional rigidity of shaft.)

The inertia moment of any cross sectional element dx is equal to

$$\frac{I_p \gamma}{g} \frac{\partial^2 \theta}{\partial t^2} dx$$

where γ = weight per unit volume of the shaft.

An equation can be written for the rotatory motion of an element of the shaft by considering the twisting moment and the inertia moment,

$$GI_p \frac{\partial^2 \theta}{\partial x^2} = \frac{\gamma I_p}{g} \frac{\partial^2 \theta}{\partial t^2} \qquad \text{or} \qquad \frac{\partial^2 \theta}{\partial t^2} = a^2 \frac{\partial^2 \theta}{\partial x^2}$$

where $a^2 = Gg/\gamma$.

This equation of motion is seen to have the same form as the equation of motion previously obtained for the longitudinal vibration of uniform bars, i.e.,

$$\frac{\partial^2 u}{\partial t^2} = a^2 \frac{\partial^2 u}{\partial x^2}, \quad a^2 = E/\rho \qquad \text{(longitudinal vibration of bars)}$$

$$\frac{\partial^2 \theta}{\partial t^2} = a^2 \frac{\partial^2 \theta}{\partial x^2}, \quad a^2 = G/\rho \qquad \text{(torsional vibration of shafts)}$$

It is evident then that any of the problems already solved in longitudinal vibration may be applied to the shaft in torsional vibration.

32. Derive the frequency equation for the torsional vibration of a uniform circular shaft with rotors attached rigidly at the ends as shown in Fig. 5-28.

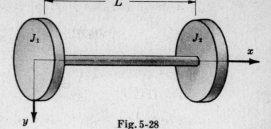

Fig. 5-28

Because the equations of motion for torsional vibration of circular shafts and for longitudinal vibration of uniform bars are identical, the general solution for the torsional vibration of circular shafts can be expressed in the following form:

$$\theta(x,t) \ = \ \sum_{i=1,2,\ldots}^{\infty} (A_i \cos p_i t + B_i \sin p_i t)(C_i \cos p_i x/a + D_i \sin p_i x/a)$$

where $a^2 = G/\rho$ and p_i = natural frequencies of the shaft.

The twisting of the shaft at both ends is produced by the inertia forces of the rotors, i.e. the boundary conditions are:

at $x = 0$, $J_1(\partial^2\theta/\partial t^2) \ = \ GI_p(\partial\theta/\partial x)$ (1)

at $x = L$, $J_2(\partial^2\theta/\partial t^2) \ = \ -GI_p(\partial\theta/\partial x)$ (2)

where G = shear modulus of elasticity, I_p = polar moment of inertia. From (1), we have

$$p_i^2 J_1 C_i + (p_i GI_p/a)D_i \ = \ 0$$

and from (2)

$$\left(p_i^2 J_2 \cos p_i L/a + \frac{p_i GI_p}{a} \sin p_i L/a\right) C_i \ + \ \left(p_i^2 J_2 \sin p_i L/a - \frac{p_i GI_p}{a} \cos p_i L/a\right) D_i \ = \ 0$$

The frequency equation, obtained by equating to zero the determinant of the coefficients of C_i and D_i, is

$$p_i^2 \left(\cos p_i L/a - \frac{p_i a J_1}{GI_p} \sin p_i L/a\right) J_2 \ + \ \frac{p_i GI_p}{a}\left(\sin p_i L/a + \frac{p_i a J_1}{GI_p} \cos p_i L/a\right) \ = \ 0$$

33. A uniform shaft of length L is fixed at one end and free at the other end as shown in Fig. 5-29. Find the free torsional vibration of the shaft.

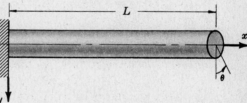

Fig. 5-29

The differential equation of motion for free torsional vibration of the shaft is

$$\frac{\partial^2\theta}{\partial t^2} \ = \ a^2 \frac{\partial^2\theta}{\partial x^2}$$

where θ = angular displacement, and $a^2 = G/\rho$. The general solution is

$$\theta(x,t) \ = \ \sum_{i=1,2,\ldots}^{\infty} (A_i \cos p_i t + B_i \sin p_i t)(C_i \cos p_i x/a + D_i \sin p_i x/a)$$

The boundary conditions are:

at $x = 0$, $\theta(0,t) \ = \ 0$ (1)

at $x = L$, $GI_p(\partial\theta/\partial x) \ = \ 0$ (2)

where I_p is the polar moment of inertia of the shaft. Using equation (1), we have

$$\theta(0,t) \ = \ \sum_{i=1,2,\ldots}^{\infty} C_i(A_i \cos p_i t + B_i \sin p_i t) \ = \ 0 \qquad \text{or} \qquad C_i = 0$$

and from boundary condition (2),

$$\theta(x,t) \ = \ \sum_{i=1,2,\ldots}^{\infty} (\sin p_i x/a)(A_i \cos p_i t + B_i \sin p_i t)$$

$$(\partial\theta/\partial x)_{x=L} \ = \ \sum_{i=1,2,\ldots}^{\infty} \frac{p_i}{a}(\cos p_i L/a)(A_i \cos p_i t + B_i \sin p_i t) \ = \ 0$$

or $\cos p_i L/a = 0, \ \ p_i = i\pi a/2L, \quad \text{where } i = 1,3,5,\ldots$

Hence the torsional vibration of the shaft is

$$\theta(x, t) = \sum_{i=1,3,\ldots}^{\infty} \sin\frac{i\pi x}{2L}\left(A_i \cos\frac{i\pi at}{2L} + B_i \sin\frac{i\pi at}{2L}\right)$$

where A_i and B_i are constants determined by initial conditions of the problem.

34. A pulley of moment of inertia J is rigidly attached to the free end of a uniform shaft of length L as shown in Fig. 5-30. Find the frequency equation for torsional vibration.

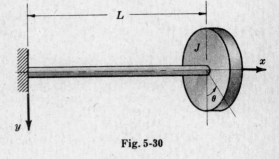

Fig. 5-30

The differential equation of motion for free torsional vibration of the shaft and the corresponding general solution are given by

$$\frac{\partial^2\theta}{\partial t^2} = a^2\frac{\partial^2\theta}{\partial x^2}$$

$$\theta(x, t) = \sum_{i=1,2,\ldots}^{\infty} (A_i \cos p_i t + B_i \sin p_i t)(C_i \cos p_i x/a + D_i \sin p_i x/a)$$

where $a^2 = G/\rho$ and p_i = natural frequencies.

The boundary conditions for this problem are

$$\theta(0, t) = 0 \tag{1}$$

$$-GI_p(\partial\theta/\partial x)_{x=L} = J(\partial^2\theta/\partial t^2) \tag{2}$$

i.e. the angular displacement of the shaft at the fixed end is equal to zero, and the restoring torque of the shaft at the free end is equal to the inertia moment of the pulley.

From boundary condition (1), $C_i = 0$; and from (2),

$$-\frac{GI_p p_i}{a}\cos\frac{p_i L}{a} = -Jp_i^2 \sin\frac{p_i L}{a} \qquad \text{or} \qquad \tan\frac{p_i L}{a} = \frac{GI_p}{aJp_i}$$

which is the frequency equation.

35. An external torque $T_0 \sin \omega t$ is applied to the free end of a uniform shaft of length L as shown in Fig. 5-31. Find the steady state vibration of the shaft.

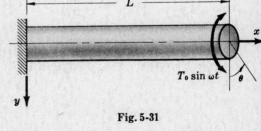

Fig. 5-31

The differential equation of motion for free torsional vibration of the shaft is

$$\frac{\partial^2\theta}{\partial t^2} = a^2\frac{\partial^2\theta}{\partial x^2} \tag{1}$$

where θ = the twist of the shaft, $a^2 = G/\rho$. Let the steady state vibration be

$$\theta = X(x)\sin\omega t \tag{2}$$

Substitute equation (2) into (1) and obtain

$$a^2\frac{d^2X}{dx^2}\sin\omega t + X\omega^2\sin\omega t = 0 \qquad \text{or} \qquad \frac{d^2X}{dx^2} + \frac{\omega^2}{a^2}X = 0$$

Therefore

$$X(x) = A\cos\omega x/a + B\sin\omega x/a$$

The boundary conditions of the problem are: the twist of the shaft at the fixed end is equal to zero, and the restoring torque of the shaft at the free end is equal to the applied torque, i.e.,

$$X(0) = 0, \qquad GI_p(\partial\theta/\partial x)_{x=L} = T_0\sin\omega t \tag{3}$$

where G = shear modulus of elasticity, I_p = polar moment of inertia of the shaft.

Using the first equation in (3), $A = 0$; and from the second equation in (3),

$$\frac{GI_pB\omega}{a}\cos\frac{\omega L}{a}\sin\omega t = T_0\sin\omega t \qquad \text{or} \qquad B = \frac{T_0a}{GI_p\omega}\sec\frac{\omega L}{a}$$

Hence the steady state vibration is

$$\theta(x,t) = \frac{T_0a}{GI_p\omega}\sec\frac{\omega L}{a}\sin\frac{\omega x}{a}\sin\omega t$$

Supplementary Problems

36. Determine the period of the fundamental mode of vibration of a steel rod of length 1000 ft and specific weight 0.28 lb/in³, if the rod is considered as a bar with both ends free.   *Ans.* $T = 0.12$ sec

37. A uniform bar of length L is fixed at the upper end and free at the lower end as shown in Fig. 5-32. Prove that a suddenly applied force at the free end produces twice as great a deflection as one gradually applied.

38. Calculate the deflection of the lower end of a uniform vertical bar subjected to its own weight as the load.

    Ans. Deflection $= wL^2/2E$, where $w =$ specific weight of bar, $L =$ length of bar, $E =$ modulus of elasticity.

39. Show that the differential equation of motion for the free longitudinal vibration of a bar with variable cross section is

$$\frac{\partial^2 u}{\partial x^2} + \frac{1}{A}\frac{\partial A}{\partial x}\frac{\partial u}{\partial x} = \frac{\rho}{E}\frac{\partial^2 u}{\partial t^2}$$

    where $A = A(x)$ is the area of the cross section of the bar.

Fig. 5-32

40. An initial velocity V_0 is given to the entire length of a simply supported beam of length L. Derive the expression for the free transverse vibration of the beam.

    Ans. $y(x,t) = \dfrac{4V_0}{\pi}\displaystyle\sum_{i=1,3,\ldots}^{\infty}\frac{1}{ip_i}\sin i\pi x/L\sin p_i t$

41. A uniform bar of length L is acted upon by a constant axial force F_0 at its center as shown in Fig. 5-33. Find the resulting vibration if this force is suddenly removed.

    Ans. $u(x,t) = \dfrac{2F_0L}{AE\pi^2}\displaystyle\sum_{i=1,3,\ldots}^{\infty}(-1)^{(i-1)/2}\left[\frac{1}{i^2}\sin\frac{i\pi x}{L}\cos\frac{i\pi a}{L}t\right]$

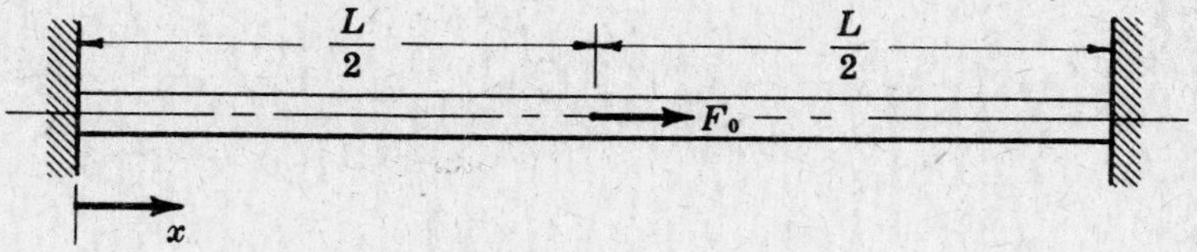

Fig. 5-33

42. In the lifting of a long drill stem, a force F_0 is suddenly applied to one end of the stem. Determine the resulting longitudinal vibration.

    Ans. $u(x,t) = \dfrac{gt^2F_0}{2\gamma LA} + \dfrac{2gLF_0}{\pi^2a^2A}\displaystyle\sum_{i=1,2,\ldots}^{\infty}\frac{(-1)^i}{i^2}\cos\frac{i\pi x}{L}\left(1-\cos\frac{i\pi a}{L}t\right)$

43. Investigate the effect of a constant longitudinal force on the natural frequency of a uniform bar undergoing longitudinal vibration.   *Ans.* No effect.

44. A uniform bar of length L is moving in a horizontal plane with velocity V_0. If the bar hits a solid wall with one end and stops, what will be the vibration of the bar?

$$Ans.\quad u(x,t) \;=\; \frac{8V_0 L}{\pi^2 a}\sum_{i=1,3,\ldots}^{\infty}\frac{1}{i^2}\sin\frac{i\pi x}{2L}\sin\frac{i\pi a}{2L}t$$

45. Show that the effect of lateral contraction on the longitudinal vibration of a uniform bar is

$$\frac{\partial^2 u}{\partial t^2} \;=\; \mu k^2 \frac{\partial^4 u}{\partial x^2\,\partial t^2} + a^2 \frac{\partial^2 u}{\partial x^2}$$

where μ = Poisson's ratio, $k^2 = \int r^2\,dA$.

46. A uniform bar of length L is acted upon by a forcing function $F_0 \sin \omega t$ at the end $x = 0$ as shown in Fig. 5-34. If both ends of the bar are free, find the steady state response of the bar.

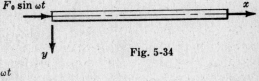

Fig. 5-34

$$Ans.\quad u(x,t) \;=\; \frac{F_0 L}{AE}(a/\omega L)\,\operatorname{cosec}\frac{\omega L}{a}\cos\!\left[\frac{\omega}{a}(L-x)\right]\sin\omega t$$

47. Derive the frequency equation for the transverse vibration of a uniform beam of length L, if one end is fixed and the other end free. $Ans.\ \cos kL \cosh kL = -1$

48. Derive the frequency equation for the transverse vibration of a uniform beam of length L, if both ends are fixed. $Ans.\ \cos kL \cosh kL = 1$

49. What is the effect of a constant longitudinal (tension and compression) force on the natural frequency of a uniform bar having transverse vibration.

$Ans.$ Tension: frequency is increased. Compression: frequency is decreased.

50. Derive the frequency equation for the longitudinal vibration of a rod of two different cross sectional areas A_1 and A_2 as shown in Fig. 5-35.
$Ans.\ \tan p_i L/a_1 \tan p_i L/a_2 = A_1 a_1 \rho_1 / A_2 a_2 \rho_2$

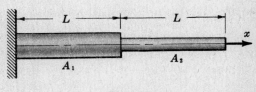

51. Show that the term $\dfrac{\gamma I}{g}\dfrac{\partial^4 y}{\partial x^2\,\partial t^2}$ represents the effect of rotatory inertia of the beam in the differential equation of motion for the transverse vibrations of beam.

Fig. 5-35

52. Determine the steady state vibration of a simply supported beam of length L acted upon by a concentrated forcing function $F_0 \sin \omega t$ as shown in Fig. 5-36.

$$Ans.\quad y(x,t) \;=\; \frac{2F_0 L^3}{\pi^4 EI}\sum_{i=1,2,\ldots}^{\infty}\frac{1}{i^4}\frac{1}{1-(\omega/p_i)^2}\sin i\pi a/L \,\sin i\pi x/L \,\sin\omega t$$

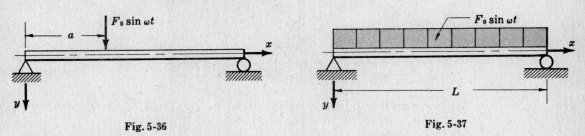

Fig. 5-36   Fig. 5-37

53. Prove that the following expression must be true in order to have zero deflection at the center of the beam as shown in Fig. 5-37: $\sec\left(\sqrt{\omega/a}\,L/2\right) + \operatorname{sech}\left(\sqrt{\omega/a}\,L/2\right) = 0$.

54. Show that the differential equation of motion for the torsional vibration of a circular shaft with variable diameter is

$$\rho(x)\,I_p(x)\,\frac{\partial^2 \theta}{\partial t^2} \;=\; G\left[I_p \frac{\partial^2 \theta}{\partial x^2} + \frac{\partial I_p}{\partial x}\frac{\partial \theta}{\partial x}\right]$$

where $\rho(x)$ is the mass density of the material of the shaft.

Chapter 6

Nonlinear Vibrations

INTRODUCTION

Many vibration problems in engineering are nonlinear in nature, i.e. the restoring forces are not proportional to the displacements and the damping forces are not proportional to the first power of the velocity. This is also true when motions of appreciable magnitude in linear vibrating systems are of concern. In such cases, the usual linear analysis may not be too desirable for many applications, and so more accurate analysis involving nonlinear differential equations are to be used.

An essential difference in the study of nonlinear systems is that a general solution cannot be obtained by superposition as in the case of linear systems. Moreover, in many instances, new phenomena occur in nonlinear systems which cannot occur in linear systems.

In general, advanced mathematics is required for the analysis and solution of nonlinear systems due to their complicated configurations and nonlinear differential equations of motion. A few simple examples are presented here.

FREE UNDAMPED VIBRATIONS WITH NONLINEAR RESTORING FORCES

Free vibrations of undamped systems with nonlinear restoring forces very often occur in actual practice. The differential equation of motion has the following form

$$\frac{d^2x}{dt^2} + f(x) = 0$$

where x is the displacement and $f(x)$ is the nonlinear restoring force as a function of x. The solution for the above equation can be expressed in elliptic integrals. (See Problem 3.)

FORCED UNDAMPED VIBRATIONS WITH NONLINEAR RESTORING FORCES

The differential equation of motion is given by

$$\frac{d^2x}{dt^2} + f(x) = F_0 \cos \omega t$$

where x = displacement,
$f(x)$ = nonlinear restoring force as a function of x,
$F_0 \cos \omega t$ = external forcing function.

The solution for the above equation can be obtained by the Iteration Method. (See Prob. 5.)

SELF-EXCITED VIBRATIONS

Self-excited vibrations are *self-governed vibrations* which draw their energy from external sources by their own periodic motions. The exciting force, in other words, is a function of the displacement, the velocity, or the acceleration of the mass of the system.

The amount of energy in a system having self-excited vibration will continue to grow until it is balanced by the same amount of energy dissipated by the system due to damping. The self-excited vibration then settles down to a steady state vibration with a frequency close to the natural frequency of the system itself. On the other hand, if the rate of energy dissipated is smaller than the rate of energy absorbed by the system, the motion of the system itself will continue to increase the total energy present in the system. The system will break down.

STABILITY

If the amplitude of vibration decreases with time, the system is said to be *stable*. If the transient increases indefinitely with time, the system is said to be *unstable*.

Some systems may appear to be unstable for small values of time, but actually are stable for longer periods of time. In particular, nonlinear systems may be unstable for small magnitudes of motion but stabilize themselves on a *limit cycle* for some larger values of motion. When the amplitudes of vibration become large, linear systems will respond nonlinearly and will also go into a stable limit cycle.

For stable systems, the total energy in the system decreases with time. The loss of energy is usually dissipated as heat due to friction. Therefore damping for stable systems must be positive.

For unstable systems, energy must have been kept adding to the system because there is continuous increase in amplitude of vibration. Work is therefore done on the system by the damping force. Hence damping for unstable systems is negative.

Thus a system will be unstable if the real part of any one of the roots for the characteristic equation of a damped system is positive, i.e.,

$$s_1 = r_1 + id_1$$

$$s_2 = r_1 - id_1$$

$$\cdots \cdots \cdots \cdots$$

$$x(t) = e^{r_1 t}(A \cos d_1 t + B \sin d_1 t) + \cdots$$

where $s_1, s_2, \ldots$ are the roots of the characteristic equation, and $i = \sqrt{-1}$.

For conservative systems, the principle of *Minimum Potential Energy* can be used to test the stability of a system. A system will be stable at an equilibrium position if the potential energy of the system is a minimum for that position, i.e.,

$$\frac{d(\text{P.E.})}{dq} = 0, \qquad \frac{d^2(\text{P.E.})}{dq^2} > 0$$

where   P.E. = potential energy of the system,
$\quad q$ = generalized coordinates.

Solved Problems

1. Two vibratory systems are given in Fig. 6-1. Show that these systems have variable spring stiffness.

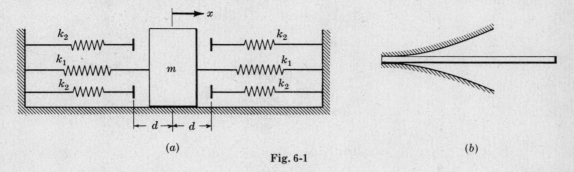

(a)

Fig. 6-1

(b)

(a) For the simple spring-mass system as shown in Fig. 6-1(a), springs k_1 are connected to m all the time, while springs k_2 come into action only when the displacement of the mass m is greater than d. Since the springs are assumed linear and the contacting surfaces smooth, we have

$$F(t) = 2k_1 x \qquad \text{(when displacement is less than } d\text{)}$$

$$\begin{aligned} F(t) &= 2k_1 x + 2k_2(x - d) \\ &= (2k_1 + 2k_2)x - 2k_2 d \qquad \text{(when displacement is greater than } d\text{)} \end{aligned}$$

i.e. when the displacement of mass m is greater than d, springs k_2 come into action, thereby increasing the total spring stiffness of the system. The variation of spring force with displacement is shown in Fig. 6-1(c).

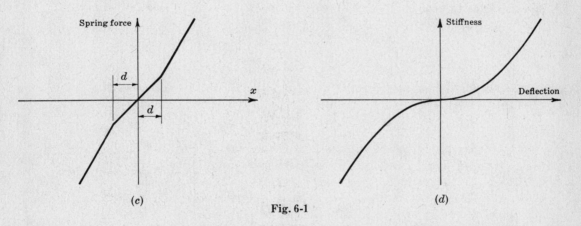

(c)

Fig. 6-1

(d)

(b) For the cantilever as shown in Fig. 6-1(b), the fixed end will contact the curved guide as it is deflected. When the deflection becomes large, the contact surface also increases. Thus deflection decreases the effective length of the cantilever, i.e. the larger the deflection, the shorter the effective length of the cantilever. As the stiffness of the cantilever depends on its effective length, deflection of the cantilever increases its stiffness. In short, the system has a variable spring stiffness as shown in Fig. 6-1(d).

2. A mass m is attached to a stretched wire with initial tension T as shown in Fig. 6-2 below. Show that the governing differential equation of motion is nonlinear.

The analysis of a similar problem with the assumption of constant tension in the wire is presented in Problem 5 of Chapter 1. Here the effect of variation in tension is taken into consideration. Then for a vertical displacement y of the mass, the corresponding strain ϵ in the wire is

$$\epsilon = \frac{\sqrt{L^2/4 + y^2} - L/2}{L/2}$$

where L is the length of the wire. Hence the tension in the wire after the displacement is $(T + AE\epsilon)$, where A is the cross section area and E is the modulus of elasticity.

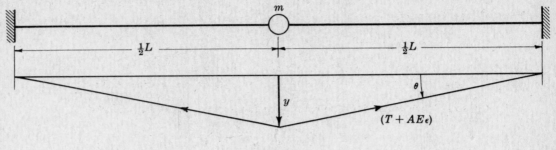

Fig. 6-2

As shown in Fig. 6-2, the restoring force on the mass m in the y-direction is $2(T + AE\epsilon) \sin \theta$, where

$$\sin \theta = \frac{y}{\sqrt{L^2/4 + y^2}}$$

Assuming that the displacement y is small compared with L, we have

$$\frac{1}{\sqrt{(L/2)^2 + y^2}} = \frac{1}{(L/2)\sqrt{1 + (2y/L)^2}} = (2/L)[1 - \tfrac{1}{2}(2y/L)^2 + \cdots]$$

or

$$\sin \theta = 2y/L - 4y^3/L^3$$

Since this is free vibration without damping, the differential equation of motion is given by

$$m\ddot{y} + 2y(T + AE\epsilon) \sin \theta = 0$$

Substitute the expressions for ϵ and $\sin \theta$ and obtain

$$m\ddot{y} + \frac{4T}{L} y + \frac{8AE}{L^3}\left(1 - \frac{T}{AE}\right) y^3 = 0$$

or

$$m\ddot{y} + \frac{4T}{L} y + \frac{8AE}{L^3} y^3 = 0$$

as the term (T/AE) represents the initial strain in the wire and would be small compared with unity. Thus the equation of motion is nonlinear.

3. A simple pendulum with no friction, as shown in Fig. 6-3, is released from some initial angle θ_0 with zero angular velocity. (a) Calculate the natural frequency for small angles of oscillation. (b) Use a two-term power series approximation of $\sin \theta$ and find the natural frequency by the use of elliptic integrals for $\theta_0 = 90°$. (c) Using the exact form of the differential equation of motion of the pendulum, determine the natural frequency.

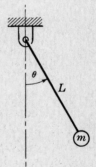

(a) For small angles of oscillation, the differential equation of motion becomes

$$mL^2 \ddot{\theta} = -mgL \sin \theta$$

or

$$\ddot{\theta} + (g/L)\theta = 0$$

where $\sin \theta$ is replaced by θ. Then

$$\omega_n = \sqrt{g/L} \text{ rad/sec}, \qquad T = \frac{6.28}{\sqrt{g/L}} \text{ sec}$$

Fig. 6-3

(b) Replacing $\sin \theta$ by $(\theta - \theta^3/6)$, the equation of motion becomes

$$\ddot{\theta} + (g/L)\theta - (g/6L)\theta^3 = 0$$

Letting $\dfrac{d\theta}{dt} = \omega$ and $\dfrac{d^2\theta}{dt^2} = \omega \dfrac{d\omega}{d\theta}$, we obtain

$$\int_{\omega_0}^{\omega} \omega \, d\omega \;\; = \;\; -\int_{\theta_0}^{\theta} \frac{g\theta}{L} d\theta \; + \; \int_{\theta_0}^{\theta} \frac{g\theta^3}{6L} d\theta$$

which yields

$$\omega^2 \;=\; \frac{g}{L}(\theta_0^2 - \theta^2) - \frac{g}{12L}(\theta_0^4 - \theta^4) \qquad \text{or} \qquad \omega \;=\; -\sqrt{\frac{g}{L}(\theta_0^2 - \theta^2)}\left[\sqrt{1 + \frac{1}{12}(\theta_0^2 + \theta^2)}\right] \;=\; \frac{d\theta}{dt}$$

Let $\theta = \theta_0 \cos \phi$; then

$$\sqrt{\theta_0^2 - \theta^2} \;=\; \theta_0 \sin \phi, \qquad \theta_0^2 + \theta^2 \;=\; \theta_0^2(1 + \cos^2 \phi),$$

$$d\theta \;=\; -\,\theta_0 \sin \phi \, d\phi, \qquad \omega \;=\; d\theta/dt \;=\; -\,\theta_0 \sin \phi \, d\phi/dt$$

Substituting these values into the expression for ω and integrating, we have

$$t \;=\; \frac{1}{\sqrt{g/L}} \int_0^{\phi} \frac{d\phi}{\sqrt{1 + \frac{1}{12}\theta_0^2(1 + \cos^2 \phi)}}$$

Let $k_1^2 = -\dfrac{g\theta_0^2}{12L}(g/L - g\theta_0^2/6L)$. Then

$$\frac{g}{L}\left[1 + \frac{\theta_0^2}{12}(1 + \cos^2 \phi)\right] \;=\; [g/L + (g/6L)(\theta_0)^2](1 - k_1^2 \sin^2 \phi)$$

and

$$t \;=\; \frac{1}{\sqrt{g/L - g\theta_0^2/6L}} \int_0^{\phi} \frac{d\phi}{\sqrt{1 - k_1^2 \sin^2 \phi}}$$

Now $\theta_0 = 90°$ or $\pi/2$, and

$$t \;=\; \frac{1}{1.19\sqrt{g/L}} \int_0^{\phi} \frac{d\phi}{\sqrt{1 - 0.386 \sin^2 \phi}}$$

which is an incomplete elliptic integral of the first kind. Hence

$$T \;=\; \frac{4}{1.19\sqrt{g/L}} \int_0^{\pi/2} \frac{d\theta}{\sqrt{1 - 0.386 \sin^2 \theta}} \;=\; \frac{7.32}{\sqrt{g/L}} \text{ sec}$$

(c) Using the exact form of the differential equation of motion, this also leads to an expression for the oscillation period which is an elliptic integral of the first kind:

$$t \;=\; 4 \int_0^{\theta_0} \frac{d\theta}{\sqrt{1 - \dfrac{2g}{L}(\cos \theta_0 - \cos \theta)}}$$

and $\quad T \;=\; \dfrac{4}{\sqrt{g/L}} \displaystyle\int_0^{\pi/2} \frac{d\theta}{\sqrt{1 - \sin^2(\theta_0/2)\sin^2 \theta}} \;=\; \dfrac{4}{\sqrt{g/L}} \displaystyle\int_0^{\pi/2} \frac{d\theta}{\sqrt{1 - 0.49 \sin^2 \theta}} \;=\; \dfrac{7.41}{\sqrt{g/L}} \text{ sec}$

Note that the results obtained are not the same. Actually, the linear analysis in part (a) is accurate for amplitudes of oscillation less than 20°. For an angle of 160°, the period is almost twice as great as for a very small amplitude. When the amplitude approaches 180°, the period of oscillation will approach infinity.

4. A homogeneous solid cylinder of radius r and mass m rolls without slipping on a horizontal plane under the action of a linear spring k with free length L_0, as shown in Fig. 6-4. Derive the equation of motion according to the first nonlinear approximation.

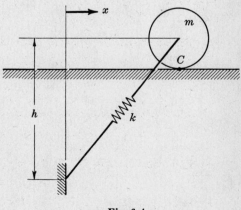

At any instant under consideration, the cylinder is rotating around point C, the point of contact or the instantaneous center of zero velocity. Hence the equation of motion is $\Sigma M = J_C \ddot{\theta}$, where $J_C = J_0 + mr^2$ is the moment of inertia of the cylinder with respect to point C. Thus

$$(\tfrac{1}{2}mr^2 + mr^2)\,\ddot{\theta} \;=\; -F_x r$$

where $F_x = $ horizontal component of the spring force.

Fig. 6-4

From the geometry of the system, the extension of the spring is

$$dL = \sqrt{h^2 + x^2} - L_0$$

and the horizontal component of the spring force is therefore given by

$$F_x = [k(\sqrt{h^2 + x^2} - L_0)]\frac{x}{\sqrt{h^2 + x^2}}$$

According to the first nonlinear approximation,

$$\frac{x}{\sqrt{h^2 + x^2}} = \frac{x}{h\sqrt{1 + (x/h)^2}} = \frac{x}{h}[1 - \tfrac{1}{2}(x/h)^2 + \cdots]$$

and so the equation of motion becomes

$$\frac{3}{2}mr^2\ddot{\theta} + krx\left(1 - \frac{L_0}{\sqrt{h^2 + x^2}}\right) = 0$$

Since there is no slipping, $\theta = x/r$ and $\ddot{\theta} = \ddot{x}/r$; then

$$\frac{3}{2}m\ddot{x} + \left\{1 - \frac{L_0}{h}[1 - \tfrac{1}{2}(x/h)^2 + \cdots]\right\}kx = 0$$

or

$$\frac{d^2x}{dt^2} + \frac{2k}{3m}\left[\left(1 - \frac{L_0}{h}\right)x + \frac{L_0}{2h^3}x^3\right] = 0$$

which represents a system with nonlinear restoring force.

5. A mechanical system with nonlinear restoring force is acted upon by an excitation $F_0 \cos \omega t$ as represented by the following equation:

$$\ddot{x} + \alpha x + \beta x^3 = F_0 \cos \omega t$$

Determine the approximate steady state vibration.

The nonlinear restoring force $\alpha x + \beta x^3$ is physically represented by a spring whose stiffness varies with displacement. If β is positive, the stiffness increases with displacement and the spring is called a *hard spring*. If β is negative, the stiffness decreases with displacement; the spring is said to be *soft*.

Considering only small values of β and F_0, we have

$$\ddot{x} = -\alpha x - \beta x^3 + F_0 \cos \omega t \qquad (1)$$

Assuming the first approximate steady state solution as

$$x_1 = A \cos \omega t \qquad (2)$$

the equation of motion then becomes

$$\ddot{x}_2 = -\alpha A \cos \omega t - \beta A^3 \cos^3 \omega t + F_0 \cos \omega t \qquad (3)$$

Now

$$\cos^3 \omega t = \tfrac{3}{4}\cos \omega t + \tfrac{1}{4}\cos 3\omega t \qquad (4)$$

and so equation (3) can be rewritten as

$$\ddot{x}_2 = -(\alpha A + \tfrac{3}{4}\beta A^3 - F_0)\cos \omega t - \tfrac{1}{4}\beta A^3 \cos 3\omega t \qquad (5)$$

Integrating equation (5) twice yields

$$x_2(t) = \frac{1}{\omega^2}(\alpha A + \tfrac{3}{4}\beta A^3 - F_0)\cos \omega t + \frac{\beta A^3}{36\omega^2}\cos 3\omega t$$

where the integration constants are taken to be zero to ensure x_1 and x_2 are periodic. This is then the second approximate steady state vibration.

As mentioned earlier, new phenomena occur in nonlinear systems which cannot occur in linear systems. Therefore it should be pointed out here that for a system with nonlinear restoring force resonance does not occur in the same manner as in the case of linear restoring force. Here the amplitude of vibration can never become very large for a driving force of any given frequency. This is because the natural frequency of the system for small amplitudes of vibration is different from the natural frequency for large amplitudes of vibration. It can be shown that for hard springs the natural frequency increases with amplitude of vibration while the opposite is true for soft springs.

Moreover, the amplitude of vibration may suddenly increase or decrease as the excitation frequency ω is increasing or decreasing. This is shown in Fig. 6-5 which includes viscous damping $c\dot{x}$.

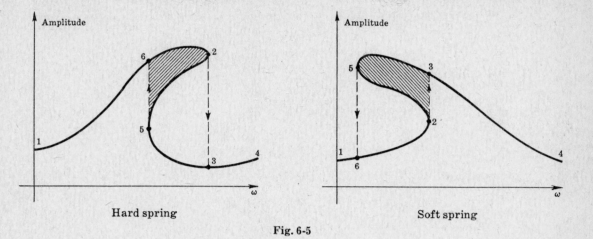

Hard spring   Soft spring

Fig. 6-5

For constant magnitude of F_0, the amplitude of vibration will increase along points 1, 2, 3, 4 on the curve when the excitation frequency ω is slowly increased. The amplitude of vibration jumps from point 2 to 3 on the curve. When the excitation frequency ω is slowly decreased, the amplitude of vibration will follow points 4, 5, 6, 1 on the curve and make a similar jump at point 5 to point 6. This is known as the *jump phenomena*.

It is clear that there are two different values for the amplitude of vibration for a given excitation frequency as shown in the shaded regions of the curves (i.e. the amplitude of vibration is not a single-valued function of excitation frequency ω, a fact contrary to linear systems).

Also, forced vibration of systems with nonlinear restoring force may have frequencies lower than the excitation frequency ω. This is known as *subharmonic vibration*. In systems with linear restoring force, the frequencies of forced vibration are always equal to the excitation frequencies.

SELF-EXCITED VIBRATION

6. A uniform beam of length 100 in. and cross section area 0.1 in² is pivoted freely at one end. A 10 lb weight is attached firmly to the other end of the beam which rests on the rim of a wheel having radius 10 in. and rotating at constant speed of 1200 rev/min. Let the coefficient of friction between the beam and wheel be $(0.2 - 0.001v)$ where v is the relative velocity at the line of contact between the beam and wheel. Determine the vibration of the weight at the end of the beam. Take $E = 10(10)^6$ lb/in² for the beam material.

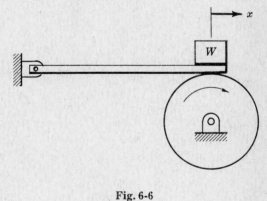

Fig. 6-6

This is typically a self-excited vibration due to dry friction between surfaces. The friction force decreases in magnitude as the relative velocity increases, giving rise to a negative damping to sustain the vibration of the system.

The velocity of a point on the rim of the wheel is $2\pi(10)(1200/60) = 1256$ in/sec. Hence the relative velocity at the point of contact is

$$v = 1256 - \dot{x}$$

where $\dot{x}$ is the velocity of the weight at the end of the beam.

The friction force acting on the beam is

$$F = \mu N = 10[0.2 - 0.001(1256 - \dot{x})] = 0.01\dot{x} - 10.56$$

The restoring force is given by the elastic action of the beam which acts as a linear spring for small extension, i.e.,

$$AE(x/L) = 0.1(10)(10)^6 x/100 = 10,000x$$

Thus the equation of motion for the weight W is given by

$$(10/g)\ddot{x} + 10,000x = 0.01\dot{x} - 10.56$$

or

$$\ddot{x} - 0.386\dot{x} + 386,000x = -407$$

the solution of which is of the form

$$x(t) = Ae^{-(c/2m)t}\sin(\omega_d t + \phi) + B$$

in which $\omega_d = \sqrt{k/m - (c/2m)^2} = \sqrt{386,000 - 0.386^2/4} = 624$ and $B = -407/386,000 = -0.0011$. Hence the vibration of the weight is

$$x(t) = Ae^{0.193t}\sin(624t + \phi) - 0.0011$$

The amplitude of vibration is seen to increase with time. At the same time, the relative velocity of the point of contact also increases; hence the magnitude of the friction force is reduced. As a result, the amplitude of vibration will decrease, and so will the relative velocity of the point of contact. This will increase the magnitude of friction force and eventually increase the amplitude of vibration. The whole cycle repeats itself.

STABILITY

7. A simple spring-mass system with damping is excited by a force proportional to the velocity of the mass m. Investigate the stability of the system.

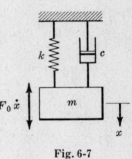

Applying $\Sigma F = ma$, the differential equation of motion is

$$m\ddot{x} + c\dot{x} + kx = F_0\dot{x} \qquad (1)$$

Rearranging,

$$m\ddot{x} + kx = (F_0 - c)\dot{x} \qquad (2)$$

Equation (2) can now be rewritten as

$$\frac{d}{dt}(\tfrac{1}{2}m\dot{x}^2 + \tfrac{1}{2}kx^2) = (F_0 - c)\dot{x}^2 \qquad (3)$$

Fig. 6-7

where the left-hand side is the rate of change of the total energy of the system (i.e. kinetic energy of the mass m and potential energy of the spring).

If $F_0 < c$, the rate of change of the total energy is negative. This means that the amount of energy possessed by the system is decreasing due to the presence of damping represented here by the dashpot c. Therefore the system is oscillating with diminishing amplitude. The system is stable.

If $F_0 > c$, we have negative damping, and the rate of change of energy is positive. This means that the amount of energy possessed by the system is increasing. As a result, the motion of the system itself tends to increase the energy or the amplitude of vibration of the system. The system is therefore unstable.

Also, equation (2) can be rewritten as

$$\ddot{x} + \frac{(c - F_0)}{m}\dot{x} + \frac{k}{m}x = 0 \qquad (4)$$

Substituting $x = e^{rt}$ into equation (4), we obtain the characteristic equation for the system as

$$r^2 + \frac{(c - F_0)}{m}r + \frac{k}{m} = 0 \qquad (5)$$

and

$$r_{1,2} = -\frac{(c - F_0)}{2m} \pm i\sqrt{\frac{k}{m} - \left[\frac{(c - F_0)}{2m}\right]^2}$$

where $i = \sqrt{-1}$. Hence

$$x(t) = Ce^{-\zeta\omega_n t}\sin(\sqrt{1 - \zeta^2}\,\omega_n t + \phi)$$

where $\zeta = (c - F_0)/2m\omega_n$ is the damping factor.

If $F_0 > c$, the damping present in the system is negative and ζ is negative. The motion of the system $x(t)$ will increase without bound and the system is unstable.

8. A simple spring-mass system with damping is acted upon by a force proportional to the displacement of the mass m. Investigate the stability of the system.

Using $\Sigma F = ma$, the differential equation of motion is

$$m\ddot{x} + c\dot{x} + kx = F_0 x \qquad (1)$$

Rearranging,

$$\ddot{x} + \frac{c}{m}\dot{x} + \frac{k-F_0}{m}x = 0 \qquad (2)$$

Substituting $x = e^{rt}$ into equation (2), we obtain the characteristic equation for the system as

$$r^2 + \frac{c}{m}r + \frac{k-F_0}{m} = 0 \qquad (3)$$

and

$$r_{1,2} = -\frac{c}{2m} \pm \sqrt{\left(\frac{c}{2m}\right)^2 - \frac{k-F_0}{m}} \qquad (4)$$

Thus

$$x(t) = Ae^{r_1 t} + Be^{r_2 t} \qquad (5)$$

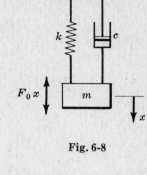

Fig. 6-8

where A and B are arbitrary constants to be evaluated by the two initial conditions of the problem, and r_1 and r_2 are given by (4).

When $F_0 > k$, r_1 is positive and r_2 is negative, that is,

$$x(t) = Ae^{r_1 t} + Be^{-r_2 t} \qquad (6)$$

where the magnitude of the first term increases with time, and the magnitude of the second term decreases with time. The motion of the system is therefore divergent and unstable.

The damped frequency is

$$\omega_d = \sqrt{\frac{k-F_0}{m} - \left(\frac{c}{2m}\right)^2}$$

and if F_0 is greater than k, the quantity under the square root bracket is negative. The motion of the system is non-oscillatory.

9. A mass m is attached to one end of a weightless stiff rod which is fixed to another rod AB. Rod AB is pivoted freely at its midpoint and restrained at its ends by springs of stiffness k as shown in Fig. 6-9. Restricting motion to the plane of the paper, investigate the stability of the system.

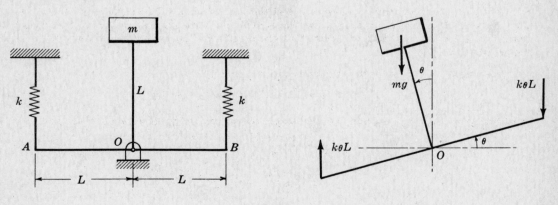

Fig. 6-9

The differential equation of motion is given by $\Sigma T_0 = I\ddot{\theta}$,

$$mL^2\ddot{\theta} = mgL\sin\theta - 2k\theta L^2\cos\theta$$

or

$$\ddot{\theta}L - g\sin\theta + 2kL\theta\cos\theta/m = 0$$

Assuming small angles of oscillation, $\sin \theta \doteq \theta$, $\cos \theta \doteq 1$, and the equation of motion becomes

$$L\ddot{\theta} + (2kL/m - g)\theta = 0 \qquad \text{and so} \qquad \omega_n = \sqrt{2k/m - g/L} \text{ rad/sec}$$

If $g < 2kL/m$, ω_n is positive. The system at this position is stable, i.e. $\theta(t) = A \cos \omega_n t + B \sin \omega_n t$.

If $g > 2kL/m$, ω_n is negative. The system at this position is unstable, i.e. $\theta(t) = C \cosh \omega_n t + D \sinh \omega_n t$.

We will test the stability of the system by the principle of minimum potential energy. The potential energy of the system is

$$\text{P.E.} = kL^2\theta^2 - mgL(1 - \cos \theta)$$

where $kL^2\theta^2$ is the energy of the springs, and $mgL(1 - \cos \theta)$ is the energy loss due to the position of mass m. Now

$$\frac{d(\text{P.E.})}{d\theta} = 2kL^2\theta - mgL \sin \theta$$

$$\frac{d^2(\text{P.E.})}{d\theta^2} = 2kL^2 - mgL \cos \theta$$

Thus
$$\frac{d(\text{P.E.})}{d\theta}\bigg]_{\theta=0} = 0$$

and for
$$\frac{d^2(\text{P.E.})}{d\theta^2}\bigg]_{\theta=0} > 0, \qquad 2kL^2 > mgL \quad \text{or} \quad 2kL/m > g$$

Hence the equilibrium position at $\theta = 0$ will be stable if $2kL/m > g$, or $k > mg/2L$.

10. A homogeneous wooden plank of length L, thickness t, and weight W is balanced on top of a semicircular cylinder of radius R as shown in Fig. 6-10(a). If the plank is slightly tipped, what is the condition for stable equilibrium at this position?

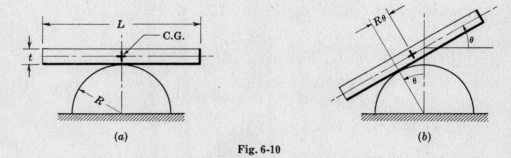

$$(a) \qquad\qquad\qquad\qquad (b)$$

Fig. 6-10

Assume that the plank rocks without slipping on the cylinder when it is slightly tipped. We will test the stability of this position of the plank using the principle of minimum potential energy.

For the displaced position of the plank as shown in Fig. 6-10(b), the potential energy of the system is due entirely to the location of the center of gravity of the plank. Thus we have

$$\text{P.E.} = W[(R + t/2) \cos \theta + R\theta \sin \theta]$$

and the first and second derivatives of P.E. with respect to θ are

$$(1) \quad \frac{d(\text{P.E.})}{d\theta} = W[-R \sin \theta - (t/2) \sin \theta + R \sin \theta + R\theta \cos \theta]$$

$$(2) \quad \frac{d^2(\text{P.E.})}{d\theta^2} = W[-(t/2) \cos \theta + R \cos \theta - R\theta \sin \theta]$$

Employing the principle of minimum potential energy, the top of the cylinder will be a stable equilibrium position if

$$\frac{d(\text{P.E.})}{d\theta} = 0, \qquad \frac{d^2(\text{P.E.})}{d\theta^2} > 0 \qquad \text{at} \quad \theta = 0$$

From (1), $d(\text{P.E.})/d\theta = 0$ when $\theta = 0$; hence $\theta = 0$ is an equilibrium position.

From (2), $d^2(\text{P.E.})/d\theta^2 > 0$ at $\theta = 0$ when $R > t/2$; hence the condition for stable equilibrium of the plank at $\theta = 0$ is $R > t/2$.

Supplementary Problems

11. Fig. 6-11 below shows a mass m sliding on a smooth horizontal plane between linear springs having stiffness k_1 and k_2 respectively. The springs are unstressed when the mass is in the equilibrium

    position. Find the natural frequency. *Ans.* $\omega_n = \dfrac{\sqrt{k_1 k_2}}{\sqrt{m}\,[\sqrt{k_2} + \sqrt{k_1}\,]}$ rad/sec

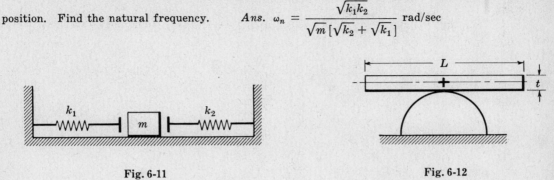

Fig. 6-11   Fig. 6-12

12. Show that the principle of superposition does not hold for a nonlinear differential equation like the following
$$\ddot{x} + \beta x^2 = 0$$

13. A homogeneous wooden plank of length L and thickness t rests on a semicircular support of radius R as shown in Fig. 6-12 above. If the plank is tipped slightly, it will rock without slipping on the support. Find the equation of motion and show that it is nonlinear.
    Ans. $\ddot{\theta}\,[(R\theta)^2 + (t/2)^2 + (t^2 + L^2)/12] + [(R\dot{\theta})^2 + Rg\cos\theta]\theta - \dfrac{3}{2}g\sin\theta = 0$

14. Determine the period of oscillation for the nonlinear equation of motion $\ddot{x} + x^3 = 0$ with initial conditions $x(0) = 1$ and $\dot{x}(0) = 0$. *Ans.* $T = 7.42$ sec

15. Show that the period T of a simple pendulum approaches infinity when the initial angle θ_0 approaches $180°$.

16. A homogeneous slender rod of length L is pinned at one end so that it is free to swing in a vertical plane as shown in Fig. 6-13 below. If the rod is released from rest in a horizontal position, calculate the angular velocity when it passes through the vertical position and the time required to reach this position. *Ans.* $\omega = \sqrt{3g/L}$ rad/sec, time required $= 1.52\sqrt{L/g}$ sec

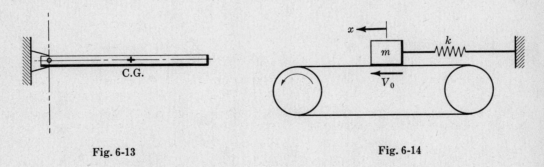

Fig. 6-13   Fig. 6-14

17. Determine the first approximation $x_1(t)$ to the solution of the following equation:
$$\ddot{x} + \alpha x + \beta x^5 = \beta F_0 \cos \omega t$$
    What is the frequency-amplitude relationship?

    Ans. $x_1(t) = A\cos\omega t + \dfrac{5\beta A^5}{128\omega^2}\cos 3\omega t + \dfrac{\beta A^5}{384\omega^2}\cos 5\omega t$

    $\omega^2 = \alpha + \dfrac{5}{8}\beta A^4 - \dfrac{\beta F_0}{A}$

18. A mass m is resting on a conveyor belt which moves at a constant velocity V_0 as shown in Fig. 6-14 above. If the coefficient of friction between the contacting surfaces decreases slightly when the relative velocity between the surfaces increases, show that self-excited vibration develops for mass m from the rest position under the slightest disturbance.

19. Show that the self-excited vibration of mass m for the system as shown in Fig. 6-15 below is unstable.

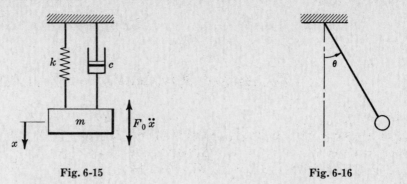

Fig. 6-15   Fig. 6-16

20. Test the stability of the following systems whose characteristic equations are given by

(a) $r^2 - 4r + 5 = 0$,   (b) $r^3 + 5r^2 + 3r + 2 = 0$

Ans. (a) Unstable, (b) Stable

21. Using the principle of minimum potential energy, show that $\theta = \pi$ position is unstable and $\theta = 0$ is stable for the simple pendulum as shown in Fig. 6-16 above.

22. A uniform stiff rod of length L and weight W is pivoted freely at one end and connected by a spring of constant k at the other end as shown in Fig. 6-17 below. Find the stable condition of the system.

Ans. $kL^2 > WL/2$

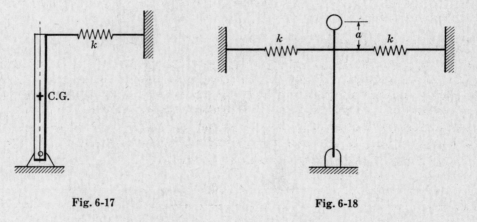

Fig. 6-17   Fig. 6-18

23. An inverted-pendulum of length L and mass m is connected by two springs of stiffness k as shown in Fig. 6-18 above. Investigate the stability of the system.

Ans. System is stable if $\dfrac{2k}{m}\left(1 - \dfrac{a}{L}\right) > (g/L)$

Chapter 7

Electrical Analogies

INTRODUCTION

Mechanical systems can be represented and studied by their equivalent electrical circuits which are more easily constructed than models of the corresponding mechanical systems and from which experimental results are more conveniently taken than from the mechanical models.

The equivalent electrical circuits are obtained by comparing the equations of motion for both systems. The mechanical and electrical systems are analogous if their differential equations of motion are mathematically the same. When this happens, the corresponding terms in the differential equations of motion are analogous to one another. The equivalent electrical circuits can then be constructed using Kirchhoff's laws.

KIRCHHOFF'S LAWS

Kirchhoff's Voltage Law: The algebraic sum of all the voltages around any closed circuit is equal to zero in any network.

Kirchhoff's Current Law: The algebraic sum of all the currents flowing toward any point is equal to zero in any circuit.

ELECTRICAL ANALOGIES

There are two electrical analogies for mechanical systems: (1) the *voltage-force* or *mass-inductance analogy*, (2) the *current-force* or *mass-capacitance analogy*. The voltage-force analogy is very useful for most systems. It is difficult to use for complicated systems. The current-force is a physical analogy rather than an analogy of similar equations of motion, and is easy to apply. It has the advantage that both the electrical circuit and the mechanical circuit are of the same form. The following table shows both the voltage-force and current-force analogies for mechanical systems.

Table 7-1

Mechanical System		Electrical System	
		Voltage-force Analogy	Current-force Analogy
	D'Alembert's principle	Kirchhoff's voltage law	Kirchhoff's current law
	Degree of freedom	Loop	Node
	Force applied	Switch closed	Switch closed
F	Force (lb)	v Voltage (volt)	i Current (ampere)
m	Mass (lb-sec²/in)	L Inductance (henry)	C Capacitance (farad)
x	Displacement (in)	q Charge (coulomb)	$\phi = \int v\, dt$
$\dot{x}$	Velocity (in/sec)	i Loop current (ampere)	v Node voltage (volt)
c	Damping (lb-sec/in)	R Resistance (ohm)	$1/R$ Conductance (mho)
k	Spring (lb/in)	$1/C$ 1/Capacitance	$1/L$ 1/Inductance
	Coupling element	Element common to two loops	Element between nodes

In general, in laying out equivalent electric circuits for mechanical systems, the following rule is observed. If the forces act in series in the mechanical system, the electric elements representing these forces are put in parallel. Forces in parallel are represented by elements in series in electric circuits.

DIMENSIONLESS NUMBERS

In order that the electrical analogue be completely equivalent to the mechanical system in question, dimensional analysis is used to obtain the right scale factors so that the two systems are identical with each other. The following dimensionless numbers can be obtained from dimensional analysis:

$$m_1/m_2 = L_1/L_2, \quad k_1/k_2 = C_2/C_1, \quad \omega\sqrt{m/k} = \omega_e\sqrt{LC}, \quad F/kx = vC/q, \quad c^2/km = R^2C/L$$

Solved Problems

1. Investigate the electrical analogues of the single-degree-of-freedom vibratory system as shown in Fig. 7-1(a).

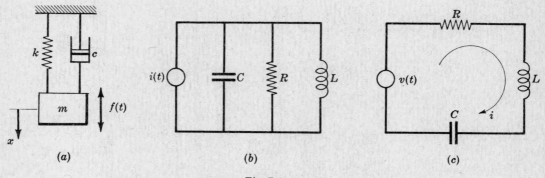

(a) (b) (c)

Fig. 7-1

Employing Newton's law of motion, the differential equation of motion is given by

$$m\frac{d^2x}{dt^2} + c\frac{dx}{dt} + kx = f(t) \tag{1}$$

For an electrical network as shown in Fig. 7-1(b), an equation of the following form can be written:

$$C\frac{d^2v}{dt^2} + \frac{1}{R}\frac{dv}{dt} + \frac{1}{L}v = \frac{di(t)}{dt} \tag{2}$$

where C = capacitance; $\left(i = C\frac{dv}{dt} \right)$,

     R = resistance; $(i = v/R)$,

     L = inductance; $\left[i = \frac{1}{L}\int v\,dt + i(0) \right]$,

   $i(t)$ = current source,

     v = voltage.

Since equations (1) and (2) are of the same form, i.e. they are identical mathematically, the two systems represented by these two equations are analogous.

Using *Kirchhoff's voltage law*, the voltage equation for the electrical network as shown in Fig. 7-1(c) is given by

$$L\frac{di}{dt} + Ri + \frac{1}{C}\int i\,dt = v(t) \tag{3}$$

Rewrite equation (1) as

$$m\frac{d\dot{x}}{dt} + c\dot{x} + k\int \dot{x}\,dt = f(t) \tag{4}$$

where dx/dt is replaced by $\dot{x}$, and x by $\int \dot{x}\,dt$. Now equations (3) and (4) are of the same form, which means that the two systems represented by these two equations are analogous. In other words, the excitation voltage $v(t)$ is analogous to the excitation force $f(t)$, the loop current i is analogous to the mass velocity $\dot{x}$, and so on. This is known as the *mass-inductance* or *voltage-force analogy*.

Integrating equation (2) once with respect to time, we obtain the current equation for the network shown in Fig. 7-1(b):

$$C\frac{dv}{dt} + \frac{v}{R} + \frac{1}{L}\int v\,dt = i(t) \tag{5}$$

(Equation (5) can also be obtained by *Kirchhoff's current law*.)

Now equations (4) and (5) are of the same form; which means that the two systems represented by these two equations are analogous. Hence the excitation current $i(t)$ is analogous to the excitation force $f(t)$, the network voltage v is analogous to the mass velocity $\dot{x}$, and so on. This is known as the *mass-capacitance* or *current-force analogy*.

2. An electrical circuit contains a capacitor C, an inductor L, and a switch in series as shown in Fig. 7-2(a). The capacitor has initially a charge q_0 and the switch is open at time $t < 0$. If the switch is closed at $t = 0$, find the subsequent charge on the capacitor.

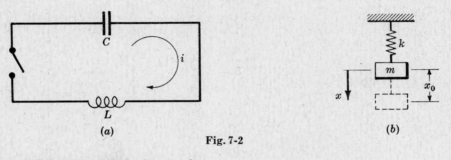

Fig. 7-2

Employing Kirchhoff's voltage law, we have

$$\frac{q}{C} + L\frac{di}{dt} = 0 \tag{1}$$

where q is the charge on capacitor C. Since $dq/dt = i$, equation (1) becomes

$$L\frac{d^2q}{dt^2} + \frac{q}{C} = 0 \tag{2}$$

or

$$\frac{d^2q}{dt^2} + \frac{q}{LC} = 0 \tag{3}$$

Hence

$$q(t) = A\sin\sqrt{1/LC}\,t + B\cos\sqrt{1/LC}\,t \tag{4}$$

At $t = 0$, $q = q_0$ and thus $B = q_0$. At $t = 0$, $\dot{q} = i = 0$ and so $A = 0$. Then

$$q(t) = q_0\cos\sqrt{1/LC}\,t = q_0\cos\omega_n t \tag{5}$$

where $\omega_n = 1/\sqrt{LC}$ is the natural frequency of the system.

Compare this electrical circuit with a simple single-degree-of-freedom spring-mass system as shown in Fig. 7-2(b). The equation of motion for this mechanical system is

$$m\frac{d^2x}{dt^2} + kx = 0$$

with solution

$$x(t) = x_0\cos\sqrt{k/m}\,t = x_0\cos\omega_n t$$

where x_0 is the initial displacement of the mass m from the static equilibrium position.

Thus the two systems are analogous to each other with L corresponding to m, q to x, $1/C$ to k, and $\omega_n = 1/\sqrt{LC}$ to $\omega_n = \sqrt{k/m}$.

3. A two-degree-of-freedom spring-mass system is shown in Fig. 7-3(a). Use both the voltage-force and current-force analogy to set up the equivalent electrical circuits for the system.

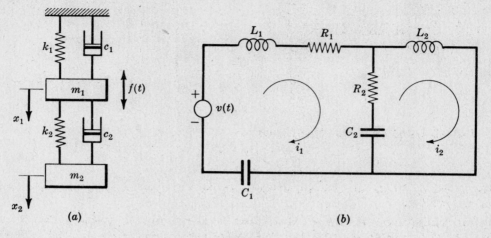

(a)   (b)

Fig. 7-3

The equations of motion given by $\Sigma F = ma$ are

$$m_1 \frac{d^2x_1}{dt^2} + (c_1 + c_2)\frac{dx_1}{dt} + (k_1 + k_2)x_1 - c_2\frac{dx_2}{dt} - k_2 x_2 = f(t)$$

$$m_2 \frac{d^2x_2}{dt^2} + c_2\frac{dx_2}{dt} + k_2 x_2 - c_2\frac{dx_1}{dt} - k_2 x_1 = 0$$

Using the voltage-force analogy given in Table 7-1, the analogous electrical equations are

$$L_1 \frac{di_1}{dt} + (R_1 + R_2)i_1 + \left[\frac{1}{C_1} + \frac{1}{C_2}\right]\int i_1\,dt - R_2 i_2 - \frac{1}{C_2}\int i_2\,dt = v(t)$$

$$L_2 \frac{di_2}{dt} + R_2 i_2 + \frac{1}{C_2}\int i_2\,dt - R_2 i_1 - \frac{1}{C_2}\int i_1\,dt = 0$$

and the analogous electrical circuit is shown in Fig. 7-3(b).

Using the current-force analogy as shown in Table 7-1, the analogous electrical equations are

$$C_1 \frac{dv_1}{dt} + \left[\frac{1}{R_1} + \frac{1}{R_2}\right]v_1 + \left[\frac{1}{L_1} + \frac{1}{L_2}\right]\int v_1\,dt - \frac{v_2}{R_2} - \frac{1}{L_2}\int v_2\,dt = i(t)$$

$$C_2 \frac{dv_2}{dt} + \frac{v_2}{R_2} + \frac{1}{L_2}\int v_2\,dt - \frac{v_1}{R_2} - \frac{1}{L_2}\int v_1\,dt = 0$$

and the analogous electrical circuit is shown in Fig. 7-3(c).

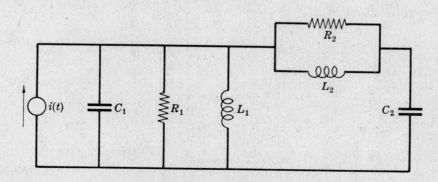

Fig. 7-3(c)

4. A simple torque system with damping is shown in Fig. 7-4(a). Using the voltage-torque analogy, draw the electrical-analogue diagram for the system.

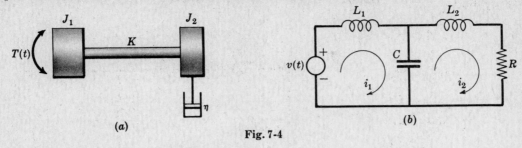

(a)   (b)

Fig. 7-4

The differential equations of motion are

$$J_1 \ddot{\theta}_1 + K(\theta_1 - \theta_2) = T(t)$$
$$J_2 \ddot{\theta}_2 + \eta \dot{\theta}_2 + K(\theta_2 - \theta_1) = 0$$

where θ_1 and θ_2 are the angular displacements of the rotors J_1 and J_2 respectively, and η is the damping coefficient.

The voltage-torque analogy is the extension of the voltage-force analogy. This is true because there is complete analogy between linear and rotational systems. Replacing torque by force, angular displacement by linear displacement, and so on, the electrical-analogue equations for the torque system are therefore given by

$$L_1 \frac{di_1}{dt} + \frac{1}{C} \int (i_1 - i_2)\, dt = v(t)$$

$$L_2 \frac{di_2}{dt} + Ri_2 + \frac{1}{C} \int (i_2 - i_1)\, dt = 0$$

and the corresponding electrical-analogue diagram is shown in Fig. 7-4(b).

5. A mechanical system is represented by the following equations:

$$A \ddot{\theta}_1 + B \dot{\theta}_1 + D\theta_1 - E \ddot{\theta}_2 = 0$$
$$a \ddot{\theta}_2 + b \dot{\theta}_2 + d\theta_2 - e \ddot{\theta}_1 = 0$$

Determine the electrical-analogue circuit for the given system.

Rewrite the given equations of motion as

$$A \ddot{\theta}_1 + B \dot{\theta}_1 + D\theta_1 = E \ddot{\theta}_2 \tag{1}$$
$$a \ddot{\theta}_2 + b \dot{\theta}_2 + d\theta_2 = e \ddot{\theta}_1 \tag{2}$$

The corresponding analogous electrical equations are given by

$$L_1 \frac{di}{dt} + R_1 i + \frac{1}{C_1} \int i\, dt = v(t) \tag{3}$$

$$C_2 \frac{dv}{dt} + \frac{v}{R_2} + \frac{1}{L_2} \int v\, dt = i(t) \tag{4}$$

Equation (3) is obtained by using the voltage-force analogy, and equation (4) by the current-force analogy.

The electrical networks for equations (3) and (4) are shown in Fig. 7-5(a) and these two separate networks can be combined into a single loop as shown in Fig. 7-5(b).

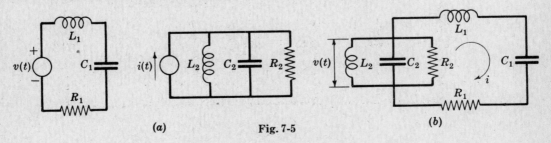

(a)   Fig. 7-5   (b)

6. Using both the voltage-force and current-force analogy, draw the electrical analogues for the mechanical system as shown in Fig. 7-6(a) and from them deduce the differential equations of motion.

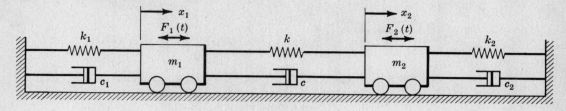

Fig. 7-6(a)

The equivalent electrical circuits are given in Fig. 7-6(b) and (c).

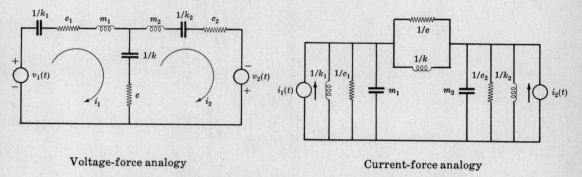

Voltage-force analogy

Fig. 7-6(b)

Current-force analogy

Fig. 7-6(c)

Hence the differential equations of motion are

$$m_1\ddot{x}_1 + c_1\dot{x}_1 + k_1x_1 + k(x_1 - x_2) + c(\dot{x}_1 - \dot{x}_2) = F_1(t)$$
$$m_2\ddot{x}_2 + c_2\dot{x}_2 + k_2x_2 + k(x_2 - x_1) + c(\dot{x}_2 - \dot{x}_1) = F_2(t)$$

7. Study the behavior of the mechanical system as shown in Fig. 7-7(a) by its electrical analogue. The mechanical elements are: $k = 50$ lb/in, $c = 0.1$ lb-sec/in, $m = 0.05$ lb-sec^2/in, $F_0 = 5$ lb, and $\omega = 10$ rad/sec. An inductance L of 0.1 henry and an alternating voltage source of frequency 100 rad/sec are available.

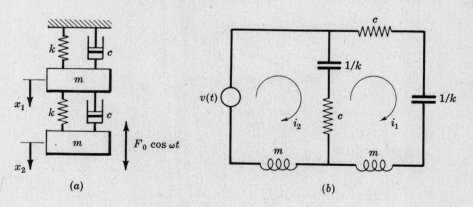

Fig. 7-7

Employing the voltage-force analogy, the electrical analogue for the above mechanical system is shown in Fig. 7-7(b).

From dimensional analysis, we have the following relations between the mechanical system and its electrical analogue:

$$\omega\sqrt{m/k} = \omega_e\sqrt{LC} \qquad (1)$$

$$c^2/km = R^2C/L \qquad (2)$$

$$F/kx = vC/q \qquad (3)$$

Using (1), we obtain

$$10\sqrt{0.05/50} = 100\sqrt{0.1C} \quad \text{or} \quad C = 100 \text{ microfarads}$$

From (2),

$$c^2/km = 0.1^2/(50)(0.05) = 100(10)^{-6}R^2/0.1 \quad \text{or} \quad R = 2 \text{ ohms}$$

and from (3),

$$F/kx = 5/50x = vC/q \quad \text{or} \quad x = (0.1/v)(q/C)$$

where v is the impressed voltage. Now the expression for x can be written as

$$x(t) = (0.1/v)v_c$$

where v_c is the voltage drop across the capacitor C.

It is clear that the displacements of the masses of the mechanical system under consideration can be obtained both qualitatively and quantitatively by measuring the voltage drop across the corresponding capacitors. Thus, the behavior of a complex mechanical system can be conveniently and accurately studied by its electrical analogue which is inexpensive, easily varied, and readily obtainable.

8. A three-degree-of-freedom spring-mass system is shown in Fig. 7-8(a). Use the voltage-force analogy to obtain the electrical-analogue network.

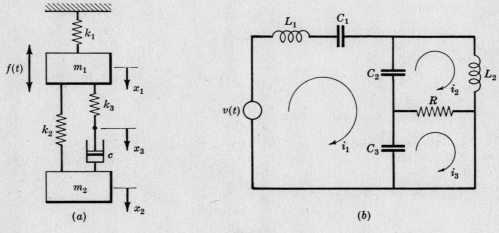

Fig. 7-8

The differential equations of motion are

$$m_1\ddot{x}_1 + k_1x_1 + k_2(x_1 - x_2) + k_3(x_1 - x_3) = f(t)$$

$$m_2\ddot{x}_2 + k_2(x_2 - x_1) + c(\dot{x}_2 - \dot{x}_3) = 0$$

$$k_3(x_3 - x_1) = c(\dot{x}_2 - \dot{x}_3)$$

Using the voltage-force analogy, the electrical-analogue equations are given by

$$L_1\frac{di_1}{dt} + \frac{1}{C_1}\int i_1\,dt + \frac{1}{C_2}\int (i_1 - i_2)\,dt + \frac{1}{C_3}\int (i_1 - i_3)\,dt = v(t)$$

$$L_2\frac{di_2}{dt} + (i_2 - i_3)R + \frac{1}{C_2}\int (i_2 - i_1)\,dt = 0$$

$$R(i_3 - i_2) + \frac{1}{C_3}\int (i_3 - i_1)\,dt = 0$$

and the corresponding electrical-analogue network is shown in Fig. 7-8(b).

9. Using the voltage-force analogy, set up the electrical-analogue circuit for the torsional vibration system as shown in Fig. 7-9(a).

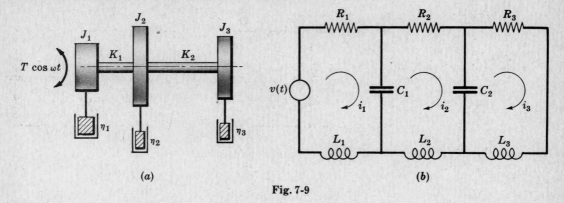

(a)   (b)

Fig. 7-9

The differential equations of motion are

$$J_1\ddot{\theta}_1 + \eta_1\dot{\theta}_1 + K_1(\theta_1 - \theta_2) = T\cos\omega t$$

$$J_2\ddot{\theta}_2 + \eta_2\dot{\theta}_2 + K_1(\theta_2 - \theta_1) + K_2(\theta_2 - \theta_3) = 0$$

$$J_3\ddot{\theta}_3 + \eta_3\dot{\theta}_3 + K_2(\theta_3 - \theta_2) = 0$$

where θ's are the angular displacements of the rotors, η's the damping coefficients, K's the coefficients of torsional stiffness, and $T\cos\omega t$ the applied torque.

Using the voltage-force analogy, the electrical-analogue equations are given by

$$L_1\frac{di_1}{dt} + R_1 i_1 + \frac{1}{C_1}\int (i_1 - i_2)\,dt = v\cos\omega t$$

$$L_2\frac{di_2}{dt} + R_2 i_2 + \frac{1}{C_1}\int (i_2 - i_1)\,dt + \frac{1}{C_2}\int (i_2 - i_3)\,dt = 0$$

$$L_3\frac{di_3}{dt} + R_3 i_3 + \frac{1}{C_2}\int (i_3 - i_2)\,dt = 0$$

and the corresponding electrical-analogue circuit is shown in Fig. 7-9(b).

Supplementary Problems

10. Using the voltage-force analogy, draw the electrical analogue of the mechanical system as shown in Fig. 7-10 and from it deduce the differential equations of motion.

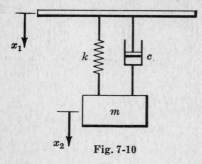

Fig. 7-10

Ans.

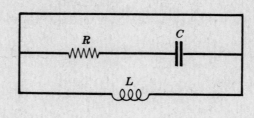

$$m\ddot{x}_2 + c(\dot{x}_2 - \dot{x}_1) + k(x_2 - x_1) = 0$$

$$c(\dot{x}_1 - \dot{x}_2) + k(x_1 - x_2) = 0$$

11. Show that the electrical circuits shown in Fig. 7-11 are analogous to each other as well as to the mechanical system.

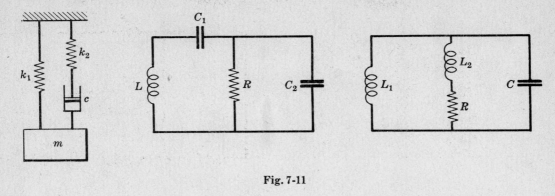

Fig. 7-11

12. Show that the electrical system as shown in Fig. 7-12(*b*) or (*c*) is equivalent to the mechanical system in Fig. 7-12(*a*).

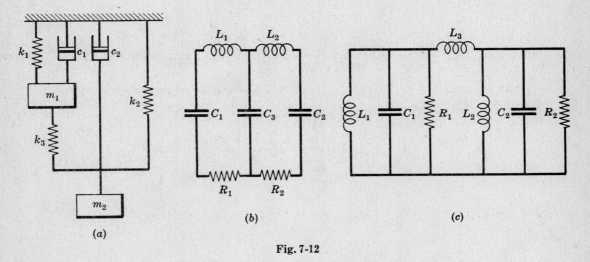

Fig. 7-12

13. Show that the electrical systems shown in Fig. 7-13(*b*) and Fig. 7-13(*c*) are equivalent to the mechanical system shown in Fig. 7-13(*a*).

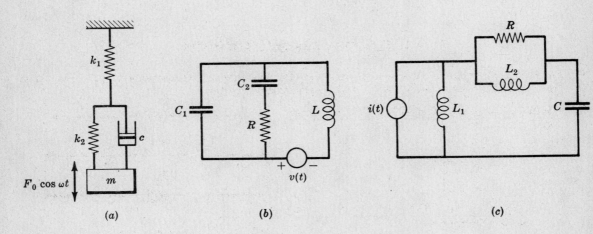

Fig. 7-13

14. Show that the electrical system shown in Fig. 7-14(b) is equivalent to the mechanical system in Fig. 7-14(a).

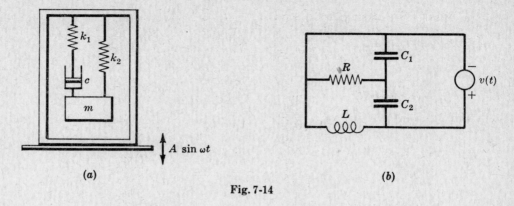

(a)   (b)

Fig. 7-14

15. Draw the electrical analogue for the branched-torsional system shown in Fig. 7-15(a).
    Ans. See Fig. 7-15(b).

(a)   (b)

Fig. 7-15

16. In viscoelasticity, the behavior of some materials is represented by the Voigt model as shown in Fig. 7-16. Using the current-force analogy, draw the electrical analogue for the representation of the Voigt model.
    Ans.

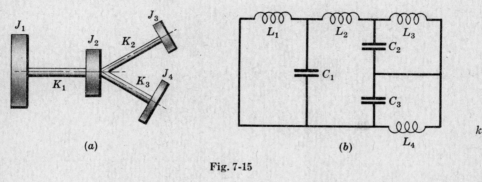

17. Fig. 7-17(a) shows the electrical circuit for the Maxwell model representation of the behavior of a solid. Draw the equivalent mechanical representation.
    Ans. See Fig. 7-17(b).

Fig. 7-16

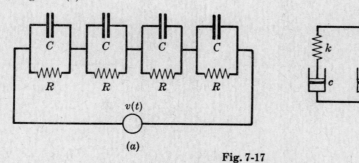

(a)   (b)

Fig. 7-17

18. The electrical circuit shown in Fig. 7-18 contains an inductor L, capacitor C, and resistor R which are connected in series with an alternating voltage source $v(t) = E_0 \sin \omega t$. Find an expression for $q(t)$. Compare the transient and steady response of the system shown in Fig. 7-19.

Ans. $q(t) = e^{-(R/2L)t} [A \sin \sqrt{1/LC - (R/2L)^2}t + B \cos \sqrt{1/LC - (R/2L)^2}t] + D \sin(\omega t - \phi)$

where $D = \dfrac{E_0/L}{\sqrt{[1/LC - \omega^2]^2 + (R\omega/L)^2}}$, $\phi = \tan^{-1}\dfrac{R\omega/L}{(1/LC - \omega^2)}$

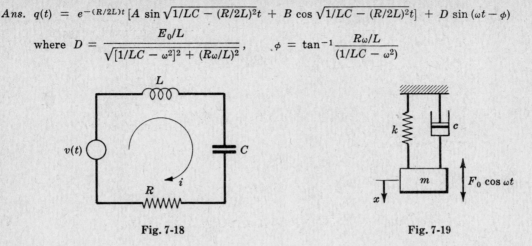

Fig. 7-18   Fig. 7-19

19. Use dimensional analysis to derive the following dimensionless quantities relating mechanical systems and their electrical analogues:

$$\omega\sqrt{m/k} = \omega_e\sqrt{LC}, \qquad c^2/km = R^2C/L, \qquad F/kx = vC/q$$

20. The mechanical elements for the system shown in Fig. 7-20 are $k = 100$ lb/in, $W = 10$ lb, and $c = 0.01$ lb-sec/in. If inductance $L = 0.01$ henry is used and the electric frequencies are 100 times the mechanical frequencies, what should be the values for the remaining elements in the equivalent electrical circuit?   Ans. $C = 0.26(10)^{-5}$ farad, $R = 0.39$ ohm

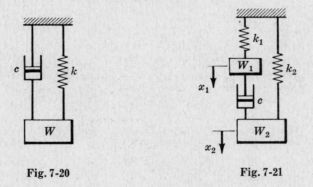

Fig. 7-20   Fig. 7-21

21. Determine the elements of the equivalent electrical circuit for the mechanical system shown in Fig. 7-21. Use the voltage-force analogy. The mechanical elements are $k_1 = 10$ lb/in, $k_2 = 100$ lb/in, $c = 0.1$ lb-sec/in, $W_1 = 10$, $W_2 = 100$ lb, $F_0 = 10$ lb, $\omega = 10$ rad/sec. Some of the electrical elements are $C_1 = 10^{-6}$, $C_2 = 2(10)^{-6}$ farad, $\omega_e = 100\omega$. $F_0 \cos \omega t$ acts on both W_1 and W_2.

Ans. $L_1 = 0.26$, $L_2 = 0.13$ henry

$\quad\ \ R = 100$ ohm

$\quad\ \ x_1 = v_{C_1}/v$, $x_2 = (v_{C_2}/v)(0.1)$

Chapter 8

Analog Computer

INTRODUCTION

The electronic analog computer is an important tool for the solution of differential equations, the simulation of physical problems, and the control of physical processes. It is widely used for vibration analysis. All information is handled in a continuous fashion by the computer with graphical output; and both a-c and d-c can be used for the computer. D-c has no phase shift, however, and tends to be more accurate than a-c.

The basic elements used in the computer are the *resistor*, the *capacitor*, and the *d-c amplifier* with high gain. The two variables in the computer are the voltage and the current.

BASIC OPERATIONS

The basic operations of an analog computer are sign inversion, summation, multiplication by a constant, and integration with respect to time.

Sign inversion of an input is accomplished by having equal values for both the input and output resistors connected to the amplifier as shown. The output voltage e_0 is equal to the negative of the input voltage e_i.

$$e_0 = -e_i$$

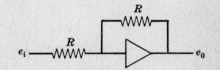

Sign inversion operation

Summation is simply adding various inputs to the same amplifier. Each input goes through its own input resistor while all input resistors are connected to the same amplifier as shown.

$$e_0 = -\left[\frac{R_0}{R_1}e_1 + \frac{R_0}{R_2}e_2 + \frac{R_0}{R_3}e_3\right]$$

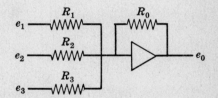

Summation operation

Integration of an input is obtained by replacing the output resistor with a capacitor and considering some initial condition. The units of the input resistor should be in *megohms* and the capacitor in *microfarads* in order to have time unit of *seconds* for the constant $1/RC$. The unit of the output voltage will then be *volts*. The output voltage e_0 can be shown to have the value

$$e_0 = -\frac{1}{RC}\int e_i\,dt + E_0$$

where E_0 is the initial condition.

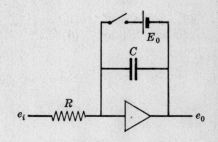

Integration operation

Multiplication of an input by a constant can either be obtained through the ratio of the output resistor to the input resistor or by the use of a *potentiometer*. Sometimes the combined use of both methods is the only way to obtain a fractional multiplication.

$$e_0 = -(R_0/R_i)e_i$$
$$e_0 = -ke_i$$
$$e_0 = -k(R_0/R_i)e_i$$

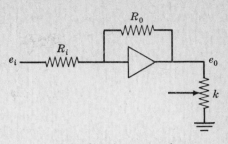

Multiplication operation

SCALING

Since most analog computers have voltage ranges ± 100 volts, the variables of any problem (represented here by the output voltages in the computer) should not exceed these values. For high accuracy, the output voltages of all amplifiers should lie between ± 10 and ± 100 volts. Therefore *magnitude scaling* is sometimes required to ensure that the output voltages lie within these given ranges.

In general, *magnitude scaling* is a process of adjusting the parameters of the differential equation, either multiplying or dividing the equation through by a certain constant, so that the output voltages will stay within the given ranges.

On the other hand, *time scaling* is a process either to speed up or slow down a particular physical problem on the computer so that observation or recording can be properly carried out.

During this process, the real time of the problem is changed to what is known as the "machine time". When a time change is made, it is important that the change must be made throughout the entire problem. To slow a problem down, let

$$t = T/a \quad \text{and in general} \quad \frac{d^n}{dt^n} = a^n \frac{d^n}{dT^n}$$

where t = real time of the problem, T = machine time, a = magnification factor. To speed up a problem, let

$$t = aT \quad \text{and in general} \quad \frac{d^n}{dt^n} = \frac{1}{a^n} \frac{d^n}{dT^n}$$

Solved Problems

1. Set up the computer circuit for the solution of the following differential equation with initial condition $x(0) = x_0$:

$$\frac{dx}{dt} + Ax = 0$$

Solve for the highest derivative and integrate the resulting equation once with respect to time:

$$\frac{dx}{dt} = -Ax \quad \text{or} \quad dx = -Ax\,dt$$

and

$$x = -\int Ax\,dt + x_0$$

Compare this equation with the equation for the integration operation of the computer,

$$e_0 = -\frac{1}{RC}\int e_i\,dt + E_0$$

We see that the two equations have the same form. Thus the output voltage e_0 can be used to represent displacement x, i.e.,

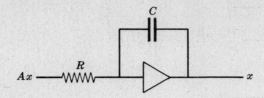

where $R = 1$ megohm, $C = 1$ microfarad, $e_i = Ax$, $e_0 = x$.

The input Ax to the amplifier can be obtained from the output x of the same amplifier by the use of a potentiometer as shown below. The initial condition can be imparted to the circuit because a capacitor is capable of retaining a voltage. This can be done by a battery-relay circuit. When time t is greater than zero, the relay is opened. When t is less than zero, the relay is closed and $x(0) = x_0$. The complete circuit is

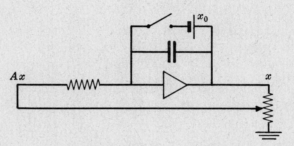

where $R = 1$ megohm and $C = 1$ microfarad as before.

2. Set up the computer circuit for the vibratory system as shown in Fig. 8-1(a). The initial conditions are $\dot{x}(0) = \dot{x}_0$ and $x(0) = x_0$.

The differential equation of motion for the system is

$$m\frac{d^2x}{dt^2} + c\frac{dx}{dt} + kx = 0$$

Rewrite the equation of motion for the computer as

$$\ddot{x} = -\frac{c}{m}\dot{x} - \frac{k}{m}x$$

Integrating the above equation once with respect to time, we have

$$\dot{x} = -\int \left[\frac{c}{m}\dot{x} + \frac{k}{m}x\right] dt + \dot{x}_0$$

and the corresponding computer circuit is

Fig. 8-1(a)

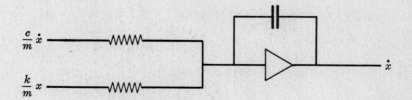

where $R = 1$ megohm, $C = 1$ microfarad.

The inputs $(c/m)\dot{x}$ and $(k/m)x$ in the above circuit are made available from the outputs of the amplifiers as shown in Fig. 8-1(b) below through potentiometers and $(1/RC)$ ratios. The initial conditions $\dot{x}(0)$ and $x(0)$ are set across amplifier 1 and 2 respectively. Amplifier 3 acts as a sign inverter.

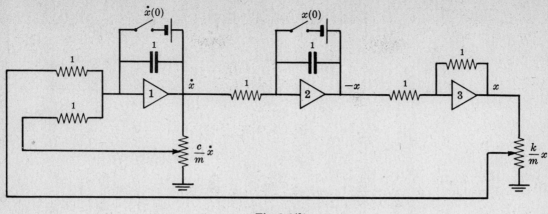

Fig. 8-1(b)

3. Set up the computer circuit for the system having forced vibration as shown in Fig. 8-2.

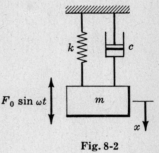

The differential equation of motion for the system is

$$m\frac{d^2x}{dt^2} + c\frac{dx}{dt} + kx = F_0 \sin \omega t$$

Rewrite the equation of motion for the computer as

$$\ddot{x} = -\frac{c}{m}\dot{x} - \frac{k}{m}x + (F_0/m) \sin \omega t$$

Integrate the above equation once with respect to time and obtain

$$\dot{x} = -\int \left[\frac{c}{m}\dot{x} + \frac{k}{m}x - (F_0/m) \sin \omega t\right] dt$$

Fig. 8-2

Compare this equation with the corresponding equation in Problem 2 for damped free vibration, and set up a similar computer circuit with an additional input $-(F_0/m) \sin \omega t$ to amplifier 1 as shown below. The forcing function is usually obtained externally from a low-frequency function generator.

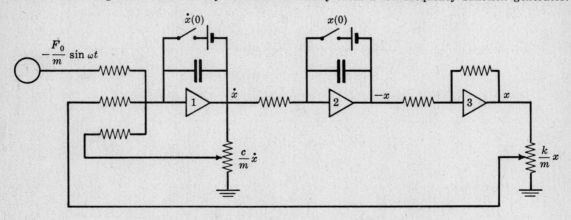

4. Set up the computer circuit to solve the following simultaneous differential equations:

$$\frac{d^2x}{dt^2} + A\frac{dx}{dt} + Bx - C\frac{d^2y}{dt^2} = 0$$

$$\frac{d^2y}{dt^2} + D\frac{dy}{dt} + Ey - F\frac{d^2x}{dt^2} = 0$$

Rewrite the given differential equations for the computer as

$$\ddot{x} = -A\dot{x} - Bx + C\ddot{y} \qquad \qquad \ddot{x} = -(A\dot{x} + Bx - C\ddot{y})$$
$$\ddot{y} = -D\dot{y} - Ey + F\ddot{x} \qquad \text{or} \qquad \ddot{y} = -(D\dot{y} + Ey - F\ddot{x})$$

which can be represented by two summation operations as

The inputs in the above summation amplifiers are obtained as follow:

(1) $\dot{x}$ obtained from integration of $\ddot{x}$, and $A\dot{x}$ from $\dot{x}$ by the use of a potentiometer or the ratio of the two resistors (R_0/R_i) or $(1/RC)$ ratio.

(2) x is obtained from integration of $\dot{x}$, and Bx from x by the use of a potentiometer or the ratio $(1/RC)$. $D\dot{y}$ and Ey can be obtained in the same way.

The circuit thus consists of two separate closed loops, each satisfying the corresponding equation. These two loops are then interconnected through amplifiers in terms of $-C\ddot{y}$ and $-F\ddot{x}$ as shown below.

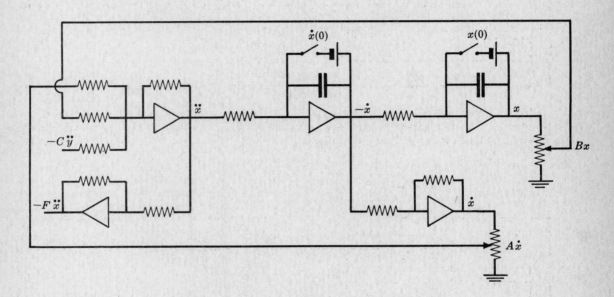

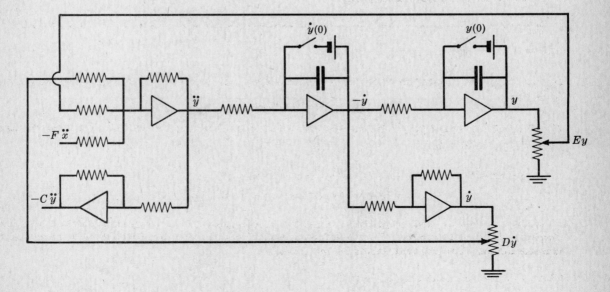

5. Set up the computer circuit to solve the following differential equation:

$$\ddot{x} + 10\dot{x} + 100x = 0$$

with initial conditions $\dot{x}(0) = 0$, $x(0) = 10$.

Rewrite the given equation for the computer as

$$\ddot{x} = -10\dot{x} - 100x$$

and integrate once with respect to time to obtain

$$\dot{x} = -\int (10\dot{x} + 100x)\, dt$$

The computer circuit for this equation is

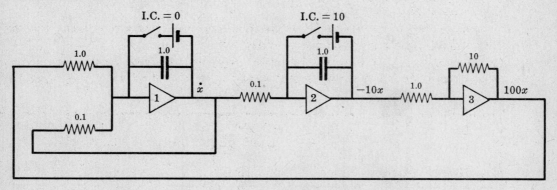

Now we have 100 volts as maximum output on amplifier 2 and 1000 volts as maximum output on amplifier 3. But the maximum output of an amplifier is limited to 100 volts. Hence this is modified by delaying the multiplication by 10 of the quantity $10x$ up to the input of amplifier 1 as shown below.

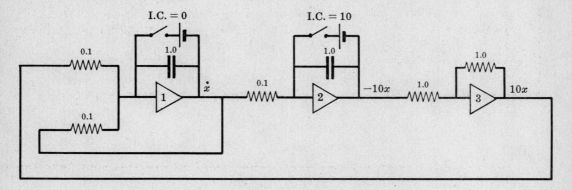

6. Study the free undamped vibration of a system represented by the following equation of motion

$$\ddot{x} + 10,000x = 0$$

by the use of the electronic analog computer. Initial conditions of the system are $\dot{x}(0) = 100$, $x(0) = 0$.

The natural frequency of this system is given by

$$\omega_n = \sqrt{k/m} = 100 \text{ rad/sec} \quad \text{or} \quad 100/2\pi = 16 \text{ cyc/sec}$$

Assume that we wish to observe the motion of the system with a recording device which has maximum speed of 10 cyc/sec. It is therefore necessary to slow down the problem with a time scale transformation. Let this time factor be 10. Thus

$$t = T/10$$

where T is the "machine time". Then

$$\frac{d}{dt} = \frac{d}{dT/10} = 10\frac{d}{dT}$$

and

$$\frac{d^2}{dt^2} = \frac{d}{dt}\left(10\frac{d}{dT}\right) = 10^2\left(\frac{d^2}{dT^2}\right) = 100\left[\frac{d^2}{dT^2}\right]$$

Substituting the expression for d^2/dt^2 into the equation of motion, we obtain

$$100\frac{d^2x}{dT^2} + 10{,}000x = 0 \qquad\text{or}\qquad \frac{d^2x}{dT^2} + 100x = 0$$

The initial conditions become

$$\frac{dx}{dT} = \frac{dx/dt}{10} \qquad\text{or}\qquad \frac{dx}{dT}(0) = 10, \qquad x(0) = 0,$$

Hence the natural frequency of the system in "machine time" T is

$$\omega_n = \sqrt{100} = 10 \text{ rad/sec} \quad\text{or}\quad 1.6 \text{ cyc/sec}$$

The computer circuit should then be drawn according to

$$\dot{x} = -\int (100x)\, dT + 10$$

and the circuit is given below.

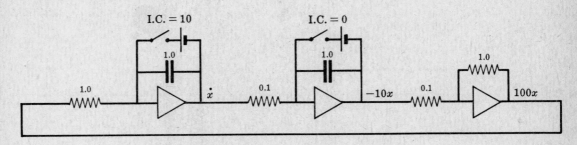

The results given by this circuit are in terms of "machine time" and should be converted to real time, i.e. 10 real seconds = 1 machine second.

7. Set up the computer circuit for the free damped vibration of a vibratory system described by the following differential equation of motion

$$\frac{d^2x}{dt^2} + 10\frac{dx}{dt} + 100x = 0$$

with initial conditions $\dot{x}(0) = 10$, $x(0) = 0$.

The damped natural frequency of the system is

$$\omega_d = \sqrt{k/m - (c/2m)^2} = \sqrt{100 - 25} = 8.5 \text{ rad/sec}$$

and for small amount of damping,

$$x = A\sin\omega_d t \qquad\text{and}\qquad \dot{x} = \omega_d A\cos\omega_d t$$

Now $\dot{x}(0) = 10 = 8.5A$; then $x_{\max} = A = 10/8.5 = 1.2$ in. and $\dot{x}_{\max} = \omega_d A = 1.2(8.5) = 10$ in/sec.

The scale factors from the maximum estimated values are

$$S_1 = \frac{100}{\dot{x}_{\max} = 10} = 10\frac{\text{volt}}{\text{in/sec}}, \qquad S_2 = \frac{100}{x_{\max} = 1.2} = 83.5 \text{ volt/in}$$

where the maximum allowable output voltage for all amplifiers is 100 volts.

Rewrite the equation of motion for the computer as

$$\ddot{x} = -10\dot{x} - 100x$$

Integrating the above equation once with respect to time,

$$\dot{x} = -\int (10\dot{x} + 100x)\, dt + 10 \tag{1}$$

and the computer circuit for equation (1) takes the following general form:

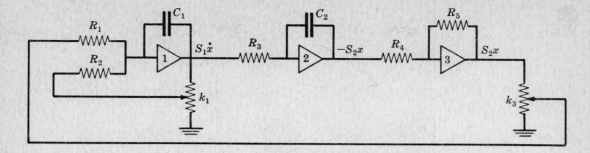

Let e_1, e_2, e_3 be the output voltages from amplifiers $1, 2, 3$ respectively. The scale factor for $\dot{x}$ is S_1, so that the voltage at the output of amplifier 1 becomes $S_1\dot{x}$. Similarly, the voltage at the output of amplifier 2 is $-S_2x$, and at amplifier 3 is S_2x.

Now the input and output voltages at amplifier 1 are related as follows:

$$e_1 = -\int \left[\frac{e_3 k_3}{R_1 C_1} + \frac{e_1 k_1}{R_2 C_1} \right] dt + E_1 = S_1\dot{x} \tag{2}$$

From equation (1),

$$S_1\dot{x} = -\int (100 S_2 x + 10 S_1\dot{x}) \, dt + 10 S_1 \tag{3}$$

Comparing (2) and (3),

$$\frac{e_3 k_3}{R_1 C_1} = \frac{(S_2 x)k_3}{R_1 C_1} = 100 S_2 x \tag{4}$$

$$\frac{e_1 k_1}{R_2 C_1} = \frac{k_1(S_1\dot{x})}{R_2 C_1} = 10 S_1\dot{x} \tag{5}$$

$$E_1 = 10 S_1 = 100 \text{ volts}$$

From equation (4), $R_1 C_1 = k_3/100$; and if $C_1 = 0.01$ microfarad and $R_1 C_1 = 0.005$, then $R_1 = 0.5$ megohm and $k_3 = 0.5$.

From equation (5), $R_2 C_1 = k_1/10$; and if $R_2 = 5$ megohms, then $R_2 C_1 = 0.05$ and $k_1 = 0.5$.

Now the input and output voltages at amplifier 2 are related as follows:

$$e_2 = -\int \frac{e_1}{R_3 C_2} dt = -S_2 x = -\int S_2 \dot{x} \, dt$$

from which

$$\frac{e_1}{R_3 C_2} = S_2 \dot{x} \tag{6}$$

Substituting $\dot{x} = e_1/S_1$ into (6), we have

$$\frac{e_1}{R_3 C_2} = \frac{S_2 e_1}{S_1} \quad \text{or} \quad R_3 C_2 = S_1/S_2 = 0.12$$

If $C_2 = 0.1$ microfarad, then $R_3 = 0.12/(10)^{-7} = 1.2$ megohms.

The potentiometer k_1 controls the amount of damping presented in the system. Thus the maximum displacement and velocity for various degrees of damping can be achieved by changing the potentiometer setting k_1.

The displacement x of the system is the voltage from the output of amplifier 3 divided by the scale factor S_2. The velocity of the system is obtained from the output of amplifier 1. These output voltages can be conveniently recorded by pen and ink recorder or oscilloscope.

8. Set up the computer circuit to solve the following nonlinear differential equation:

$$\frac{d^2 x}{dt^2} + \frac{dx}{dt} + 0.1 x^3 = 0$$

Rewrite the given equation for the computer as

$$\ddot{x} = -\dot{x} - 0.1 x^3$$

Integrating the above equation once with respect to time,

$$\dot{x} = -\int (\dot{x} + 0.1x^3)\, dt + \dot{x}(0)$$

and the computer circuit is

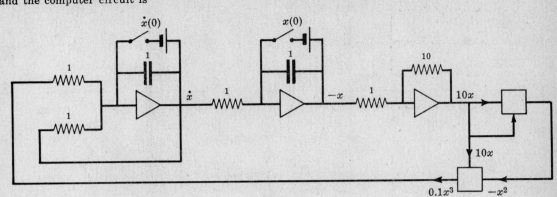

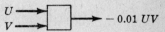

where the function of the multiplier, which is usually built into the computer, is to form the product of two inputs with coefficient (-0.01) as shown in the adjacent figure.

$$U \longrightarrow \boxed{} \longrightarrow -0.01\,UV$$
$$V \longrightarrow$$

Supplementary Problems

9. Write the equation for the computer circuit shown in Fig. 8-3.

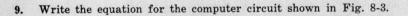

Fig. 8-3

Ans. $y = 2x_1 - 5x_2$

10. Write the equation for the computer circuit shown in Fig. 8-4.

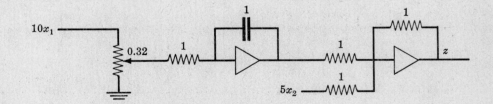

Fig. 8-4

Ans. $z = \int 3.2x_1\, dt - 0.2x_2$

11. Write the equation for the computer circuit shown in Fig. 8-5.

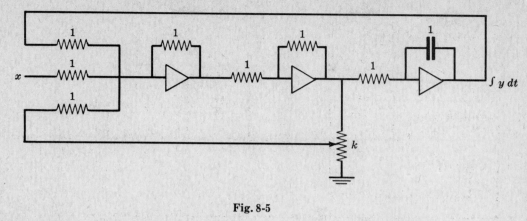

Fig. 8-5

Ans. $y = -dx/dt$ as $k \to 1$.

12. Write the equation for the computer circuit shown in Fig. 8-6.

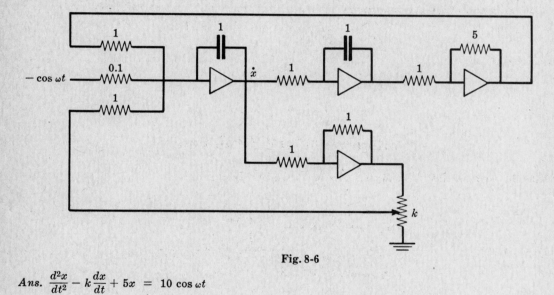

Fig. 8-6

Ans. $\dfrac{d^2x}{dt^2} - k\dfrac{dx}{dt} + 5x = 10 \cos \omega t$

13. The computer circuit as shown in Fig. 8-7 has a broken wire, find the resulting equation solved by this circuit.

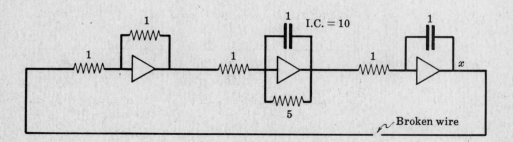

Fig. 8-7

Ans. $x = -50\, e^{-(t/5)} + 50$

14. Find the equation solved by the computer circuit as shown in Fig. 8-8. All resistors have equal value of 1 megohm and all capacitors have equal value of 1 microfarad.

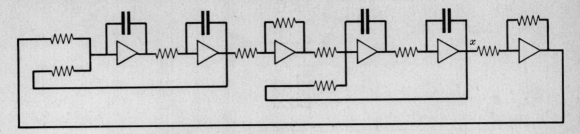

Fig. 8-8

Ans. $\dfrac{d^4x}{dt^4} = 2\dfrac{d^2x}{dt^2}$

15. Set up the computer circuit for the following differential equation:

$$\frac{d^2x}{dt^2} + \frac{dx}{dt} + x = A + Bt$$

Ans.

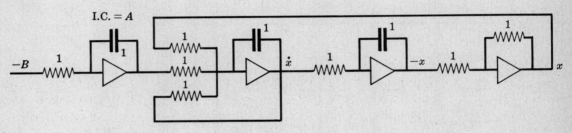

16. Set up the computer circuit to solve the following differential equation for the simple pendulum:

$$\frac{d^2\theta}{dt^2} + \sin\theta = 0$$

Ans.

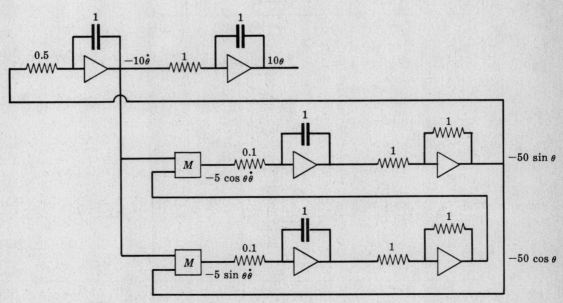

Chapter 9

Vibration and Sound

INTRODUCTION

Sound may be defined as audible vibration of air. Sound generators, like musical instruments, set into motion one or more of the normal modes of vibration of air. These travel as *longitudinal sound waves* in all directions, reflected, scattered, or absorbed.

The *pitch* is determined by the frequency of vibration, and the *intensity* of sound is governed by the rate at which energy is being transmitted along the wave. In short, the analysis of sound is an analysis of vibration.

THE VIBRATING STRING

The study of the vibrating string plays an important role in the analysis of sound. The vibrating string has its mass spread uniformly along its length and is the simplest case of a system with an infinite number of frequencies of vibration. In comparison with vibration of bars, where stiffness is the most important property, vibration of flexible strings is an extreme case. Here, tension force comes first.

VIBRATION OF STRINGS

The general differential equation of motion is

$$\frac{\partial^2 y}{\partial t^2} = a^2 \frac{\partial^2 y}{\partial x^2}$$

where y = deflection of the string,

x = coordinate along the longitudinal axis of the string,

$a^2 = T/\rho$,

T = tension in the string,

ρ = mass per unit length of the string.

The general solution is given by

$$y(x, t) = \sum_{i=1,2,\ldots}^{\infty} \left(A_i \sin \frac{p_i}{a} x + B_i \cos \frac{p_i}{a} x \right)(C_i \sin p_i t + D_i \cos p_i t)$$

where A_i and B_i are constants to be evaluated by boundary conditions, C_i and D_i are constants to be evaluated by initial conditions, and p_i the natural frequencies of the system.

The general solution can also be expressed as

$$y(x, t) = f_1(x - at) + f_2(x + at)$$

where f_1 and f_2 are arbitrary functions. The first part $f_1(x - at)$ represents a wave of arbitrary shape traveling in the positive x-direction with velocity a, while $f_2(x + at)$ represents a wave of arbitrary shape traveling in the negative x-direction with velocity a.

Solved Problems

1. Investigate the transverse vibration of a stretched string of length L in a plane, assuming the tension in the string remains constant.

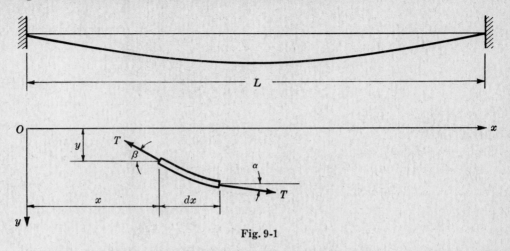

Fig. 9-1

In general, it can be assumed that the flexible string offers no resistance to bending nor to shear, and its tension is constant for small displacements.

The differential equation of motion for an infinitesimal element of the string as shown in Fig. 9-1 can be written as

$$\Sigma F = m\ddot{y}$$

or

$$(\rho\,\Delta x)\frac{\partial^2 y}{\partial t^2} = -T\sin\beta + T\sin\alpha$$

where ρ is the mass per unit length of the string and T is the tension in the string. Partial derivatives are used because there are two independent variables, x and t.

But $\left(\dfrac{\partial y}{\partial x}\right)_{x=x} = \tan\beta$, $\left(\dfrac{\partial y}{\partial x}\right)_{x=x+\Delta x} = \tan\alpha$. And for small displacements, $\sin\beta \doteq \tan\beta$ and $\sin\alpha \doteq \tan\alpha$. Hence

$$(\rho\,\Delta x)\frac{\partial^2 y}{\partial t^2} = -T\left(\frac{\partial y}{\partial x}\right)_{x=x} + T\left(\frac{\partial y}{\partial x}\right)_{x=x+\Delta x}$$

or

$$\frac{\partial^2 y}{\partial t^2} = \frac{(T/\rho)[(\partial y/\partial x)_{x+\Delta x} - (\partial y/\partial x)_x]}{\Delta x}$$

$$\frac{\partial^2 y}{\partial t^2} = \frac{T}{\rho}\frac{\partial^2 y}{\partial x^2}$$

which is generally known as the *wave equation*, and is usually written in the form

$$\frac{\partial^2 y}{\partial t^2} = a^2\frac{\partial^2 y}{\partial x^2}$$

replacing T/ρ by the constant a^2.

The solution of this wave equation can be obtained by the "separate variables" method. Since y is a function of x and t, it can be represented as

$$y(x, t) = X(x)\cdot T(t)$$

Then

$$\frac{\partial^2 y}{\partial x^2} = T\frac{d^2 X}{dx^2}, \qquad \frac{\partial^2 y}{\partial t^2} = X\frac{d^2 T}{dt^2}$$

and the wave equation becomes

$$X\frac{d^2 T}{dt^2} = a^2 T\frac{d^2 X}{dx^2}$$

Separating the variables,

$$\frac{d^2 T/dt^2}{T} = a^2\frac{d^2 X/dx^2}{X}$$

As X and T are independent of each other, the above expression must equal a certain constant. Let this constant be $-p^2$. This then leads to two ordinary differential equations,

$$\frac{d^2T}{dt^2} + p^2T = 0 \qquad \text{and} \qquad \frac{d^2X}{dx^2} + \frac{p^2}{a^2}X = 0$$

and the solution is

$$y(x, t) = \left(A \sin\frac{px}{a} + B \cos\frac{px}{a}\right)(C \sin pt + D \cos pt)$$

With both ends of the string fixed, the boundary conditions are

$$y(0, t) = 0 \tag{1}$$
$$y(L, t) = 0 \tag{2}$$

From condition (1),

$$0 = B(C \sin pt + D \cos pt) \qquad \text{or} \qquad B = 0$$

and from condition (2),

$$0 = (A \sin pL/a)(C \sin pt + D \cos pt)$$

Because A cannot equal zero all the time, $\sin pL/a$ must equal zero. Therefore the frequency equation is

$$\sin pL/a = 0$$

and the natural frequencies of the string are given by

$$p_i = i\pi a/L \qquad \text{where} \quad i = 1, 2, 3, \ldots$$

It is clear that there are an infinite number of natural frequencies; this is in agreement with the fact that all continuous systems are composed of an infinite number of mass particles.

For this particular configuration of the vibrating string, i.e. with both ends fixed, the normal function $X(x)$ is therefore given by

$$X_i(x) = \sin i\pi x/L$$

and

$$y(x, t) = (A \sin px/a)(C \sin pt + D \cos pt)$$

In general, the expression for the vibrating string is given by

$$y(x, t) = \sum_{i=1,2,\ldots}^{\infty} \left(\sin\frac{i\pi x}{L}\right)(C_i \sin p_i t + D_i \cos p_i t)$$

in which the principle of superposition is used to represent the many natural modes of vibration of the string.

2. A uniform string of length L and high initial tension is statically displaced h units from the center and released as shown in Fig. 9-2. Find its subsequent displacements.

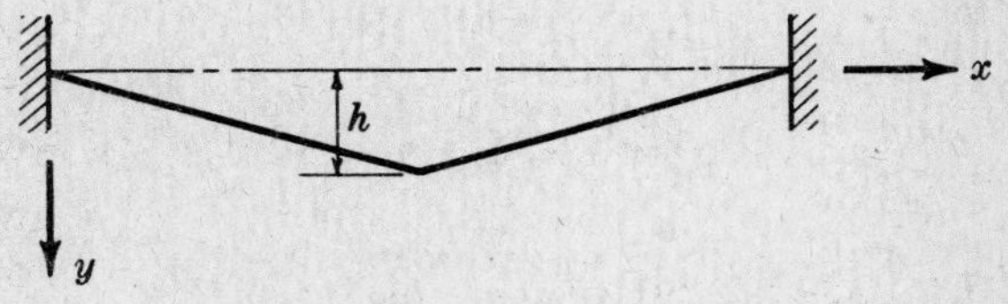

Fig. 9-2

The general expression for the free vibration of a string fixed at both ends is

$$y(x, t) = \sum_{i=1,2,\ldots}^{\infty} \left(\sin\frac{i\pi x}{L}\right)(A_i \sin p_i t + B_i \cos p_i t)$$

The initial conditions are

$$\dot{y}(x, 0) = 0, \qquad y(x, 0) = \begin{cases} 2hx/L, & 0 \le x \le L/2 \\ 2h(1 - x/L), & L/2 \le x \le L \end{cases}$$

which are equal to

$$y(x, 0) = \sum_{i=1,2,\ldots}^{\infty} B_i \sin\frac{i\pi x}{L}, \qquad \dot{y}(x, 0) = \sum_{i=1,2,\ldots}^{\infty} A_i p_i \sin\frac{i\pi x}{L}$$

Hence $A_i = 0$, and

$$\sum_{i=1,2,\ldots}^{\infty} B_i \sin\frac{i\pi x}{L} = \begin{cases} 2hx/L, & 0 \leq x \leq L/2 \\ 2h(1 - x/L), & L/2 \leq x \leq L \end{cases}$$

Multiplying both sides of the above equation by $\sin i\pi x/L$ and integrating between the limits $x = 0$ and $x = L$, we obtain

$$\int_0^L B_i \sin\frac{i\pi x}{L} \sin\frac{i\pi x}{L}\, dx = \int_0^{L/2} \frac{2hx}{L} \sin\frac{i\pi x}{L}\, dx + \int_{L/2}^L 2h\left(1 - \frac{x}{L}\right) \sin\frac{i\pi x}{L}\, dx$$

or

$$LB_i/2 = \frac{2h}{L}\left[\int_0^{L/2} x \sin\frac{i\pi x}{L}\, dx + \int_{L/2}^L (L - x) \sin\frac{i\pi x}{L}\, dx\right]$$

and thus

$$B_i = (-1)^{(i-1)/2} \frac{8h}{i^2\pi^2} \qquad \text{where } i = 1, 3, \ldots$$

The natural frequencies are given by

$$p_i = \frac{i\pi a}{L} \qquad \text{or} \qquad p_1 = \frac{\pi a}{L},\ p_2 = \frac{3\pi a}{L},\ p_3 = \frac{5\pi a}{L},\ \ldots$$

Therefore the expression for the displacement of the string is

$$y(x,t) = \frac{8h}{\pi^2}\left(\sin\frac{\pi x}{L}\cos\frac{\pi a}{L}t - \frac{1}{9}\sin\frac{3\pi x}{L}\cos\frac{3\pi a}{L}t + \frac{1}{25}\sin\frac{5\pi x}{L}\cos\frac{5\pi a}{L}t - \cdots\right)$$

where $a = \sqrt{T/\rho}$ and ρ is the mass per unit length of the string.

3. Find the velocity of propagation of waves along a steel wire. The modulus of elasticity for the steel is $30(10)^6$ lb/in², and the steel weighs 0.283 lb/in³.

The velocity of propagation of waves along a steel wire is given by $a = \sqrt{E/\rho}$, where E is the modulus of elasticity and ρ the density of the steel. Hence

$$a = \sqrt{\frac{Eg}{W}} = \sqrt{\frac{30(10)^6(32.2)(12)}{0.283}} = 16{,}900 \text{ ft/sec}$$

4. Derive an expression for the potential energy of a uniform vibrating string of length L, considering that the tension is not constant.

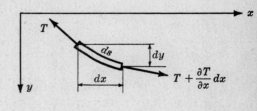

Consider the element ds of the string as shown in Fig. 9-3. In the equilibrium state this element ds is simply equal to dx. When it is displaced as shown, the amount of potential energy stored in this element ds of the string is

$$d(\text{P.E.}) = (ds - dx)\left[\frac{2T + (\partial T/\partial x)\, dx}{2}\right]$$

Fig. 9-3

which is equal to zero when the string is unstretched. Now

$$(ds)^2 = (dx)^2 + (dy)^2 \qquad \text{or} \qquad ds = \sqrt{1 + (dy/dx)^2}\, dx$$

Expanding by the binomial theorem, we obtain

$$ds = [1 + \tfrac{1}{2}(dy/dx)^2 + \cdots]\, dx$$

As deflection y is a function of x and t, ds can be expressed in partial derivatives as

$$ds = [1 + \tfrac{1}{2}(\partial y/\partial x)^2 + \cdots]\, dx$$

Hence

$$d(\text{P.E.}) = \frac{1}{2}T\left(\frac{\partial y}{\partial x}\right)^2 dx + \frac{1}{4}\frac{\partial T}{\partial x}\left(\frac{\partial y}{\partial x}\right)^2 (dx)^2 = \frac{1}{2}T\left(\frac{\partial y}{\partial x}\right)^2 dx$$

and the potential energy in the string is $\dfrac{1}{2}\displaystyle\int_0^L T\left(\dfrac{\partial y}{\partial x}\right)^2 dx$.

5. A uniform string of length L is fixed at both ends. If the tension in the string is constant, determine the normal coordinates of the system.

The expression for the potential energy of a vibrating string under constant tension T is

$$\text{P.E.} = \tfrac{1}{2}T \int_0^L (\partial y/\partial x)^2\, dx$$

and the kinetic energy expression is

$$\text{K.E.} = \tfrac{1}{2}\rho \int_0^L (\partial y/\partial t)^2\, dx$$

where ρ is the mass per unit length of the string.

The motion of the string fixed at both ends is

$$y(x,t) = \sum_{i=1,2,\ldots}^{\infty} \left(\sin\frac{i\pi x}{L} \right)\left(A_i \sin\frac{i\pi a}{L}t + B_i \cos\frac{i\pi a}{L}t \right)$$

where $a^2 = T/\rho$.

Rewrite the above equation as $\quad y(x,t) = \sum_{i=1,2,\ldots}^{\infty} \phi_i \sin\frac{i\pi x}{L}$. Now

$$\frac{\partial y}{\partial x} = \frac{\pi}{L}\sum_{i=1,2,\ldots}^{\infty} i\phi_i \cos\frac{i\pi x}{L}, \qquad \frac{\partial y}{\partial t} = \sum_{i=1,2,\ldots}^{\infty} \dot{\phi}_i \sin\frac{i\pi x}{L}$$

Hence

$$\text{P.E.} = \frac{\pi^2 T}{2L^2}\sum_{i=1,2,\ldots}^{\infty}\sum_{i=1,2,\ldots}^{\infty}\left(i^2\phi_i^2 \int_0^L \cos\frac{i\pi x}{L}\cos\frac{i\pi x}{L}\, dx \right) = \frac{\pi^2 T}{4L}\sum_{i=1,2,\ldots}^{\infty}\sum_{i=1,2,\ldots}^{\infty} i^2\phi_i^2$$

$$\text{K.E.} = \frac{\rho}{2}\sum_{i=1,2,\ldots}^{\infty}\sum_{i=1,2,\ldots}^{\infty}\left(\dot{\phi}_i^2 \int_0^L \sin\frac{i\pi x}{L}\sin\frac{i\pi x}{L}\, dx \right) = \frac{\rho L}{4}\sum_{i=1,2,\ldots}^{\infty}\sum_{i=1,2,\ldots}^{\infty} \dot{\phi}_i^2$$

The Lagrange equation for ϕ_i is

$$\frac{d}{dt}\frac{\partial(\text{K.E.})}{\partial \dot{\phi}_i} - \frac{\partial(\text{K.E.})}{\partial \phi_i} + \frac{\partial(\text{P.E.})}{\partial \phi_i} = 0$$

or

$$\ddot{\phi}_i + (i^2 T\pi^2/\rho L^2)\phi_i = 0$$

This is a second-order linear differential equation with constant coefficients and one dependent variable, and so can be solved independently. Clearly, the function ϕ_i satisfies all requirements for normal coordinates. Hence the normal coordinates of the system are

$$\left(A_i \sin\frac{i\pi a}{L}t + B_i \cos\frac{i\pi a}{L}t \right)$$

6. A tightly stretched string is initially straight when a hammer blow results in an initial velocity to that part of the string of length s struck by the hammer. The rest of the string is initially undisturbed. If the length of the string is L and the tension T in the string remains constant, find the expression $y(x,t)$ for the subsequent motion of the string.

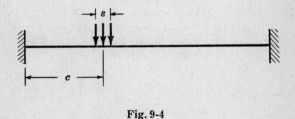

Fig. 9-4

The general expression for free vibration of uniform strings with both ends fixed is

$$y(x,t) = \sum_{i=1,2,\ldots}^{\infty} \left(\sin\frac{i\pi x}{L} \right)\left(A_i \cos\frac{i\pi a}{L}t + B_i \sin\frac{i\pi a}{L}t \right)$$

where $a^2 = T/\rho$, and ρ is the mass of the string per unit length. A_i and B_i are arbitrary constants to be evaluated by the initial conditions, and T is the tension in the string. Substituting $t=0$ in the above expression and in the derivative of the above expression with respect to t, we obtain

$$y(x,0) = \sum_{i=1,2,\ldots}^{\infty} A_i \sin\frac{i\pi x}{L}, \qquad \dot{y}(x,0) = \sum_{i=1,2,\ldots}^{\infty} \frac{i\pi a}{L}B_i \sin\frac{i\pi x}{L}$$

Now the initial conditions are

$$y(x,0) = 0$$
$$\dot{y}(x,0) = V_0 \quad \text{from } x = c - s/2 \text{ to } x = c + s/2$$

where V_0 is the initial velocity. From the first initial condition, we have $A_i = 0$; and from the second initial condition, we obtain

$$\sum_{i=1,2,\ldots}^{\infty} \frac{i\pi a}{L} B_i \sin\frac{i\pi x}{L} \;=\; V_0$$

Multiplying both sides of the above expression by $\sin i\pi x/L$ and integrating from $x = 0$ to $x = L$, we have

$$\int_0^L \frac{i\pi a}{L} B_i \sin\frac{i\pi x}{L} \sin\frac{i\pi x}{L}\, dx \;=\; \int_0^L V_0 \sin\frac{i\pi x}{L}\, dx$$

Since $\displaystyle\int_0^L \sin\frac{i\pi x}{L} \sin\frac{i\pi x}{L}\, dx = \frac{L}{2}$, the above expression becomes

$$B_i \;=\; \frac{2}{i\pi a} \int_{c-s/2}^{c+s/2} V_0 \sin\frac{i\pi x}{L}\, dx \;=\; \frac{2V_0}{i\pi a}\left\{\frac{-L}{i\pi}\cos\frac{i\pi x}{L}\right]_{c-s/2}^{c+s/2}\right\}$$

$$=\; \frac{-2V_0 L}{i^2\pi^2 a}\left(-2\sin\frac{i\pi c}{L}\sin\frac{i\pi s}{2L}\right) \;=\; \frac{4V_0 L}{i^2\pi^2 a}\sin\frac{i\pi c}{L}\sin\frac{i\pi s}{2L}$$

Hence

$$y(x,t) \;=\; \frac{4V_0 L}{\pi^2 a}\sum_{i=1,2,\ldots}^{\infty} \frac{1}{i^2}\sin\frac{i\pi c}{L}\sin\frac{i\pi s}{L}\sin\frac{i\pi x}{L}\sin\frac{i\pi a}{L}t$$

7. A uniform string of length L is fixed at both ends, and a damping force proportional to the velocity of the string acts upon all points of the string. Find the transient vibration of the string.

The free undamped vibration of a string fixed at both ends is given by

$$y(x,t) \;=\; \sum_{i=1,2,\ldots}^{\infty} \left(\sin\frac{i\pi x}{L}\right)(A_i \sin p_i t + B_i \cos p_i t) \tag{1}$$

where $p_i = i\pi a/L$ and $a^2 = T/\rho$. Equation (1) can be rewritten as

$$y(x,t) \;=\; \sum_{i=1,2,\ldots}^{\infty} \phi_i \sin\frac{i\pi x}{L} \tag{2}$$

The damping force on an element dx of the string may be expressed as

$$-c\frac{\partial y}{\partial t}\, dx$$

where c is the damping coefficient per unit length of the string. The negative sign indicates that the force is opposite to the velocity. Then the wave equation of the string becomes

$$\rho\frac{\partial^2 y}{\partial t^2} + c\frac{\partial y}{\partial t} \;=\; T\frac{\partial^2 y}{\partial x^2} \tag{3}$$

where ρ is the mass per unit length of the string and T is the tension in the string.

Substituting equation (2) into equation (3), we obtain

$$\sum_{i=1,2,\ldots}^{\infty}\left[\left(\ddot\phi_i + \frac{c}{\rho}\dot\phi_i + \frac{i^2\pi^2 T}{L^2\rho}\phi_i\right)\sin\frac{i\pi x}{L}\right] \;=\; 0$$

and solving for ϕ_i,

$$\phi_i \;=\; e^{-(c/2\rho)t}(A_i' \sin p_i' t + B_i' \cos p_i' t)$$

where A_i' and B_i' are constants of integration to be evaluated by the initial conditions. The damped natural frequencies are

$$p_i' \;=\; \sqrt{\frac{i^2\pi^2 a^2}{L^2} - \frac{c^2}{4\rho^2}}$$

The transient vibration of the string is therefore given by

$$y(x,t) \;=\; \sum_{i=1,2,\ldots}^{\infty} \sin\frac{i\pi x}{L}\, e^{-(c/2\rho)t}(A_i' \sin p_i' t + B_i' \cos p_i' t)$$

which diminishes with time due to the decay factor $e^{-(c/2\rho)t}$.

8. A taut uniform string of length L is fixed at both ends and is acted upon by a uniformly distributed sinusoidal force $F_0 \cos \omega t$ as shown in Fig. 9-5. Determine the steady state vibration of the string.

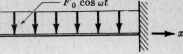

Fig. 9-5

The differential equation of motion for free vibration of uniform strings is

$$\rho \frac{\partial^2 y}{\partial t^2} - T \frac{\partial^2 y}{\partial x^2} = 0$$

where ρ is the mass per unit length of the string and T is the constant tension in the string.

The equation of motion for forced vibration of uniform strings is then

$$\frac{\partial^2 y}{\partial t^2} = a^2 \frac{\partial^2 y}{\partial x^2} + \frac{F_0}{\rho} \cos \omega t$$

where $a^2 = T/\rho$.

Let the general expression for steady state vibration be

$$y(x, t) = Y(x) \cos \omega t$$

and thus
$$\frac{\partial^2 y}{\partial t^2} = -\omega^2 Y(x) \cos \omega t, \qquad \frac{\partial^2 y}{\partial x^2} = \frac{d^2 Y}{dx^2} \cos \omega t$$

The equation of motion for forced vibration then becomes

$$\frac{d^2 Y}{dx^2} + \frac{\omega^2}{a^2} Y = -\frac{F_0}{\rho a^2}$$

The complete solution of this equation is

$$Y(x) = A \cos \frac{\omega}{a} x + B \sin \frac{\omega}{a} x - \frac{F_0}{\rho \omega^2}$$

where A, B are arbitrary constants to be evaluated by the boundary conditions.

Here the deflections of the string at both ends are equal to zero, i.e. the boundary conditions are given by

$$Y(0) = 0 \qquad \text{and} \qquad Y(L) = 0$$

or
$$A - F_0/\rho\omega^2 = 0 \qquad \text{and} \qquad A \cos \frac{\omega}{a} L + B \sin \frac{\omega}{a} L - F_0/\rho\omega^2 = 0$$

Thus
$$A = F_0/\rho\omega^2 \qquad \text{and} \qquad B = \frac{F_0 (1 - \cos \omega L/a)}{\rho\omega^2 \sin \omega L/a}$$

Since $\tan \frac{1}{2} x = \dfrac{1 - \cos x}{\sin x}$, we have $B = \dfrac{F_0}{\rho\omega^2} \tan \dfrac{\omega L}{2a}$. Hence the forced vibration of the string is

$$y(x, t) = \frac{F_0}{\rho\omega^2} \left(\cos \frac{\omega}{a} x + \tan \frac{\omega L}{2a} \sin \frac{\omega}{a} x - 1 \right) \cos \omega t$$

9. Show that it is possible to express the solution for the wave equation

$$\frac{\partial^2 y}{\partial t^2} = \frac{T}{\rho} \frac{\partial^2 y}{\partial x^2}$$

by the following traveling waves equation

$$y = f_1(x - at) + f_2(x + at)$$

where f_1 and f_2 are arbitrary functions and $a^2 = T/\rho$.

The traveling wave equation has two parts: the first part $f_1(x - at)$ represents a wave of arbitrary shape traveling in the positive x-direction with velocity a, and the second part $f_2(x + at)$ represents a wave of arbitrary shape traveling in the negative x-direction with velocity a.

Let $(x - at) = g$ and $(x + at) = h$. Then

$$y = f_1(g) + f_2(h)$$

Now
$$\frac{\partial y}{\partial t} = \frac{df_1}{dg} \frac{\partial g}{\partial t} + \frac{df_2}{dh} \frac{\partial h}{\partial t} = a \left(\frac{df_2}{dh} - \frac{df_1}{dg} \right)$$

and so
$$\frac{\partial^2 y}{\partial t^2} = a^2 \left(\frac{d^2 f_1}{dg^2} + \frac{d^2 f_2}{dh^2} \right)$$

Similarly,

$$\frac{\partial^2 y}{\partial x^2} = \frac{d^2 f_1}{dg^2} + \frac{d^2 f_2}{dh^2}$$

Substituting these expressions into the wave equation yields

$$a^2 \left(\frac{d^2 f_1}{dg^2} + \frac{d^2 f_2}{dh^2} \right) = \frac{T}{\rho} \left(\frac{d^2 f_1}{dg^2} + \frac{d^2 f_2}{dh^2} \right)$$

Thus the traveling wave solution is a possible solution to the wave equation if $a^2 = T/\rho$.

10. A uniform string of length L and fixed at both ends is released at zero initial velocity from the displaced position as shown in Fig. 9-6(a). By means of the wave-travel method, sketch the shape of the string at time intervals of $L/8a$ for one half cycle of the motion of the string.

As shown in the following figures, solid lines represent the actual shape of the string, and dotted lines the traveling waves in opposite directions. At any time under consideration, the shape of the string is the resultant configuration of the traveling waves.

The shape of the traveling wave is determined by the initial displacement of the string. Here, as shown in Fig. 9-6(b), it is the shape of a triangle of height $h/2$. The initial configuration of the string is made up of two identical traveling waves on top of each other but traveling in opposite directions.

At the end of the first time interval $L/8a$ (where a is the velocity of the traveling waves), the traveling waves have moved a distance of $L/8$, one to the right and the other to the left. The configuration of the string at this moment is the resultant of the two traveling waves and is shown in Fig. 9-6(c).

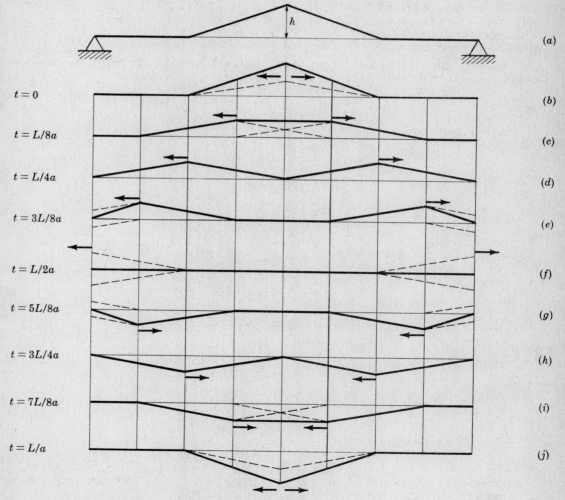

Fig. 9-6

When the traveling waves reach the fixed ends of the string as shown in Fig. 9-6(e), they reflect and change sign. Then the traveling waves just keep moving as shown in the following figures. This procedure goes on for the rest of the cycle. At the end of the cycle, i.e. when $t = 2L/a$, the cycle repeats itself.

In the absence of damping, this procedure will continue indefinitely and the amplitudes as well as the shapes of the traveling waves will remain the same.

The traveling wave representation of the vibration of a string, however, becomes very involved if the initial velocity is not equal to zero.

11. Investigate the wave motion and energy transmission of a compound string. **Refer** to Fig. 9-7.

To account for the change of phase and mass density of the string, we use the complex exponential rather than the sine or cosine in the following forms:

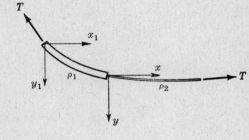

$$y_1(x, t) = A_1 e^{i\omega(t - x/a_1)} + A_2 e^{i\omega(t + x/a_1)} \quad (1)$$

$$y_2(x, t) = B e^{i\omega(t - x/a_2)} \quad\quad\quad (2)$$

where $a_1^2 = T/\rho_1$, $a_2^2 = T/\rho_2$; T is the tension in the string and ρ is the mass per unit length of the string. In the right hand side of equation (1), the first term refers to the incident wave traveling in the positive x-direction with velocity a_1 while the second term refers to the reflected wave traveling in the negative x-direction with velocity a_1. $y_2(x, t)$ represents the transmitted wave traveling in the positive x-direction with velocity a_2.

Fig. 9-7

At the junction of the string, the displacement as well as the force given by the two expressions y_1 and y_2 should be the same, i.e.,

$$(y_1)_{x=0} = (y_2)_{x=0} \quad\quad\quad (3)$$

$$T(\partial y_1/\partial x)_{x=0} = T(\partial y_2/\partial x)_{x=0} \quad\quad\quad (4)$$

Substituting equations (1) and (2) into (3) and (4) respectively, we obtain

$$A_1 e^{i\omega t} + A_2 e^{i\omega t} = B e^{i\omega t}$$

or

$$A_1 + A_2 = B \quad\quad\quad (5)$$

and

$$-i\frac{\omega}{a_1}A_1 e^{i\omega t} + i\frac{\omega}{a_1}A_2 e^{i\omega t} = -i\frac{\omega}{a_2}B e^{i\omega t}$$

or

$$\frac{A_1 - A_2}{a_1} = \frac{B}{a_2} \quad\quad\quad (6)$$

Solving equations (5) and (6) simultaneously yields

$$\frac{A_2}{A_1} = \frac{a_1 - a_2}{a_1 + a_2} \quad \text{and} \quad \frac{B}{A_1} = \frac{2a_1}{a_1 + a_2}$$

Putting $a_1 = \sqrt{T/\rho_1}$ and $a_2 = \sqrt{T/\rho_2}$, the above expressions become

$$\frac{A_2}{A_1} = \frac{\sqrt{\rho_1} - \sqrt{\rho_2}}{\sqrt{\rho_1} + \sqrt{\rho_2}} \quad\quad\quad (7)$$

$$\frac{B}{A_1} = \frac{2\sqrt{\rho_1}}{\sqrt{\rho_1} + \sqrt{\rho_2}} \quad\quad\quad (8)$$

If ρ_2 is very large (for fixed end, $\rho = \infty$), equation (7) gives

$$A_2/A_1 = -1$$

The reflected wave A_2 is equal to the incident wave A_1 except for the negative sign. This means reflection with reversal.

If $\rho_2 = \rho_1$ (for uniform string), equation (8) gives

$$B/A_1 = 1$$

The transmitted wave B is exactly the same as the incident wave A_1.

If $\rho_2 > \rho_1$ (for non-uniform string), equation (8) gives

$$B < A_1$$

The amplitude of the transmitted wave B is smaller than the amplitude of the incident wave A_1 and vice versa.

If ρ_2 is very small (for free end, $\rho = 0$), we have

$$A_2 = A_1$$

The reflected wave A_2 is exactly the same as the incident wave A_1.

The energy per unit length of the string for each of the different waves is given by

$$\text{incident energy} \quad = \tfrac{1}{2}\rho_1 A_1^2 \omega^2$$
$$\text{reflected energy} \quad = \tfrac{1}{2}\rho_1 A_2^2 \omega^2$$
$$\text{transmitted energy} = \tfrac{1}{2}\rho_2 B^2 \omega^2$$

From the principle of conservation of energy, the rate of energy approaching the junction must equal the rate of energy leaving the junction. Thus

$$\tfrac{1}{2}\rho_1 A_1^2 \omega^2 a_1 = \tfrac{1}{2}\rho_1 A_2^2 \omega^2 a_1 + \tfrac{1}{2}\rho_2 B^2 \omega^2 a_2$$

or
$$Z_1 A_1^2 = Z_1 A_2^2 + Z_2 B^2 \tag{9}$$

where $Z = \rho a$ is called the mechanical impedance.

From equations (6) and (9), we obtain

$$\frac{\text{reflected energy}}{\text{incident energy}} = \frac{(Z_1 - Z_2)^2}{(Z_1 + Z_2)^2}, \qquad \frac{\text{transmitted energy}}{\text{incident energy}} = \frac{4 Z_1 Z_2}{(Z_1 + Z_2)^2}$$

In order to obtain maximum transmission of energy, the two impedances must match each other. In other words, when $Z_1 = Z_2$ there is no reflected energy, and transmitted energy is equal to incident energy.

12. Determine the frequency equation of a compound string fixed at both ends as shown in Fig. 9-8.

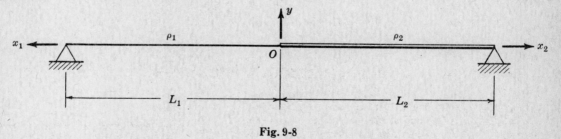

Fig. 9-8

The general solution for vibrating strings can be expressed in complex exponential form as

$$y_1(x, t) = A_1 e^{i(\omega t - k_1 x_1)} + B_1 e^{i(\omega t + k_1 x_1)}$$
$$y_2(x, t) = A_2 e^{i(\omega t - k_2 x_2)} + B_2 e^{i(\omega t + k_2 x_2)}$$

where $k_1 = \omega/a_1$ and $k_2 = \omega/a_2$.

At the junction O, the two expressions given for the deflection and slope should be equal, i.e.,

$$(y_1)_{x_1=0} = (y_2)_{x_2=0} \tag{1}$$
$$(dy_1/dx_1)_{x_1=0} = -(dy_2/dx_2)_{x_2=0} \tag{2}$$

From equation (1),

$$A_1 + B_1 = A_2 + B_2 \tag{3}$$

and from (2),

$$-A_1 k_1 + B_1 k_1 = A_2 k_2 - B_2 k_2 \tag{4}$$

At the fixed ends of the string, the deflections are equal to zero, i.e. $(y_1)_{x_1=L_1} = (y_2)_{x_2=L_2} = 0$, or

$$A_1 e^{-ik_1L_1} + B_1 e^{ik_1L_1} = 0 \tag{5}$$

$$A_2 e^{-ik_2L_2} + B_2 e^{ik_2L_2} = 0 \tag{6}$$

From equations (5) and (6), we have

$$A_1 = \frac{-B_1 e^{ik_1L_1}}{e^{-ik_1L_1}} \tag{7}$$

$$A_2 = \frac{-B_2 e^{ik_2L_2}}{e^{-ik_2L_2}} \tag{8}$$

Substituting equations (7) and (8) into equations (3) and (4), we obtain

$$B_1 \left(1 - \frac{e^{ik_1L_1}}{e^{-ik_1L_1}}\right) = B_2 \left(1 - \frac{e^{ik_2L_2}}{e^{-ik_2L_2}}\right) \tag{9}$$

$$B_1 k_1 \left(1 + \frac{e^{ik_1L_1}}{e^{-ik_1L_1}}\right) = -B_2 k_2 \left(1 + \frac{e^{ik_2L_2}}{e^{-ik_2L_2}}\right) \tag{10}$$

Dividing equation (9) by (10) yields

$$\frac{(e^{-ik_1L_1} - e^{ik_1L_1})}{k_1(e^{-ik_1L_1} + e^{ik_1L_1})} = \frac{-(e^{-ik_2L_2} - e^{ik_2L_2})}{k_2(e^{-ik_2L_2} + e^{ik_2L_2})}$$

Since $e^{ix} = \cos x + i \sin x$ and $e^{-ix} = \cos x - i \sin x$, where $i = \sqrt{-1}$, the above equation can be simplified to

$$\frac{-2i \sin k_1L_1}{2k_1 \cos k_1L_1} = \frac{2i \sin k_2L_2}{2k_2 \cos k_2L_2} \qquad \text{or} \qquad \frac{\tan k_1L_1}{k_1} = -\frac{\tan k_2L_2}{k_2}$$

which finally becomes

$$\frac{1}{a_1} \tan (\omega L_1/a_1) + \frac{1}{a_2} \tan (\omega L_2/a_2) = 0$$

where $a_1 = \sqrt{T/\rho_1}$, $a_2 = \sqrt{T/\rho_2}$, T is the tension in the string, and ρ_1 and ρ_2 are the masses per unit length of the two parts of the string as shown in Fig. 9-8.

Supplementary Problems

13. A uniform wire of length 10 ft is under a tensile pull of 100,000 lb at the free end as shown in Fig. 9-9. The wire weighs 3.22 lb/ft. What are the natural frequencies of the transverse vibration of the wire?

    Ans. $\omega_n = 5.00i$ cyc/sec, $i = 1, 2, 3, \ldots$

    Fig. 9-9

14. Employ energy considerations to derive the wave equation $\partial^2 y/\partial t^2 = a^2(\partial^2 y/\partial x^2)$ for a flexible string stretched to a constant tension.

15. Find the motion of a uniform string of length L and fixed at both ends if initially it is displaced a distance h at the center. Express the solution in terms of traveling waves.

    Ans. $y(x,t) = \dfrac{4h}{\pi^2} \left\{ \left[\sin \dfrac{\pi a}{L}\left(\dfrac{x}{a} - t\right) + \sin \dfrac{a\pi}{L}\left(\dfrac{x}{a} + t\right) \right] \right.$

    $\left. - \dfrac{1}{9}\left[\sin \dfrac{3\pi a}{L}\left(\dfrac{x}{a} - t\right) + \sin \dfrac{3\pi a}{L}\left(\dfrac{x}{a} + t\right) \right] + \cdots \right\}$

16. A uniform taut string of length L is plucked a distance h at $x = L/4$ and then released as shown in Fig. 9-10. Find the resulting free vibration.

Ans. $y(x, t) = \dfrac{32h}{3\pi^2} \displaystyle\sum_{i=1,2,\ldots}^{\infty} \dfrac{1}{i^2} \sin\dfrac{i\pi}{4} \sin\dfrac{i\pi x}{L} \cos\dfrac{i\pi a}{L} t$

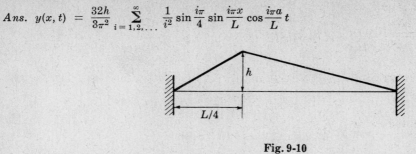

Fig. 9-10

17. A uniform taut string fixed at both ends is struck at the center so as to obtain an initial velocity which varies linearly from zero at the ends to V_0 at the center. What is the resulting free vibration? The length of the string is L.

Ans. $y(x, t) = \dfrac{8V_0 L}{a\pi^3} \displaystyle\sum_{i=1,2,\ldots}^{\infty} \dfrac{1}{i^3} \sin\dfrac{i\pi}{2} \sin\dfrac{i\pi x}{L} \sin\dfrac{i\pi a}{L} t$

18. A uniform string of length L is held firmly at its ends with a high initial tension. If an initial velocity V_0 is given to each point along the string, determine the resultant motion.

Ans. $y(x, t) = \dfrac{4V_0 L}{\pi^2 a} \displaystyle\sum_{i=1,2,\ldots}^{\infty} \dfrac{1}{(2i-1)^2} \sin\dfrac{(2i-1)\pi x}{L} \sin\dfrac{(2i-1)\pi a}{L} t$

19. Determine the fundamental natural frequency of a uniform string of length L and fixed at both ends. Use the complex exponential approach.

Ans. $\omega = (\pi/L)\sqrt{T/\rho}$ rad/sec

20. A uniform string of length L, held firmly at its ends with a high initial tension, is plucked at $x = L/3$ with an amplitude h and released. Determine the resultant motion.

Ans. $y(x, t) = \dfrac{9h}{\pi^3} \displaystyle\sum_{i=1,2,\ldots}^{\infty} \dfrac{1}{i^2} \sin\dfrac{i\pi}{3} \sin\dfrac{i\pi x}{L} \cos\dfrac{i\pi a}{L} t$

21. A uniform string of length L is fixed at both ends. If the string is given an initial displacement $y_0 \sin \pi x/L$, find its subsequent displacements.

Ans. $y(x, t) = y_0 \sin\dfrac{\pi x}{L} \cos\dfrac{\pi a}{L} t$

22. A uniform string of length L is fixed at both ends as shown in Fig. 9-11. The right support is mounted on a shake table which generates a steady sinusoidal motion $y = A \sin \omega t$. Determine the steady state response of the string.

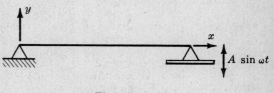

Ans. $y(x, t) = \dfrac{A \sin \omega x/a}{\sin \omega L/a} \sin \omega t$

Fig. 9-11

23. A uniform string of length L is stretched between two rigid supports. It is driven by a force $F_0 e^{-i\omega t}$ concentrated at its midpoint. Show that the amplitude of vibration of the midpoint is $(F_0/2\rho a\omega) \tan \omega L/2a$.

24. A uniform string of length L is stretched between two rigid supports. A force $F_0 \cos \omega t$ is applied at a single point $x = d$ of the string. Find the steady state vibration of the string.

Ans. $y(x, t) = \dfrac{2F_0 L}{(i^2\pi^2 a^2 - L^2\omega^2)} \sin\dfrac{i\pi d}{L} \cos \omega t$

25. A uniform cable of length L is stretched to an initial tension T at its two fixed supports. The cable is submerged in water with damping coefficient per unit length of the cable equal to c. A force $F_0 \sin \omega t$ is applied at a single point $x = d$. Find the steady state vibration.

$Ans.$ $y(x, t) = \dfrac{2F_0}{L\sqrt{\rho^2(p_i^2 - \omega^2)^2 + \omega^2 c^2}} \sin \dfrac{i\pi d}{L} \sin(\omega t - \psi),$ where $p_i = i\pi a/L,$ $\tan \psi = \dfrac{c\omega}{\rho(p_i^2 - \omega^2)}$

26. Show that the potential energy of a uniform string under non-uniform tension is

$$\text{P.E.} = -\frac{1}{2} \int_0^L \frac{d}{dx}\left(Ty \frac{dy}{dx} \right) dx$$

where T is the tension in the string.

27. A uniform taut wire of length L is initially at rest with one end fixed and the other end subjected to a given motion $y(t) = y_0 \sin \omega t$. Determine the motion of the wire.

$Ans.$ $y(x, t) = \dfrac{2y_0 \omega a}{L} \sum\limits_{i=1,2,\ldots}^{\infty} (-1)^{i+1} \left[\dfrac{\sin(i\pi a/L)t \, \sin i\pi(L-a)/L}{(\omega^2 - i^2\pi^2 a^2/L^2)} \right] + y_0 \dfrac{\sin \omega(L-x)/a}{\sin \omega L/a} \sin \omega t$

28. A uniform taut string of length L is fixed at both ends. If initial displacement is zero and initial velocity is $V_0 x(L - x)$, find the motion of the string.

$Ans.$ $y(x, t) = \dfrac{8L^3 V_0}{\pi^4 a} \sum\limits_{i=1,2,\ldots}^{\infty} \dfrac{1}{(2i-1)^4} \sin(2i-1)\dfrac{x}{L}\pi \sin\dfrac{(2i-1)\pi a}{L} t$

29. Show that the work done in displacing the center of a stretched string by an amount h equals the sum of the energies present in the various modes of vibration when the string is released.

30. A concentrated mass M is attached to the center of a uniform string of length L, fixed at both ends. If the initial tension T in the string remains constant for small angles of oscillation, determine the frequency equation of the system.

$Ans.$ $\cot \omega L/a = \frac{1}{2}\omega M/a\rho$

31. A concentrated mass M is attached to the center of an infinite uniform string of mass ρ per unit length, stretched to an initial high tension T. What is the transverse mechanical impedance of the mass M?

$Ans.$ $2\rho a + iM\omega,$ where $i = \sqrt{-1}$

32. A uniform taut string of length L is fixed at both ends. Initial displacement is

$$y = \begin{cases} 0, & (0 \leq x \leq L/3) \\ \sin 3\pi x/L, & (L/3 \leq x \leq 2L/3) \\ 0, & (2L/3 \leq x \leq L) \end{cases}$$

while initial velocity is zero. Find the motion of the string.

$Ans.$ $y(x, t) = \dfrac{6}{\pi} \sum\limits_{i=1,2,\ldots}^{\infty} \dfrac{1}{i^2 - 9} \sin\dfrac{i\pi x}{L} \cos\dfrac{i\pi a}{L} t \left\{ \sin 2(i+3)\dfrac{\pi}{2} - \sin(i+3)\dfrac{\pi}{2} \right\}$

33. A coil spring of free length L has stiffness k lb/in and total weight W_0 lb. It is subjected to a pulse at one end which reaches the other end in 10 seconds. The spring is then put under a tension of kL lb. What will be the pulse arrival time? $Ans.$ $t = 10$ sec

34. Prove the orthogonality principle for vibration of uniform strings:

$$\int_0^L \rho(x) \, y_i(x) \, y_j(x) \, dx = 0$$

35. Prove that the velocity of the traveling waves in strings is $a = \sqrt{T/\rho}$.

INDEX

Catalog

If you are interested in a list of SCHAUM'S
OUTLINE SERIES send your name
and address, requesting your free catalog, to:

SCHAUM'S OUTLINE SERIES, Dept. C
McGRAW-HILL BOOK COMPANY
1221 Avenue of Americas
New York, N.Y. 10020